Summer

June 1,	1:00 a.m. Local Daylight Time
July 1,	11:00 p.m. Local Daylight Time
August 1,	9:00 p.m. Local Daylight Time

NORTH

PERSEUS

CAMELOPARDALIS

ANDROMEDA

LYNX

CASSIOPEIA

Polaris

δ Cephei

CEPHEUS

URSA MINOR

Dubhe

Merak

LEO MINOR

Kochab

LEO

URSA MAJOR

Deneb

DRACO

Alcor

Mizar

CANES VENATICI

PEGASUS

Eltanin

COMA BERENICES

CYGNUS

Vega

Denebola

Albireo

LYRA

M 13

BOOTES

M 3

EQUULEUS

DELPHINUS

HERCULES

Arcturus

CORONA BOREALIS

AQUARIUS

Altair

AQUILA

SERPENS CAPUT

VIRGO

Equator

M 5

CAPRICORNUS

SERPENS CAUDA

OPHIUCHUS

Spica

CORVUS

Ecliptic

LIBRA

HYDRA

M 4

Antares

SAGITTARIUS

SCORPIUS

CORONA AUSTRALIS

EAST

SOUTH
SUMMER

Autumn

September 1, 11:30 p.m. Local Daylight Time
October 1, 9:30 p.m. Local Daylight Time
November 1, 6:30 p.m. Local Standard Time

NORTH

Merak

AURIGA
Capella
CANES VENATICI
CAMELOPARDALIS
Dubhe
URSA MAJOR
Kochab
Mizar
M 3
Pleiades
PERSEUS
URSA MINOR
Polaris
Double Cluster
BOOTES
Algol
Arcturus
CASSIOPEIA
CEPHEUS
DRACO
ANDROMEDA
Eltanin
CORONA BOREALIS
TRIANGULUM
M 31
M 13
Deneb
Vega
Alpheratz
HERCULES
CYGNUS
LYRA
PISCES
Albireo
PEGASUS

WEST

Mira
DELPHINUS
Enif
Altair
OPHIUCHUS
AQUILA
Equator
CETUS
AQUARIUS
Deneb Kaitos
Ecliptic
CAPRICORNUS
SAGITTARIUS
Fomalhaut
PISCES AUSTRINUS
CORONA AUSTRALIS

SOUTH
AUTUMN

Winter

December 1, 10:30 p.m. Local Standard Time
January 1, 8:30 p.m. Local Standard Time
February 1, 6:30 p.m. Local Standard Time

Hold the map overhead so that the map lines up with north, south, east, and west, at your viewing location. This series of maps is drawn for observers at 40° north latitude.

Spring

March 1, 12:30 a.m. Local Standard Time
April 1, 10:30 p.m. Local Standard Time
May 1, 9:30 p.m. Local Standard Time

NORTH

δ Cephei

CASSIOPEIA
Double Cluster
CEPHEUS
Algol
PERSEUS
Vega
Eltanin
CAMELOPARDALIS
Polaris
DRACO
URSA MINOR
Capella
Kochab
AURIGA
Aldebaran
HERCULES
M 13
TAURUS
Dubhe
Alcor
CORONA BOREALIS
Merak
M 35
Mizar
BOOTES
URSA MAJOR
Castor
LYNX
GEMINI
CANES VENATICI
Pollux
SERPENS CAPUT
Betelgeuse
Arcturus
LEO MINOR
M 44
ORION
COMA BERENICES
Ecliptic
CANIS MINOR
M 42
M 5
Denebola
LEO
CANCER
Procyon
Regulus
VIRGO
MONOCEROS
Equator
Sirius
LIBRA
SEXTANS
Alphard
Spica
CANIS MAJOR
CRATER
HYDRA
CORVUS
PUPPIS

WEST

SOUTH
SPRING

ASTRONOMY Magazine Star Dome Map by Adolf Schaller.

ASTRONOMY NOW

JAY M. PASACHOFF

Williams College—Hopkins Observatory
Williamstown, Massachusetts

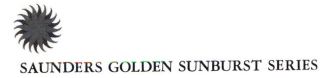

SAUNDERS GOLDEN SUNBURST SERIES

1978
W. B. SAUNDERS COMPANY · Philadelphia · London · Toronto

W. B. Saunders Company: West Washington Square
Philadelphia, PA 19105

1 St. Anne's Road
Eastbourne, East Sussex BN21 3UN, England

1 Goldthorne Avenue
Toronto, Ontario M8Z 5T9, Canada

Library of Congress Cataloging in Publication Data

Pasachoff, Jay M

  Astronomy now.

  (Saunders golden sunburst series)

  Bibliography: p.
  Includes index.

  1. Astronomy.   I. Title.

QB45.P289   520   77–84678

ISBN 0–7216–7100–4

Astronomy Now   ISBN 0-7216-7100-4

Last digit is the print number: 9   8   7   6   5   4   3   2   1

Preface

Astronomy, as a science, combines the best of new and old traditions. Every year we find something new or unexpected; quasars, molecules in interstellar space, pulsars, coronal holes, x-ray bursts, black holes, the nature of the surface of Mars, and the rings of Uranus are but a few examples from recent times. At the same time, astronomy has a rich and fascinating history.

I have attempted, in this book, to describe the state of astronomy as it is now. I have tried to fit the latest discoveries in their places alongside established results in order to give a contemporary picture of the state of our science.

ASTRONOMY NOW is written for students with no background in mathematics or physics. It discusses astronomy in non-mathematical terms. Students with mathematics or physics backgrounds, however, will still find the book valuable because of its unique coverage of a wide range of modern topics. ASTRONOMY NOW is largely a shorter version of my text, CONTEMPORARY ASTRONOMY. Many people might find it more suitable for courses that do not have the time to cover the entire longer version. Neither book uses much mathematics nor requires any as background. A mathematical version, which uses algebra and trigonometry but not calculus, is also available; this expanded version is written in collaboration with Marc L. Kutner. The overwhelmingly favorable response to CONTEMPORARY ASTRONOMY has led to my preparation of these alternative versions.

The books emphasize topics that are of current interest to astronomers and in which current research is very active. In order to gain space in the books and time in the academic year for the contemporary material, some traditional material that is discussed at length in other texts has been minimized. But I have attempted to cover all fields of astronomy, without major gaps.

A major theme of the books has been the expansion of our senses to all parts of the spectrum. We see time and again through the entire book how observations of gamma rays, x-rays, ultraviolet light, infrared light, and radio waves are fit together with observations of visible light and with theory to improve our knowledge of the universe and the objects in it. Such new results appear throughout the entire book, rather than being segregated in a few chapters. Similarly, the historical side of astronomy is also treated as the various topics come up, rather than having historical discussions segregated.

A major goal of the books has been to describe astronomy in clear, thorough, understandable, and colloquial terms. Many discussions of especially important topics have been expanded in order to have space to describe various aspects. All the steps in chains of reasoning have been included in order to make certain that the major points are clear.

An unusual aspect of the books is that background discussions of physics and astronomy appear as they are needed in the text rather than as lengthy introductory chapters. This has the advantage of making more apparent the connection between the background discussions and the discussions of modern research. It has the further advantage of allowing the student to begin with real astronomy at the very start of the course, instead of having to wait until the introductory chapters are completed. Certain background material may be briefly recapitulated in later chapters even though it may have been dealt with more thoroughly earlier. Once again, the reason for this is to stress the connection between underlying concepts and their applications.

In order to depict the widespread activity in astronomy, and to demonstrate the human aspect of astronomical research, I have sometimes named individual astronomers and institutions. Further, details that may not be important in themselves are sometimes used to embellish the central point. The ability to expand in this way has enabled me to set many matters in especially interesting contexts. But at no time should students find themselves memorizing material instead of understanding concepts. As an aid in determining just what is most important to remember, I have provided summaries of each chapter in this book, and a detailed Student Study Guide (Pasachoff, Kutner, Pasachoff and Kutner, *Student Study Guide to ASTRONOMY NOW*, W. B. Saunders Company, 1978) is also available.

My stress has continually been on how to understand what is going on, and why astronomers think or act as they do. My criterion has been to include mostly material that can be remembered for years and that provides a basic understanding of the topic, rather than facts that are merely learned for examinations and then soon forgotten. In order to give a feel of what it is like to be an astronomer, I have at several places in the text described just what an astronomer does in different circumstances, such as observing at a large telescope, observing with a satellite, or working at a solar eclipse. I have myself been fortunate to be able to observe with a variety of optical, radio, and space telescopes in this country and abroad, and I am glad to have the opportunity to describe the excitement an observer feels.

It also seems to me that after completing a course in astronomy a student should be able to carry on a reasonable conversation with an astronomer. This has led me to continue the usage of the somewhat strange units that are retained by astronomers for historical reasons rather than for reasons of logic. For example, magnitudes are used to describe the brightness of stars, though I try to make clear the relation of the magnitude scale to a direct scale of brightness. Anyway, I find it easier to remember magnitudes, which are simple numbers like 2 or 16, rather than luminosities like 2×10^4. If these units are madness, there is method in it.

In view of the importance to scientific research of governmental support, and the need for everyone to be informed about the value of such research, I have described from time to time the benefits that might accrue to society from certain types of astronomical research.

Arrangement of the book

Since stars are the main point of reference that most individuals have to astronomy, I have begun my discussion with them. After discussing basic definitions and terms, I discuss the sun as a special case of a star, and then proceed to stellar evolution, including its fascinating end-products: white dwarfs, neutron stars (which we observe as pulsars), and black holes. I then discuss another special case of objects closely associated with stars that we happen to be able to study because they are so near: the planets. After the discussions of stars and planets, we continue outward through the universe from small to large, with interstellar matter, our own galaxy, galaxies in general, quasars, and then cosmological consideration of the universe as a whole. The book thus describes astronomy from the inside out, beginning with the stars.

One topic that many professors treat in different manners is Section III on the planets. I have placed it in this book adjacent to the discussion of stars, but it does not depend on any material later than Section 2.3. The material on the planets could be taught at other places in the syllabus — including immediately following Section 2.3 — or even omitted. It is not dependent on the discussions that preceded it in Chapters 3 through 9, and neither do the sections that follow depend on it. Some professors might want to teach the planets first; others might want to defer the study of planets until the very end. Another alternative is to teach Chapter 3 on observing ahead of all or most of Chapter 2. Several alternate plans for syllabi are listed in the *Teacher's Guide.*

Some of the sections in this book have been specially marked as "optional." Individual teachers, of course, may wish to include these sections and omit others.

The *Student Study Guide to ASTRONOMY NOW*, written in collaboration with Marc L. Kutner, Naomi Pasachoff, and Nancy P. Kutner, all of the Rensselaer Polytechnic Institute, provides discussions of the material in ASTRONOMY NOW chapter by chapter, worked problems, quizzes with answers, and chapter by chapter lists of books and articles. In some cases, it will be adopted for class use along with the main text; in other cases, individual students might choose to use the book on their own.

A *Teacher's Guide* to accompany *ASTRONOMY NOW* is also available. It contains labs, tests, and answers to questions, and lists films, tapes, and other audio and visual aids for use as supplementary material. The *Teacher's Guide* also includes a card that can be used to order a yearly supplement that I shall prepare to be sent out each summer, so that no teacher need be far behind the exciting new results that we hear of all the time.

The publishers and I are making available the artwork and photographs that are included in this book in a format that can be projected in class. If you are interested in such material, please contact Mr. John J. Vondeling, Editor-in-Chief, College Department, W. B. Saunders Company, West Washington Square, Philadelphia, Pa 19105.

Acknowledgments

The publishers and I have placed a heavy premium on accuracy, and have made certain that the manuscript and proof have been read not only by students for clarity and style but also by several astronomers for their professional comment. I would like to thank, in particular, Thomas T. Arny (University of Massachusetts at Amherst), Marc L. Kutner (Rensselaer Polytechnic Institute), Richard L. Sears (University of Michigan), and Joseph F. Veverka (Cornell University) for reading the entire manuscript. I am also grateful to David E. Koltenbah (Ball State University), Karl F. Kuhn (Eastern Kentucky University), and Stephen T. Gottesman (University of Florida) for consulting on the preparation of this short edition.

I would also like to thank other astronomers and physicists who have commented on parts of the manuscript in their particular areas of expertise. They include (in the approximate order in which the topics are treated in the book):

Jean Pierre Swings (Institut d'Astrophysique, Liège, Belgium), stars and quasars; Jeffrey L. Linsky (Joint Institute for Laboratory Astrophysics, Boulder), radiative transfer; David Park (Williams College), atomic physics; James G. Baker (Harvard College Observatory), optics; Jonathan E. Grindlay (Harvard-Smithsonian Center for Astrophysics), black holes and x-ray bursts; Peter V. Foukal (Harvard-Smithsonian Center for Astrophysics), the sun; Robert F. Howard (Hale Observatories), the sun; Lawrence Cram (Fraunhofer Institute, Freiburg, Germany), opacity effects; J. Craig Wheeler (University of Texas), stellar evolution; Raymond Davis, Jr. (Brookhaven National Laboratory), solar neutrinos; Joseph H. Taylor, Jr. (University of Massachusetts, Amherst), pulsars; John A. Wheeler (University of Texas), black holes; Kenneth Brecher (Massachusetts Institute of Technology), black holes; Peter Conti (University of Colorado), stars; Bruce E. Bohannan (University of Colorado), stars; Maurice Shapiro (Naval Research Laboratory), cosmic rays;

Owen Gingerich (Harvard-Smithsonian Center for Astrophysics), history of astronomy; David D. Morrison (University of Hawaii and NASA Headquarters), the solar system; James P. Pollack (NASA Ames Research Center), the solar system; Farouk El-Baz (NASA Headquarters and National Air and Space Museum, Smithsonian Institution), the moon; Ewen A. Whitaker (Lunar and Planetary Laboratory, University of Arizona), the moon; Gerald J. Wasserburg (California Institute of Technology), lunar chronology; William R. Moomaw (Williams College), ozone; Ian Halliday (Herzberg Institute of Astrophysics, Ottawa), Neptune and Pluto; J. Donald Fernie (David Dunlap Observatory, Toronto), the Adams-Leverrier affair; P. Kenneth Seidelmann (U.S. Naval Observatory), ephemerides and planetary parameters; Brian Marsden (Harvard-Smithsonian Center for Astrophysics), comets, asteroids, and meteoroids; George D. Gatewood (Allegheny Observatory), Barnard's Star; Everett Mendelsohn (Harvard University), superstition; Robert G. Strom (Lunar and Planetary Laboratory, University of Arizona), Mercury;

Agris Kalnajs (Mt. Stromlo Observatory, Australia), spiral structure; Barry Turner (National Radio Astronomy Observatory), the latest interstellar molecules; Juri Toomre (Joint Institute for Laboratory Astrophysics), gravitational effects on galaxies; Leonid Weliachew (Observatoire de Paris, France), galaxies; George K. Miley (Leiden Observatory, the Netherlands), interferometry; Bernard J. T. Jones (Institute for Astronomy, Cambridge,

England), galaxy formation; Jerome Kristian (Hale Observatories), quasars, Seyfert and N galaxies; John Lathrop (Williams College), cosmology; and Alan T. Moffet (California Institute of Technology), radio astronomy.

Many others have read shorter bits of manuscript, cleared up particular points, or provided special illustrations, and I thank them as well. In particular, let me acknowledge the assistance of Maarten Schmidt (Hale Observatories), Hyron Spinrad (University of California at Berkeley), Richard B. Dunn (Sacramento Peak Observatory), Freeman D. Miller (University of Michigan), Peter van de Kamp (Sproul Observatory), Kenneth D. Tucker (Fordham University), Edwin C. Krupp (Griffith Observatory), Sarah Lee Lippincott (Sproul Observatory), and R. Newton Mayall (American Association of Variable Star Observers). I am grateful to Elske v. P. Smith (University of Maryland) for comments early on in the preparation of this book. The many institutions and individuals who provided photographs or other illustrations are acknowledged separately; I an grateful to all of them.

Marc L. Kutner has worked closely with me on the questions. Naomi Pasachoff worked closely with me on the glossary.

I thank many people at the W. B. Saunders Company for their efforts on this book. John J. Vondeling, Editor-in-Chief of the College Division, merits special thanks for his continued support. He has made it possible for me to include so many illustrations and photographs. Lloyd Black of the Manuscript Editorial Department has worked long and hard on the manuscript and figures. John Hackmaster, Susan O'Neill, Celeste Brennan, and Linda Downham of the Art Department, Ray Kersey and Peg Shaw of the Illustration Department, Jay Freedman of the Manuscript Editorial Department, and Suzanne Rommel and Deborah Patterson, assistants to John Vondeling, have also contributed to many phases of the book. Lorraine Battista designed the book and the cover. The detailed layout is by Joan De Lucia. The support and efforts of George Laurie of the Art Department and of Tom O'Connor and Frank Polizzano of the Production Department helped keep the book on schedule.

Most of the drawings were executed by George Kelvin of Science Graphics from my sketches. Other drawings were executed by Grant Lashbrook and John Hackmaster at Saunders.

I am grateful to Zadig Mouradian of the Observatoire de Paris and to Jean-Claude Pecker of the Collège de France for their hospitality during part of the period on which I was working on this book. I also remember fondly the influence that Donald H. Menzel of the Harvard College Observatory had on my career over a period of many years.

I am also grateful to my colleagues at the University of Massachusetts at Amherst, where I hold an adjunct appointment, and to William Irvine and others there for arranging that appointment as a supplement to my position at Williams.

Carla A. Weiss and Eudora Tharp have worked with me in Williamstown on many aspects of the manuscript, illustrations, and proof, and have been of tremendous assistance. Susan K. Waldron typed early drafts of the manuscript. I thank them for their assistance.

Various members of my family have provided a vital and valuable editorial service. In particular, my wife Naomi Pasachoff, my father Dr. Samuel S. Pasachoff, and my mother Anne T. Pasachoff have worked many hours on the manuscript and proof. My daughter Eloise came in to help type from time to time; as she was then not yet two, this did not necessarily speed things up. My daughter Deborah is also now spurring me on.

The index was put together in collaboration with Nancy P. Kutner, who also aided with the page proof.

I am extremely grateful to all of the above individuals for their assistance. Of course, it is I who have put this all together, and I alone am responsible for any errors that have crept through our sieve. I would appreciate hearing from readers, not just about typographical or other errors, but also with suggestions for presentation of topics or even with comments about certain points that need clarifying. I invite readers to write me c/o Williams College, Hopkins Observatory, Williamstown, Mass. 01267. I promise a personal response to each writer.

JAY M. PASACHOFF

Williamstown, Massachusetts

Contents

LIST OF BOXES

LIST OF TABLES

Chapter 1

The universe is a place of great variety—after all, it has everything in it! Thus astronomers have much to study. Some of the things astronomers study are of a size and scale that we humans can easily comprehend: the planets, for instance. Most astronomical objects, however, are so large and so far away that our minds have trouble grasping the sizes and distances involved.

Moreover, astronomers study the very small in addition to the very large. The radiation we receive from distant bodies is emitted by atoms, which are much too small to see with the unaided eye. Also, the properties of the large astronomical objects are often determined by changes that take place on a minuscule atomic scale. Thus the astronomer must be an expert in the study of objects the size of atoms as well as in the study of objects the size of galaxies.

It should come as no surprise to hear that the variety of objects at very different distances from us or with very different properties often must be studied with widely differing techniques. Clearly, different tests are required to analyze the properties of Martian soil from a spacecraft on the Martian surface than are required to test the composition of a quasar deep in space. There is one unifying method, however, that links much of astronomy. This method is *spectroscopy,* which involves, as we shall see, analyzing components of the light or other radiation that we receive from distant objects and studying the components in detail. Throughout this book, we shall return to spectroscopic methods time and again, to study not only visible light but also other types of radiation. Not all of astronomy is spectroscopy, of course; much important information is gained by making images—pictures—of planets, the sun, and galaxies, for example, or by studying the variation of the amount of light coming from a star or a galaxy over time.

The explosion of astronomical research in the last few decades has been fueled by our new ability to study radiation other than light—gamma rays, x-rays, ultraviolet radiation, infrared radiation, and radio waves. Astronomers' use of their new abilities to study such radiation is a major theme of this book. All the kinds of radiation together make up the *electro-*

The Whirlpool Galaxy, M51, in the constellation Canes Venatici.

magnetic spectrum, which will be discussed in Chapter 2. As we shall see there, we can think of radiation as waves, and all the types of radiation have similar properties except for the length of the waves. Still, although to a scientist x-rays and visible light may be similar, our normal experiences tell us that very different techniques are necessary to study them.

The earth's atmosphere shields us from most kinds of radiation. Not only light waves but also radio waves penetrate the atmosphere, though most other types of radiation are blocked. Over the last forty years, radio astronomy has become a major foundation of our astronomical knowledge. For the last twenty years, we have been able to send satellites into orbit outside the earth's atmosphere, and we are no longer limited to the study of radio and visible (light) radiation. Many of the fascinating discoveries of this decade—the probable observation of a black hole, for example— were made because of our newly extended senses. We shall be discussing how all parts of the spectrum are used to help us understand the universe. And besides the information that the spectrum provides, we can also get information from direct sampling of bodies in our solar system (as we shall discuss in Part III), from cosmic rays (which are particles whizzing through space and which we shall discuss in Sections 8.3 and 22.2), and perhaps even from gravitational waves (which, as we shall discuss in Section 9.7, may or may not have been observed).

1.1 A SENSE OF SCALE

Let us try to get a sense of scale of the universe, starting with sizes that are part of our experience and then expanding toward the infinitely large. In the margin, we can keep track of the size of our field of view as we expand in powers of 100: each diagram will show a square 100 times greater on a side and thus 10,000 times greater in area than the previous diagram.

Literary scholars could say that a "milli-Helen" is a unit of beauty sufficient to launch one ship.

We shall use the metric system, which is in common use by scientists. The basic unit of length is the meter, which is equivalent to 39.37 inches, slightly more than a yard. Prefixes are commonly used in conjunction with the word "meter" (see Appendix 2) to define new units. The most commonly used are "milli-," meaning 1/1000, "centi-," meaning 1/100, and "kilo-," meaning 1000 times. Thus 1 millimeter is 1/1000 of a meter, or about .04 inch, and a kilometer is 1000 meters, or about 5/8 mile. We will keep track of the powers of 10 by which we multiply 1 m by writing the number of tens we multiply together as an exponent; 1000 m, for example, is 10^3 m.

We can also keep track of distance in units that are based on the length of time that it takes light to travel. The speed of light is, according to Einstein's *special theory of relativity* (published in 1905), the greatest speed that is physically attainable. Light travels at 300,000 km/sec (186,000 miles/sec), fast enough to circle the earth 7 times in a single second. Even at that fantastic speed, we shall see that it would take years for us to reach the stars. Similarly, it has taken years for the light we see from stars to reach us, and we are thus really seeing the stars as they were years ago. In a sense, we are looking backward in time. The distance that light travels in a year is called a *light year*; note that the light year is a unit of length rather than a unit of time even though the term "year" appears in it.

Let us begin our journey through space with a view of a circle one millimeter in diameter (Fig. 1–1).

1 mm = 0.1 cm

10 cm = 100 mm

A square 100 times larger on each side is 10 centimeters × 10 centimeters (since the area of a square is the length of a side squared, the area of a 10 cm square is 10,000 times the area of a 1 mm square). The area encloses a flower (Fig. 1–2).

10 m = 1000 cm

As we move far enough away to see an area 10 meters on a side, we are seeing an area approximately that taken up by half a tennis court (Fig. 1–3).

1 km = 10³ m

A square 100 times larger on each side is now 1 kilometer square, about 250 acres. An aerial view of several square blocks in New York City shows how big an area this is (Fig. 1–4).

100 km = 10⁵ m

The next square, 100 km on a side, encloses a major city (Fig. 1–5). Note that we are still bound to the limited area of the earth, but the area we can see is increasing rapidly.

10,000 km = 10⁷ m

A square 10,000 km on a side covers nearly the entire earth (Fig. 1–6).

1 LIGHT SECOND
300,000 KM

EARTH ● MOON ●

1,000,000 km = 10⁹ m = 3 lt sec

When we have receded 100 times farther, we see a square 100 times larger in diameter: 1 million kilometers across. It encloses the orbit of the moon around the earth (Fig. 1–7). We can measure with our wristwatches the amount of time that it takes light to travel this distance. If we were carrying on a conversation by radio with someone at this distance, there would be pauses of noticeable length after we finished speaking before we heard an answer. This is because radio waves, even at the speed of light, take that amount of time to travel. Astronauts on the moon have to get used to these pauses when speaking to earth.

When we look on from 100 times farther away still, we see an area 100 million kilometers across, 2/3 the distance from the earth to the sun. We can now see the sun and the inner planets in our field of view (Fig. 1–8). This is the scale representing approximately the limit of our current ability to send spacecraft to sample the planets directly. We have now sent spacecraft to the moon, Mercury, Venus, Mars, and Jupiter; the results will be discussed at length in Part III.

VENUS

MERCURY

SUN

10¹¹ m = 5 lt min

EARTH
MARS
JUPITER
SATURN
URANUS
NEPTUNE
PLUTO

10¹³ m = 8 lt hrs

An area 10 billion kilometers across shows us the entire solar system in good perspective. It now takes light 8 hours to travel this distance. The outer planets have become visible and are receding into the distance as our journey outward continues (Fig. 1–9).

From 100 times farther away, we see little that is new. The solar system seems smaller and we see how vast is the empty space around us. We have not yet reached the scale at which another star besides the sun is in our field of view, although, of course, many stars are visible in the background (Fig. 1–10).

ORBIT OF PLUTO
5.9 BILLION KM

10¹⁵ m = 38 lt days

10^{17} m = 10 ly

As we continue to recede from the solar system, the nearest stars finally come into view. We are seeing an area 10 light years across, which contains only a few stars (Fig. 1–11). Some of these stars turn out to be systems in which several stars revolve around each other. Part II of this book discusses the properties of stars.

10^{19} m = 10^3 ly

By the time we are 100 times farther away, we can see a fragment of our galaxy (Fig. 1–12). We see not only many individual stars but also many clusters of stars and many areas of glowing, reflecting or opaque gas or dust called nebulae. There is even a lot of material between the stars that is mostly invisible to our eyes but that can be studied with radio telescopes. Part IV of this book is devoted to the study of our galaxy and its contents.

In a field of view 100 times larger in diameter, we can now see an entire galaxy. The photograph (Fig. 1–13) shows a galaxy called M74, located in the direction that we call the constellation Pisces. This galaxy shows arms wound in spiral form. Our galaxy also has spiral arms, though they are wound more tightly.

10^{21} m = 10^5 ly

10^{23} m = 10^7 ly

Next we move sufficiently far away so that we can see an area 10 million light years across at a glance (Fig. 1–14). There are 10^{25} centimeters in 10 million light years, about as many centimeters as there are grains of sand in all the beaches of the earth. Our galaxy is in a cluster of galaxies, called the Local Group, that would take up only 1/3 of our angle of vision. In this group are all types of galaxies, which we will discuss in Chapter 23. The photograph shows part of a cluster of galaxies in the constellation Leo.

Figure 1–15 Straight lines and triangles are shown on a plane, on the surface of a sphere, and on a saddle-shaped surface. These types of surfaces have a property called curvature that is different for each type.

We live not in two dimensions as on these surfaces, but rather in three dimensions or, in some sense, in four dimensions (three of space plus one of time). Yet in principle we could test the curvature of our space, though in practice this has not yet been possible to do.

If we could see a field of view 1 billion light years across, our Local Group of galaxies would appear as but one of many clusters. We are not able to observe this scale.

Before we could enlarge our field of view another 100 times we might see a supercluster—a cluster of clusters of galaxies. We would be seeing almost to the distance of the quasars, which are the topic of Chapter 24. Quasars may be explosive events in the cores of normal galaxies, or may involve phenomena very different from those astronomers normally consider. We do not know what is involved. Light from the farther quasars may have taken ten billion years to reach us on earth, and if this is so we are thus looking back to epochs billions of years ago. Since we think that the big bang that began the universe took place approximately 13 billion years ago, we are looking back almost to the beginning of time.

At this scale, we cannot accurately predict what the universe would look like. Einstein's general theory of relativity predicts that space itself could show properties of being curved, just as the surface of a saddle or the surface of a sphere is fundamentally curved (Fig. 1–15). On none of these surfaces, for example, can we draw straight lines that remain parallel out to infinity, that is, never crossing and always remaining the same distance from each other. On a saddle the lines diverge; on a sphere the lines converge. Is space in the universe curved positively like a sphere, curved negatively like a saddle, or flat? We do not know. Studies of the depths of the universe may tell us about the curvature of space. In any case, we can understand certain things about the universe as a whole, and we even think that we have detected the radiation given off at the time of the big bang itself. A combination of radio, ultraviolet, and optical studies is allowing us to explore the past and predict the future of the universe. All this is discussed in Chapter 25.

1.2 THE VALUE OF ASTRONOMY

Throughout history, observations of the heavens have led to discoveries that have had major impact on people. Even the dawn of mathematics may have followed ancient observations of the sky, made in order to keep track of seasons and seasonal floods in the fertile areas of the earth. Observations of the motions of the moon and the planets, which are free of such complicating terrestrial forces as friction and which are massive enough so that gravity dominates their motions, led to an understanding of gravity and of the forces that govern all motion.

We can consider the regions of space studied by astronomers as a cosmic laboratory for the study of matter or radiation under a variety of conditions, often under conditions that we cannot duplicate on earth.

Many of the discoveries of tomorrow—perhaps the control of nuclear fusion or the discovery of new sources of energy, or perhaps something so revolutionary that it cannot now be predicted—will undoubtedly also be based on discoveries made through such basic research as the study of astronomical systems. Considered in this sense, astronomy is an investment in our future.

The impact of astronomy on our conception of the universe has been strong through the years. Discoveries that the earth is not in the center of the universe, or that the universe has been expanding for billions of years, affect our philosophical conceptions of ourselves and our relations to space and time.

Yet most of us study astronomy not for its technological and philosophical benefits but for its grandeur and inherent interest. We must stretch our minds to understand the strange objects and events that take place in the far reaches of space. The effort broadens us and continually fascinates us all. Ultimately, we study astronomy because of its fascination and mystery.

QUESTIONS

1. Why do we say that our senses have been expanded in recent years?

2. The speed of light is 3×10^5 km/sec. Express this number in meters/sec and in cm/sec.

3. During the Apollo explorations of the moon, we had a direct demonstration of the finite speed of light when we heard ground controllers speak to the astronauts. The sound from the astronauts' earpieces was sometimes picked up by the astronauts' microphones and retransmitted to earth. We then heard the words repeated. What is the time delay between the original and the "echo"?

4. The distance to the Andromeda galaxy is 2×10^6 light years. If we could travel at one-tenth the speed of light, how long would a round trip take?

5. How long would it take to travel to Andromeda at 1000 km/hr, the speed of a jet plane?

6. List the following in order of increasing size: (a) light year, (b) distance from earth to sun, (c) size of Local Group, (d) size of a football stadium, (e) size of our galaxy, (f) distance to a quasar.

7. Of the examples of scale in this chapter, which would you characterize as part of "everyday" experience? What range of scale does this encompass? How does this range compare with the total range covered in the chapter?

8. What is the largest of the scales discussed in this chapter that could reasonably be explored in person by humans with current technology?

9. How can we tell a flat space from a curved space?

10. What is the value of astronomy to you? Save your answer for comparison with your answer to a similar question in Chapter 26 or your answer when you have completed this course.

Part II
The Stars

A solar prominence.

(*Left*) The Cygnus loop, the remnant of a supernova. The remainder of the star that explodes as a supernova contracts to become either a pulsar or a black hole, depending on how much mass is left behind.

II   The Stars . . .

Our first contact with astronomy involves the stars. When we look up at the sky at night, most of the objects we see are stars. In the daytime, we see the sun, which is itself a star. The moon, the planets, even a comet may give a beautiful show, but, however spectacular, they are only minor actors on the stage of observational astronomy.

From the center of a city, we may not be able to see very many stars, because city light scattered by the earth's atmosphere makes the light level of the sky brighter than most stars. All together, there are about 6000 stars in the sky bright enough to be seen with the naked eye under good observing conditions.

Even a cursory glance at the sky shows that there are patterns in the distribution of stars. The different areas of sky are called *the constellations,* and the names that the ancients gave to these areas, associating them with figures and objects from their myths and religions, are still used down to this day. It takes a great imagination to see Orion (the Hunter) or Draco (the Dragon) in the sky above, but it is nice to be like St. Exupéry's Little Prince (Fig. 2–1), and not let our detailed knowledge cause us to lose our sense of wonder at the sky above and the objects in space.

The first High-Energy Astronomy Observatory, HEAO-1, launched in 1977 to study x-rays and gamma rays.

The stars we see defining a constellation may be at very different distances from us. One star may appear bright because it is relatively close to us even though it is intrinsically faint; another star may appear bright even though it is far away because it is intrinsically very luminous. When we look at the constellations, we are observing only the directions of the stars and not whether the stars are physically close together (Fig. 2–3). Nor do the constellations tell us very much about the nature of stars. Astronomers tend not to be interested in the constellations because the constellations do not give us information about how the universe works. After all, the constellations merely tell the directions in which the stars lie. Most astronomers want to know *why* and *how*. Why is there a star? How does it shine? Such studies of the workings of the universe are called *astrophysics*. Almost all modern day astronomers are astrophysicists as well, for they always not only make observations but also think about their meaning.

Some properties of the stars that can be seen with the naked eye tell us about the natures of the stars themselves. For example, some stars seem blue-white in the sky, while others appear slightly reddish. From information of this nature, astronomers are able to determine the temperatures of the stars. Chapter 2 is devoted to the basic properties of stars and Chapter 3 to various methods of making astronomical observations. Some stars are not constant in brightness. In some cases, as we shall see in Chapter 4, the variations can tell us how far away these stars are, and thus give the distances to the astronomical groupings in which these stars are located. In Chapter 5, we study an average star in detail. It is the only star we can see close up—the sun. The phenomena we observe on the sun presumably take place on other stars as well.

We can't follow an individual star from cradle to grave, for that would take billions of years, but by studying many different stars we can study enough stages of development to write out a stellar biography. Chapters 6 through 9 are devoted to such life stories. Similarly, we could study the stages of human life not by watching someone's aging, but rather by studying people of all ages who are present in, say, a city on a given day.

In Chapter 6, we shall study the birth of stars, and observe some places in the sky where we expect stars to be born very soon, possibly even in our own lifetimes. We shall then consider the properties of stars during the long, stable phase in which they spend most of their histories. Then we go on to consider the ways in which stars end their lives. Stars like the sun sometimes eject shells of gas that glow beautifully; the remainders then collapse until they are as small as the earth (see Chapter 7). Sometimes when a star is newly visible in a location where no star was previously known to exist, we may be seeing the spectacular death throe of a star that had earlier been too faint to observe (see Chapter 8). The study of what happens to stars after that explosive event is currently a topic of tremendous interest to many astronomers. Some of the stars become pulsars (also discussed in Chapter 8). Other stars, we think, may even wind up as black holes, objects that are invisible and therefore difficult—though not impossible—to observe (see Chapter 9).

To do their research, astronomers still often use optical telescopes to gather light from the stars. Following how a stellar biography can be written from the clues that the light carries is like reading a wonderful detective story. But in recent years, the heavens have been studied in other ways besides observing the ordinary light that is given off by many astronomical objects. Radio waves, x-rays, gamma rays, ultraviolet light, infrared light, and cosmic rays are increasingly studied from the earth's surface or from space. Many a contemporary astronomer—even many who consider themselves observers (who mainly carry out observations) rather than theoreticians (who do not make observations, but rather construct theories)—has never looked through an optical telescope. So we mustn't grow to think that the appearance of the stars in ordinary light is the only part of astronomy. Still, studying the light from the stars has been very fruitful for astronomers, and our story begins there.

Ordinary Stars

What are these stars that we see as twinkling dots of light in the night-time sky? The stars are luminous balls of gas scattered throughout space. All the stars we see in the sky are among the 100 billion stellar members of a collection of stars and other matter called the Milky Way Galaxy (see Part IV). The sun, which is the controlling star of our planet, is so close to us that we can see detail on its surface, but it is just an ordinary star like the rest.

An important property of stars is that they generate their own energy and light. Deep inside them they generate energy by a process that we think we understand even though we cannot see it directly—nuclear fusion. It has been only fifty years since the realization that the stars shine in this way, and we are now looking to the stars to provide sources of energy for the earth in the future, either by harnessing the energy from the sun that falls upon the earth or by learning how to tame the nuclear fusion that takes place inside the sun and the other stars and re-creating it on earth.

We see only the outer layers of the gas of the stars; the interiors are hidden from our view by all the gas surrounding them. The outer layers are not generating energy by themselves, but are merely glowing because of the effects of energy that is transported outward from the stellar interiors. So when we study the light that we get from the stars we are not observing the processes of energy generation directly. We must use our studies of the exteriors to tell us what is going on inside.

What about the twinkling? It is not a property of the stars themselves, but merely an effect of our earth's atmosphere. The starlight is always being bent by moving volumes of air in our atmosphere, making the images of the stars appear to be larger than points, to dance around slightly, and to change rapidly in intensity when observed with the naked eye (Fig. 2–2). The planets, on the other hand, do not usually seem to twinkle. This is because they are close enough to earth that they appear as tiny disks large enough to be seen through telescopes though not by the naked eye. When the disk of light representing a planet moves slightly in the sky by the same amount as does a neighboring star, the new position of the disk still overlaps much of the former position. The visual effect to the naked eye is that

Figure 2–1 When asked by the Little Prince to draw a sheep, the author drew the above. "This is only his box. The sheep you asked for is inside." The Little Prince replied that the picture was exactly the way he wanted it, and asked for quiet because the sheep had gone to sleep.

From *The Little Prince* by Antoine de Saint-Ex-upéry.

Cassiopeia, from Johann Bayer's *Uranometria*, first published in 1603.

13

Figure 2–2 Stars appear to twinkle because they are point images (left). Because planets appear as disks (right), they do not usually twinkle, but planets can twinkle too under conditions in which the atmosphere is sufficiently turbulent.

Figure 2–3 The stars we see as a constellation are actually differing distances from us. In this case we see the true distances of the stars in the Big Dipper, part of the constellation Ursa Major, in the lower part of the figure. Their appearance projected on the sky is shown in the upper part.

the planets appear to shine more steadily, even though the earth's atmosphere is affecting each bit of their light in just the same manner as it affects the light from the stars.

The stars are balls of gas that are held together by the force of gravity, the same force that keeps us on the ground with our feet toward the center of the earth, no matter whether we are in the United States or half way around the globe in Australia. Long ago, gravity compressed large amounts of gas and dust into dense spheres, which became stars. Energy from this original contraction has heated up the interiors of the spheres until they were hot enough for nuclear fusion to begin (see Chapter 6).

One difference between astronomers and scientists of most other fields is that many astronomers are usually *observers* rather than *experimentalists* in that they can **observe** light from distant objects but cannot ordinarily **experiment** with the objects directly. Even though we can now sample the moon and some of the planets, we cannot seriously hope to sample the distant stars.

2.1 LIGHT FROM THE STARS

The most obvious thing we notice about the stars is that they have different brightnesses. Over two thousand years ago, the Greek astronomer Hipparchus divided the stars that he could see into classes of brightness. His measurements were extended by Claudius Ptolemy, a Greek/Egyptian astronomer in Alexandria in about A.D. 150. In their classification, the brightest stars were said to be of the first magnitude, somewhat fainter stars were of the second magnitude, and so on down to the sixth magnitude, which represented the faintest stars that could be seen with the naked eye.

In the 19th century, when astronomers became able to make quantitative measurements of the brightnesses of stars, the magnitude scale was placed on a numerical basis. It was discovered that the brightest stars that could be seen with the naked eye were about 100 times brighter than the faintest stars. This corresponded to a difference of 5 magnitudes on Hipparchus's scale, a number that was used to set up the present magnitude system.

The new system is based on the definition that a difference of five magnitudes means a ratio of intensity of exactly 100 times. Thus a star of first magnitude is exactly 100 times brighter than a star of the sixth magnitude. The brightnesses of the stars that had been classified by the Greeks were placed on this new magnitude scale (Fig. 2–4). Many of the

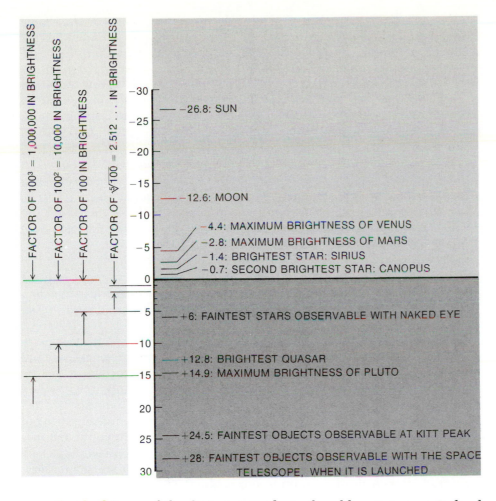

Figure 2–4 The magnitude scale.

stars that had been of the first magnitude in the old system were indeed first magnitude in the new system. But a few stars were much brighter, and thus corresponded to "zeroth" magnitude, that is, a number less than one. A couple, like Sirius, the brightest star in the sky, were brighter still. The numerical magnitude scale could easily be extended to negative numbers. On this scale, Sirius is brighter than magnitude −1. The new scale, being numerical, admits fractional magnitudes. Sirius is actually magnitude −1.4, for example. Physiologically, this type of measurement makes sense because the human eye perceives equal **ratios** of luminosity as roughly equal **intervals** of brightness (on the magnitude scale).

The scale was extended not only to stars brighter than first magnitude, but also to stars fainter than 6th magnitude. Since 5 magnitudes corresponds to a hundred times in brightness, 11th magnitude is 100 times fainter than 6th magnitude (that is, 1/100 times as bright); 16th magnitude, in turn, is 100 times fainter than 11th magnitude. The faintest stars that can be photographed with the 5-meter (200-inch) telescope on Palomar Mountain in California, the largest optical telescope in the United States and until recently the largest optical telescope in the world, are fainter than 23rd magnitude. The interval from magnitude −1.4 to magnitude 23.5 is nearly 25 magnitudes. Since each 5 magnitudes represents a multiplication of 100 times in brightness, 25 magnitudes is a factor of $100 \times 100 \times 100 \times 100 \times 100$ (since it is 100 multiplied by itself 5 times, we abbreviate this by writing 100^5), or 10,000,000,000.

½ mag = 1.585 times
1 mag = 2.512 times
2 mag = 6.310 times
3 mag = 15.85 times
4 mag = 39.81 times
5 mag = 100 times
6 mag = 251.2 times
7 mag = 631.0 times
8 mag = 1585 times
9 mag = 3981 times
10 mag = 10^4 times
15 mag = 10^6 times
20 mag = 10^8 times

In astronomy we often find ourselves writing numbers that have strings of zeros attached, so we use what is called scientific notation. *In scientific notation, which we have already employed in Chapter 1, we merely count the number of zeros, and write it as a superscript to the number 10. Thus the number 10,000,000,000, a 1 followed by 10 zeros, is written 10^{10}. If the number had been 30,000,000,000, we would have written 3×10^{10}. This method of writing numbers simplifies our writing chores considerably, and helps prevent making mistakes in copying long strings of numbers.*

We can handle numbers much less than one with negative exponents. A minus sign in the exponent of a number means that the number is actually one divided by what the quantity would be if the exponent were positive. Thus $10^{-2} = 1/10^2$. One can count the number of places by which the decimal point has to be moved until it is at the right of the first non-zero digit. Thus, for example, $.000001435 = 1.435 \times 10^{-6}$.

Scientific notation also allows simplification in calculation. The 5 in $(100)^5$ is called the exponent. *When a number with an exponent is itself raised to an exponent, as we say, then one simply multiplies the exponents. Thus, since $100 = 10^2$, we see that $(100)^5 = (10^2)^5$ and by merely multiplying 2×5 to get 10, we have 10^{10}.*

The Student Study Guide *provides practice in the use of scientific notation.*

We have seen how a difference of 5 magnitudes corresponds to a factor of 100 in brightness. What does that indicate about a difference of 1 magnitude? Since an increase of 1 in the magnitude scale corresponds to a decrease in brightness by a certain factor, we need a number that, when multiplied by itself 5 times, will equal 100. This number is just $\sqrt[5]{100}$. Its value is 2.512 . . ., with the dots representing an infinite string of other digits. For many practical purposes, it is sufficient to know that it is nearly 2.5.

Thus a second magnitude star is about 2.512 times fainter than a first magnitude star. A third magnitude star is about 2.512 times fainter than a second magnitude star. Thus, a third magnitude star is approximately $(2.5)^2$ (which is about 6) times fainter than a first magnitude star. (We could easily give more decimal places, writing 6.3 or 6.31, but the additional figures would not be meaningful. We started with only one digit being significant when we wrote down "2nd magnitude"—we say that the number was given to an accuracy of *one significant figure*—and we must be careful not to fool ourselves that we have gained in accuracy in the multiplication process.)

In similar fashion, using a factor of 2.512 for the single magnitudes and a factor of 100 for each group of 5 magnitudes, we can very simply find the ratio of intensities of stars of any brightness.

Example: What is the difference between magnitude -1 and magnitude 6, the ratio between the brightest and faintest stars that are seen with the naked eye?

Answer: The difference is 7 magnitudes, and the factor is thus $(100)(2.5)(2.5)$, which is about 600. The star of -1st magnitude is 600 times brighter than the star of 6th magnitude.

One unfortunate thing about the magnitude scale is that it operates in the opposite sense from a direct measure of brightness. Thus the brighter the star, the lower the magnitude (a second magnitude star is brighter than a third magnitude star), which is certainly a confusing convention. This kind of problem comes up occasionally in an old science like astronomy, for at each stage astronomers have made their new definitions so as to have continuity with the past.

2.2 THE SPECTRUM

It was discovered over three hundred years ago that when ordinary light is passed through a prism, a band of color like the rainbow comes out the other side. Thus "white light" is composed of all the colors of the rainbow.

These colors are always spread out in a certain order, which has traditionally been remembered by the initials of the friendly fellow ROY G. BIV, which stands for Red, Orange, Yellow, Green, Blue, Indigo, Violet. No matter what rainbow you watch, or what prism you use, the order of the colors never changes.

Modern sensibility, though, no longer sees "indigo" as a separate color of the same importance as the others listed.

We can understand why light contains the different colors if we think of light as waves of radiation. These waves are all traveling at the same speed, 3×10^{10} cm/sec (186,000 miles/sec). This speed is the speed that light would have if it were in a vacuum; it is normally called the *speed of light* (even though light normally travels at a slightly lower speed because the light is not usually in a perfect vacuum).

The distance between one crest of the wave to the next or one trough to the next, or in fact between any point on a wave and the similar point on the next wave, is called the *wavelength* (Fig. 2–5). Light of different wavelengths appears as different colors. It is as simple as that. Red light has approximately 1½ times the wavelength of blue light. Yellow light has a wavelength in between the two. Actually there is a continuous distribution of wavelengths, and one color blends subtly into the next.

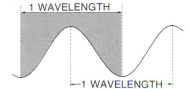

Figure 2–5 The definition of the wavelength of electromagnetic radiation.

The wavelengths of light are very short: just a few hundred-thousandths of a centimeter. Astronomers use a unit of length called an angstrom, named after the Swedish physicist A. J. Ångström. One angstrom (1 Å) is 10^{-8} cm. (Remember that this is 1 divided by 10^8, which is .000 000 01 cm.) The wavelength of violet light is approximately 4000 angstrom units, usually written 4000 angstroms (4000 Å). Yellow light is approximately 6000 Å, and red light is approximately 6500 Å in wavelength.

The human eye is not sensitive to radiation whose wavelength is much shorter than 4000 Å or much longer than 6600 Å, but other devices exist that can measure light at shorter and longer wavelengths. At wavelengths shorter than violet, the radiation is called *ultraviolet*; at wavelengths longer than red, the radiation is called *infrared*. (Note that it is not "infared," a common misspelling; "infra" is a Latin prefix that means "below.")

All these types of radiation—visible or invisible—have much in common. Many of us are familiar with the field of force produced by a magnet— a magnetic field—and the field of force produced by a charged object like a hair comb on a dry day—an electric field. It has been known for over a century that fields that vary rapidly behave in a way that no one would guess from the study of magnets and combs. In fact, light, x-rays, and radio waves are all examples of electric fields and magnetic fields that have become detached from their sources and move rapidly through space. For this reason, these radiations are referred to as *electromagnetic radiation*.

We can draw the entire *electromagnetic spectrum*, often simply called

The strength of an electric field or of a magnetic field can change in very short intervals of time.

Figure 2–6 The electromagnetic spectrum.

the *spectrum* (plural: *spectra*), ranging from radiation of wavelength shorter than 1 Å to radiation of wavelength many meters long and more (Fig. 2–6).

Note that from a scientific point of view there is no real qualitative difference among types of radiation at different wavelengths. They can all be thought of as *electromagnetic waves,* and light waves comprise but one limited range. When an electromagnetic wave has a wavelength of 1 Å, we call it an x-ray. When it has a wavelength of 5000 Å, we call it light. When it has a wavelength of 1 cm (which is 10^8 Å), we call it a radio wave. Of course, there are obvious practical differences in the methods by which we detect x-rays, light, and radio waves, but the principles that govern their existence are the same.

Note also that light occupies only a very small portion of the entire electromagnetic spectrum. It is obvious that the new ability that astronomers have to study parts of the electromagnetic spectrum other than light waves enables us to increase our knowledge of celestial objects manyfold.

Only certain parts of the electromagnetic spectrum come through the

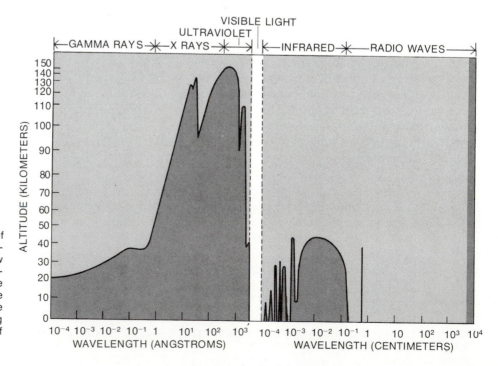

Figure 2–7 Windows of transparency in the terrestrial atmosphere allow only part of the solar spectrum to penetrate to the earth's surface. The curve specifies the altitude where the intensity of arriving radiation is reduced to half its original value.

earth's atmosphere. We call these parts of the spectrum at which the earth's atmosphere is transparent by the name "windows." One window passes what we call "light," and what astronomers more technically call *visible light*, or *the visible*, or *the optical part of the spectrum*. Another "window" falls in the radio part of the spectrum, and modern astronomy uses "radio telescopes" to detect that radiation (Fig. 2–7).

But we of earth are no longer bound to our planet's surface; balloons, rockets, and satellites carry telescopes above the atmosphere to observe in parts of the electromagnetic spectrum that do not reach the earth's surface. By now, we have made at least some observations in each of the named parts of the spectrum (of course, there is really no limit to how short or how long a wavelength can be). It seems strange, in view of the long identification of astronomy with visible observations, to realize that optical astronomers are perhaps in a minority nowadays.

2.3 SPECTRAL LINES

As early as 1666, Isaac Newton showed that sunlight was composed of all the colors of the rainbow. William Wollaston in 1804 and Joseph Fraunhofer in 1811 also studied sunlight as it was dispersed (spread out) into its rainbow of component colors (Fig. 2–8). Wollaston and Fraunhofer were able to see that at certain colors there were gaps that looked like dark lines across the spectrum at those colors. These gaps are thus called *spectral lines*; the continuous radiation in which the gaps appear is called the *continuum* (pl.: *continua*). The dark lines in the spectrum of the sun, and in the spectra of most stars, are mostly gaps representing the diminution of electromagnetic radiation at those particular wavelengths. We interpret them by supposing that originally a continuous band of color was

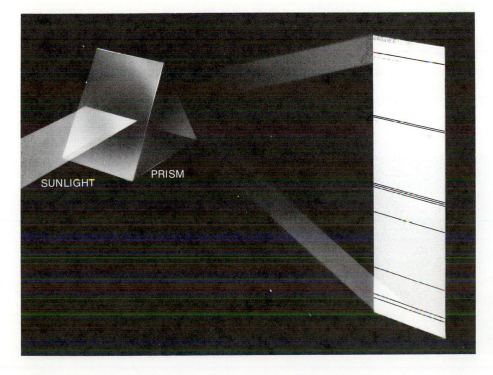

SUNLIGHT PRISM

Figure 2–8 When sunlight is dispersed by a prism, we see not only a continuous spectrum but also dark Fraunhofer lines.

radiated, and that something between the source and us absorbed the radiation at those particular wavelengths. These dark lines are thus called *absorption lines*; they are also known (for the sun, in particular) as *Fraunhofer lines*. It is also possible to have wavelengths at which there is somewhat more radiation than at neighboring wavelengths; these are called *emission lines*. We shall see that the nature of a spectrum, and whether we see emission or absorption lines, can provide considerable information about the nature of the body that was the source of the light. We say that the lines are "in emission" or "in absorption."

It was discovered in laboratories on earth that patterns of spectral lines can be explained as the absorption or emission of energy at particular wavelengths by atoms of chemical elements in gaseous form. If a vapor (the gaseous form) of any specific element is heated, it gives off a characteristic set of **emission** lines. That element, and only that element, has that specific set of lines. If on the other hand, a continuous spectrum radiated by a source of energy at a high temperature is permitted to pass through cooler vapor of any specific element, a set of **absorption** lines (Fig. 2–9) appears in the continuous spectrum at the same characteristic wavelengths of the emission lines of that element. Thus the vapor of an element through which light has passed has subtracted energy from the continuous spectrum at the set of wavelengths that is characteristic of that element (Fig. 2–10 and Color Plate 3). This was discovered by the German chemist Gustav Kirchhoff in 1859. If a continuous spectrum is directed first through the vapor of one element and then through the vapor of a second element, or through a mixture of the two gases, then the absorption spectrum that results will show the characteristic spectral absorption lines of both elements.

The same characteristic patterns of spectral lines are observed in the spectra of stars, and we conclude basically that the same elements are in the outermost layers of the stars. They absorb radiation from a continuous spectrum generated below them in the star, and thus cause the formation of absorption lines. Evidently, since each element has its own characteristic pattern of lines when it is in a certain range of conditions of temperature and density, the absorption spectrum of a star can be used to identify the chemical constituents of the star's atmosphere, in other words, the types of atoms that make up the gaseous outer layers of the star.

Where does the radiation that is absorbed go? A very basic physical law called the *law of conservation of energy* says that it cannot simply dis-

Kirchhoff is pronounced with a "hard" ch (that is, like k): kirk'-hoff. The name of the author of this book is also pronounced with a hard ch: pa'-sa-koff.

Figure 2–9 When we view a source that emits a continuum through a vapor that emits emission lines, we may see absorption lines at wavelengths that correspond to the emission lines.

SPECTRUM

SOURCE

SOURCE

CONTINUOUS RADIATION

LINE RADIATION

SPECTRUM

Figure 2-10 When continuous radiation from a source (*top*) shines through a gas that by itself has an emission line spectrum (*center*) then an absorption line spectrum results (*bottom*) whenever the gas in the middle is cooler than the source of the continuous radiation.

appear. The energy of the radiation may be taken up in a collision of the absorbing atom with another atom. Alternatively, it may be emitted again, sometimes at the same wavelength, but even so it would go off in any random direction and not necessarily toward the observer. Thus, fewer bits of energy proceed straight ahead at that particular wavelength than were originally heading in that direction.

Moreover, if one element is present in relatively great abundance, then its characteristic spectral lines will be especially strong. By observing the star's spectrum, one can therefore tell not only which chemical elements are present in a star but also their relative abundances (Appendix 4).

The relative intensities of lines in the characteristic spectrum of each element depend not only on abundance but also on the temperature of the element at the location at which it is absorbing radiation. Certain lines may be stronger than other lines at a temperature of, say, 6000 to 7000 degrees, but be less intense at either lower or higher temperatures. (We shall study the reasons for this behavior later in this chapter.) Observation of a stellar spectrum can therefore also tell us the temperature and other conditions of the stellar atmosphere.

The method of spectral analysis is a powerful tool that can be used to explore the universe from our vantage point on earth. It has many uses other than those in astronomy. For example, by analyzing the spectrum, one can determine the presence of impurities in an alloy deep inside a blast furnace in a steel plant on earth. One can measure from afar the constituents of lava erupting from a volcano. Sensitive methods developed by astronomers trying to advance our knowledge of the universe often are put to practical uses in fields unrelated to astronomy.

Figure 2–11 The intensity of radiation for different stellar temperatures, according to Planck's law. The wavelength scale is linear, i.e., equal spaces signify equal wavelength intervals. The dotted line shows how the peak of the curves shifts to shorter wavelengths as temperature increases; this is known as Wien's displacement law. Each curve is labeled with the temperature of the gas that produced it; temperature is given in the Kelvin scale (abbreviated K), which is described on page 25.

2.4 THE COLORS OF STARS

Once we understand that the different colors merely correspond to different wavelengths, as stated in Section 2.2, we proceed to measure the colors of the stars. We can graph the intensity of light from a star at a given wavelength on one axis and the wavelength on the other axis. The measurements of intensity can be obtained either by passing the light through a spectrograph (a device that measures the spectrum, as described in Chapter 3), and measuring the intensity at each wavelength, or simply by measuring the amount of light that emerges from a set of filters that pass only certain groups of wavelengths and block others.

When we examine a set of these graphs, we can see that, spectral lines aside, the radiation follows a fairly smooth curve (Fig. 2–11). Closer analysis reveals more complicated structure, which we will discuss later on. The radiation peaks in intensity at a certain wavelength, and decreases in amount more slowly on the long wavelength side of the peak than it does on the short wavelength side. (By convention, in optical astronomy longer wavelengths are always graphed to the right.)

The nature of these graphs can best be understood by first considering the radiation that represents different temperatures. When you first put an iron poker into a fire, it glows faintly red. Then it gets a brighter red, and as it gets hotter becomes yellower and then whiter. (Because the spectrum the poker emits is at its maximum intensity in the middle of the visible spectrum—we say "it peaks" there—the emission decreases to both longer and shorter wavelengths than the wavelength of the peak. The amount of radiation that is emitted at any visible wavelength is not that different from the amount emitted at any other visible wavelength. This makes the radiation seem white.) If the poker could be even hotter without melting, it would turn blue-white. White hot is hotter than red hot (Fig. 2–12).

A set of physical laws governs the sequence of events when material is heated. As the material gets hotter, the peak of the radiation shifts toward the blue. This is known as *Wien's displacement law* (Table 2–1).

Also, as the material gets hotter, the total energy of the radiation grows rapidly. The energy follows the *Stefan-Boltzmann law,* as is shown in Fig. 2–11, whereby the total energy emitted from a source of given area per second grows with the fourth power of the temperature. That is, a gas at 10,000 degrees gives off $2^4 = 2 \times 2 \times 2 \times 2 = 16$ times more energy than does the same amount of gas at 5000 degrees. (We must measure our scale of degrees from absolute zero, the minimum temperature possible. But aside from this it doesn't matter what scale of temperature we use.) We are discussing, for the moment, only the continuous part of the spectrum; it is

Figure 2–12 As a rod or poker (perhaps made of iron or carbon) is heated, it first glows red hot and eventually, when it is hotter, appears white hot.

TABLE 2-1 WIEN'S DISPLACEMENT LAW

Temperature	Wavelength of Peak	Spectral region of wavelength of peak
3 K	.97 mm	infrared-radio
3000 K	9660 Å	infrared
6000 K	4830 Å	green
12,000 K	2415 Å	ultraviolet
24,000 K	1207 Å	ultraviolet

possible to observe both a continuous spectrum and superimposed spectral lines in a single spectrum.

From Wien's displacement law, we can see that the colors of stars in the sky are telling us something about their temperatures. The reddish star Betelgeuse, thus, is a comparatively cool star, while the blue-white star Sirius is very hot. By simply measuring the intensity of a star with a set of filters at different wavelengths, we can assess the stellar temperature (Fig. 2-13).

2.5 PLANCK'S LAW AND BLACK BODIES

These laws of the distribution of energy from a heated gas are much more general than their particular application to the stars. In principle, if a gas is heated up, its continuous emission follows a certain peaked distribution called *Planck's law*, suggested in 1900 by the German physicist Max Planck (Fig. 2-14). Wien's displacement law and the Stefan-Boltzmann law, which had been discovered earlier, can be derived from Planck's law. Real gases may not, and usually don't, follow Planck's law exactly. For example, Planck's law governs only the continuous spectrum; any line emission or absorption is already a deviation from Planck's law.

Because of these difficulties, scientists choose to consider an idealization of the actual case. We call this fictional body a *black body*, since we define it as something that absorbs all radiation that falls on it. We choose the name since black things in our everyday experience do not reflect light. Of course, for a black body we are talking of an idealized black. Ordinary black paint, for example, does not absorb 100 per cent of the light that hits it. Further, something that looks black in visible light might not seem very black at all in the infrared.

An important physical law that governs the radiation from a black body says that anything that is a good absorber is a good emitter too. **The radiation from a black body follows Planck's law.** As a black body is heated, the peak of the radiation shifts towards the short wavelength end of the spectrum (Wien's displacement law) and the total amount of radiation grows rapidly (Stefan-Boltzmann law). The radiation from a black body at a higher temperature is greater at every wavelength than the radiation from a cooler black body, but as the temperature increases, the intensity of radiation at shorter wavelengths increases more rapidly than the intensity at longer wavelengths increases. Note that the curves of black bodies at different temperatures in Figure 2-11 do not cross.

In sum, black bodies are fictional, idealized objects that absorb all the radiation that hits them. They emit radiation according to Planck's law. If the temperature of a black body is sufficiently high, then it can appear quite bright. The atmospheres of stars appear, to a certain extent, as black bodies in that over a broad region of the spectrum that includes the visible, the continuous radiation from stars follows Planck's law fairly well.

Figure 2-13 The top photograph is a view of several stars in blue light, and the bottom photograph is a view of the same stars in red light. Star A appears about the same brightness in both. Star B, a relatively hot star, appears brighter in the blue photograph. Star C, a relatively cool star, appears brighter in the red photograph.

Figure 2-14 Max Planck. Planck's law is illustrated in Figure 2-11.

Figure 2–15 The staff of the Harvard College Observatory circa 1917. Henrietta S. Leavitt, who studied variable stars (see Section 4.4a), is fifth from the left and Annie Jump Cannon, who classified hundreds of thousands of stellar spectra, is fifth from the right.

2.6 SPECTRAL TYPES

In the last decades of the 19th century, spectra of thousands of stars were photographed. There were many differences among spectra from different stars, and classifications of the different types of spectra were developed.

At first, the stellar spectra were classified only by the strengths of certain absorption lines from hydrogen, and were lettered alphabetically: A for stars with the strongest hydrogen lines, B for stars with slightly weaker lines, and so on. These categories are called *spectral types* or *spectral classes* (Fig. 2–16). But it was later realized that the types of

Figure 2–16 Spectral types. Note that most stellar spectra show absorption lines. The roman numeral "V" indicates that the stars are normal ("dwarf") stars, which will be defined in Section 2.1. The numbers following the letters in the left-hand column represent a subdivision of the spectral types into tenths; the step from B8 to B9 is equal to the adjacent step from B9 to A0.

spectra varied primarily because of differing temperatures of the stellar atmospheres. The hydrogen lines were strongest in stars at a certain temperature and were weaker at both higher and lower temperatures. When we now list the spectral types of stars in order of decreasing temperature, they are no longer in alphabetical order. From hottest to coolest, the spectral types are O B A F G K M.

O stars (that is, stars of type O) are the hottest, and the temperatures of M stars are more than ten times lower. Generations of American students and teachers have remembered the spectral types by the mnemonic: Oh, Be A Fine Girl, Kiss Me. The spectral types have been subdivided into 10 subcategories each. For example, the hottest B stars are B0, followed by B1, B2, B3, and so on. Spectral type B9 is followed by spectral type A0. It is fairly easy to tell the spectral type of a star by mere inspection of its spectrum.

We must choose the units in which we will discuss temperature for the rest of this book. In America we have used the Fahrenheit temperature scale, in which water freezes at 32°F and boils at 212°F, a span of 180°F between freezing and boiling. On this scale, absolute zero (the minimum temperature theoretically possible for any material thing) is −459.4°F (Fig. 2–17). This scale is not used in any scientific work.

Most of the rest of the world uses degrees Celsius. The United States is changing over gradually to this system; most scientists already have for their professional work. On the Celsius scale, water freezes at 0°C and boils at 100°C, a span of 100°C. Thus 180 Fahrenheit degrees are equivalent to 100 Celsius degrees; simplifying the ratio shows that 9/5°F = 1°C.

On the Celsius scale, absolute zero is −273°C. For terrestrial purposes, temperatures conveniently range around zero or a few tens. But the freezing and boiling points of water don't have much relevance to the stars. Astronomers choose to use the Kelvin temperature scale, which has its zero point at absolute zero instead of at the freezing point of water. The size of a Kelvin degree (now officially called one kelvin and abbreviated K instead of °K) is the same as the size of a Celsius degree. Thus absolute zero is 0 K, water freezes at +273 K, and water boils at +373 K.

It is simple to change from Celsius degrees to kelvins; simply add 273 to the Celsius temperature. To change from Fahrenheit to Celsius is more complicated. One must first subtract 32°F, to align the freezing points, and then multiply by 5/9 to transform the smaller Fahrenheit degrees into the larger Celsius degrees. The formula is °C = 5/9 (°F − 32).

Stellar temperatures, whether expressed in Fahrenheit, Celsius, or Kelvin degrees, are very high — in the thousands or tens of thousands (Fig. 2–18).

Figure 2–17 Temperature scales.

An easy way to convert from degrees Celsius to degrees Fahrenheit is to take the Celsius number, double it, subtract 10% of the result, and add 32. Example: take 30°C, double it to get 60°, subtract 10% of 60° = 6° from 60° to get 54°, and add 32° to get the answer of 86°F.

Figure 2–18 Stellar temperature scales.

Figure 2–19 A helium atom contains two protons and two neutrons in its nucleus and two orbiting electrons.

An electron volt is one of the several units in which energy can be expressed. It takes 6×10^{11} eV to equal even a small unit of energy called an erg, and 10^9 ergs per second to light a 100 watt light bulb. Obviously, one eV is pretty small.

2.7 THE FORMATION OF SPECTRAL LINES

Spectral lines arise when there is a change in the amount of energy that any given atom has. The amount of energy that an atom can have is governed by the laws of *quantum mechanics,* a field of physics that was developed in the 1920's and whose applications to spectra were further worked out in the 1930's. According to the laws of quantum mechanics, light (and other electromagnetic radiation) has the properties of waves in some circumstances and the properties of particles under other circumstances. One cannot understand this dual set of properties of light intuitively and some of the greatest physicists of the time had difficulty adjusting to the idea.

An atom (Fig. 2–19) can be thought of as consisting of particles in its core, which is called the *nucleus,* surrounded by orbiting particles called *electrons.* Examples of nuclear particles are the *proton,* which carries one unit of positive electric charge, and the *neutron,* which has no electric charge. Electrons, which each have one unit of negative electric charge, are 1/1800 as massive as protons or neutrons.

According to quantum mechanics, an atom can exist only in a specific set of energy states, as opposed to a whole continuum (continuous range) of energy states. An atom can have only discrete values of energy; the energy cannot vary continuously. For example, the energy corresponding to one of the states might be 10.2 electron volts (eV). The next energy state allowable by the quantum mechanical rules might be 12.0 eV. We say that these are *allowed states.* The atom **simply could not have** an energy intermediate between 10.2 and 12.0 eV. In sum, the energy states are discrete; we say that they are *quantized.* We call them *energy levels.*

When an atom drops from a higher energy state to a lower energy state, without colliding with another atom, the difference in energy is sent off as a bundle of radiation. This bundle of energy, a *quantum,* is also called a *photon,* which may be thought of as a particle of electromagnetic radiation. Photons always travel at the speed of light. Each photon has a specific energy, which does not vary.

A link between the particle version of light and the wave version is seen in the equation that relates the energy of the photon with the wavelength it has: $E = hc/\lambda$. In this equation, E is the energy, h is a constant named after Planck, c is the speed of light (in a vacuum, as always), and λ (lambda) is the wavelength. Since λ is in the denominator on the right hand side of the equation, a small λ corresponds to a photon of great energy. When, on the other hand, λ is large, the photon has less energy. An x-ray photon, for example, of wavelength 1 Å, has much more energy than does a photon of visible light of wavelength 5000 Å.

When a gas is heated up so that many of its atoms are in energy states other than the lowest possible, atoms not only are being raised to higher energy states but also spontaneously drop back to their lowest energy levels, emitting photons as they do so. These photons represent energies at certain wavelengths. As there were not necessarily any photons already existing at those wavelengths, the new photons appear, when they correspond to the visible part of the spectrum, as bright wavelengths. These are the **emission** lines (Fig. 2–20).

As we saw, when we allow continuous radiation from a body at some relatively high temperature to pass through a cooler gas, the atoms in the gas can take energy out of the continuous radiation. We now see that the

Figure 2-20 When pho-
tons are emitted, we see
an emission line; a continuum may or may not also
be present. An absorption
line must absorb radiation
from something. Hence an
absorption line necessarily
appears in a continuum.

atoms in the gas are being changed into higher energy states. Less radiation
remains at certain wavelengths to come out of the gas than went in. These
wavelengths are those of the **absorption** lines (Fig. 2–20). In the visible,
these wavelengths appear dark.

Note that emission lines can appear with or without a continuum, since
they are merely the addition of energy to the radiation field. But absorption
lines must be absorbed **from** something, namely the continuum. A normal
stellar spectrum is a continuum with absorption lines. This implies that the
temperature in the outer layers of the star is decreasing with height. (See
Section 2.9 for a discussion of the formation of a normal stellar spectrum.)
As a result of this temperature trend with height, emission lines, which result from the presence of gas hotter than any background continuum, are
rarely found in stellar spectra. Emission lines can appear, however, in
spectra of stars surrounded by hot shells and in some other types of astronomical objects.

We can categorize the energy state of an atom by an amount of energy,
and think of a change from one energy state to another as the result of a
change in energy of the electrons orbiting the nucleus of the atom. (The

Figure 2-21 An atom missing one or more electrons
is *ionized*.

Figure 2–22 As the temperature of a cloud of gas rises, more and more electrons are excited to higher and higher energy levels. The *Boltzmann distribution* tells us how many atoms will be in each of its energy states for a given temperature.

production of spectral lines does not involve changes in the nucleus itself.) If a gas were at a temperature of absolute zero, then electrons in it would be in the lowest possible energy levels. The lowest energy level that an electron can be on is called the *ground state* of that atom. As we add energy to the atom, we can "excite" one or more electrons to higher energy levels; that is, the electrons are pushed to higher energy levels. If we were to add even more energy, some of the electrons would be given not only sufficient energy to be excited but also enough energy to escape entirely from the atom. We then say that the atom is *ionized*, and the remnant with less than its quota of electrons is called an *ion* (Fig. 2–21). As the remnant has more positive charges in its nucleus than it has negative charges on its electrons, its net charge is positive and it is sometimes called a *positive ion*. An atom is *singly ionized* when it has lost one electron, *doubly ionized* when it has lost two electrons, and so on.

Atoms can be excited or ionized either by collisions or by radiation. In a star the former can be the case when the density is sufficiently high that collisions of the atoms with electrons are frequent. When the gas is "heated," this simply means that the constituent atoms are given more energy. Thus in a "hotter" gas, more atoms will be excited or ionized (Fig. 2–22).

2.8   THE HYDROGEN SPECTRUM

Hydrogen emits, and absorbs, a set of spectral lines that fall across the visible spectrum in a distinctive pattern (Fig. 2–23). The strongest line is in the red, the second strongest is in the blue, the third strongest is farther in the blue, and the other lines continue this series, with the spacing between the lines getting less and less.

It is easy to visualize the hydrogen atom in the picture (Fig. 2–24) that Niels Bohr (Fig. 2–25), the Danish physicist, laid out in 1913. In the *Bohr atom* electrons can have orbits of different sizes that correspond to different energy levels. Only certain orbits were allowable, hence the quantiza-

Figure 2–23 The Balmer series, representing transitions down to or up from the second energy state of hydrogen.

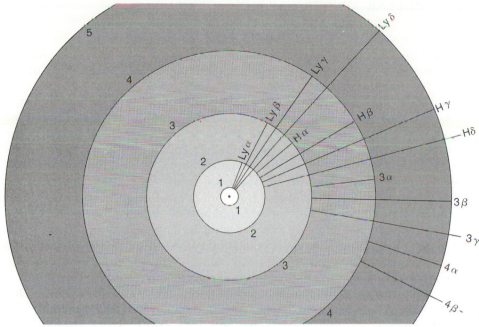

Figure 2–24 The representation of hydrogen energy levels known as the Bohr atom.

tion of energy levels. This simple picture of the atom was superseded by the development of quantum mechanics, but the notion that the energy levels are quantized remains.

The letter *n*, which labels the energy levels and which appears in the formulae for their energies, is called the *principal quantum number*. We call the energy level for n = 1 the *ground level* or *ground state,* as it is the lowest possible energy state. The series of transitions from or to the ground level is called the Lyman series, after the American physicist Theodore Lyman. The lines fall in the ultraviolet, at wavelengths far too short to pass through the earth's atmosphere. Lyman's observations of what we now call the Lyman lines were made in a laboratory in a vacuum tank. At present, we can observe Lyman lines from the sun and the other stars with telescopes aboard earth satellites.

The series with n = 2 as the lowest level is in the visible part of the hydrogen spectrum. This series is called the Balmer series, after the Swiss physicist J. J. Balmer who, in 1885, first described the mathematical formula that predicts the spacing between the lines. For example, the bright red line in the visible part of the hydrogen spectrum, known as the Hα line, arises from the transition from level 3 to level 2. The transition from the n = 4 to the n = 2 level accounts for the Hβ line, etc.

There are many other series of lines, each corresponding to transitions to a given lower level.

Since the higher energy levels have greater energy, a spectral line caused by the transition from a higher level to a lower level yields an emission line. When, on the other hand, continuous radiation falls on cool hydrogen gas, some of the atoms in the gas can be raised to higher energy levels. Absorption lines result in this case. The concept of the Bohr atom shows us why a given line of any element appears at the same wavelength no matter whether it is in emission or in absorption.

Lyman went so far as to climb Mount McKinley in 1915, hoping that from that great altitude there might happen to be a window of transparency that would pass Lyman alpha. Unfortunately, there is no such window, and his photographs of the spectrum were blank in the region where ultraviolet radiation would have appeared.

Figure 2–25 Niels Bohr.

2.9 STELLAR SPECTRA

When we observe the light from a star, we are seeing only the radiation from a thin layer of that star's atmosphere. The interior is hidden within. As we discuss stellar spectra, let us keep this fact in mind. In parts of Chapters 5 to 9, we will be discussing the processes that go on inside stars.

As we consider stars of different spectral types (Fig. 2–16) ranging from the hottest to the coolest, we can observe the effects of temperature on the nature of the spectra. For example, the hottest stars, those of type O, have temperatures that are sufficiently high so that there is enough energy to remove the outermost electron or electrons from most of the atoms. Since most of the hydrogen is ionized, there are relatively few neutral hydrogen atoms left and the hydrogen spectral lines are weak. Helium happens not to be easily ionized—it takes a large amount of energy to do so—yet even helium appears not in its neutral state but in its singly-ionized state in an O star. Temperatures range from 30,000 K to over 60,000 K in O stars. O stars are relatively rare since they have short lifetimes; none of the nearest stars to us are O stars. Only a handful of stars as hot as spectral types O3 or O4 are known at any distance; no O0, O1 or O2 stars have ever been discovered.

B stars are somewhat cooler, 10,000 K to 30,000 K. The hydrogen lines are stronger than they were in the O stars, and lines of neutral helium instead of ionized helium are present. Rigel and Spica are familiar B stars.

A stars are cooler still, 7500 K to 10,000 K. The lines of hydrogen are strongest in this spectral type. Lines of singly ionized elements like magnesium and calcium begin to appear. (All elements other than hydrogen or helium are known as *metals* for this purpose.) Sirius, Deneb and Vega are among the bright A stars in the sky. O, B, and A stars are all bluish in color.

F stars have temperatures of 6000 K to 7500 K. Hydrogen lines are weaker in F stars than they are in A stars, but the lines of singly ionized calcium are stronger. Singly ionized calcium has a pair of lines that are particularly conspicuous; they are easy to pick out and recognize in the spectrum. These lines are called H and K. Canopus, the second brightest star in the sky (not visible from the latitudes of most of the United States), and Polaris are prominent F stars.

G stars, the spectral type of our sun, are 5000 K to 6000 K. They are yellowish in color, as the peak in their spectra falls in the yellow to yellow/green part of the spectrum. The hydrogen lines are visible, but the H and K lines are the strongest in the spectrum. The H and K lines are stronger in G stars than they are in any other spectral type.

K stars are relatively cool, only 3500 K to 5000 K. The spectrum is covered with many lines from neutral metals, a strong contrast to the spectra of the hottest stars, which show few spectral lines. Arcturus and Aldebaran, both visible as reddish points in the sky, are K stars.

Box 2.1 Do not confuse:
K stars: a spectral class
K: a kelvin, a unit of temperature
K: a strong line in many stellar spectra
K: the element potassium

Figure 2–26 Neutral hydrogen, the hydrogen ion, and the negative hydrogen ion.

M stars are cooler yet, with temperatures less than 3500 K. Their atmospheres are so cool that molecules can exist without being torn apart, and the spectrum shows many molecular lines. (Hotter stars do not usually show molecular lines.) Lines from the molecule titanium oxide are particularly numerous. Betelgeuse is an example of such a reddish star of spectral class M.

We have listed the ordinary spectral types of stars, all of which show absorption spectra, that is, a continuum crossed with dark lines. There are many unusual stellar spectra, though, and additional spectral types.

Where does the continuous spectrum come from? In order for there to be an absorption line, there must be some continuous radiation to be absorbed. The particular mechanism that causes the continuous emission itself is different for different spectral types. In the sun, the continuous emission is caused by a strange type of ion: the negative hydrogen ion (Fig. 2–26). When second electrons join neutral hydrogen atoms, continuous emission results, unlike the spectral lines that are formed when electrons bound in an atom jump between two levels.

2.10 THE HERTZSPRUNG-RUSSELL DIAGRAM

In about 1910, Ejnar Hertzsprung in Denmark and Henry Norris Russell at Princeton University in the United States independently plotted a new kind of graph (Fig. 2–27). On one axis, the abscissa (x-axis), each graphed a measure of the temperature of stars. On the other axis, the ordinate (y-axis), each graphed a measure of the brightness of stars. They found that all the points that they plotted fell in limited regions of the graph rather than being widely distributed over the graph.

There are several possible ways to plot the temperature. One doesn't need to plot the temperature itself. One can plot in its place the spectral type, for example, or even some measure of the color of a star.

To plot the brightness, there are also several possible ways. The simplest is merely to plot the magnitudes of stars, with the brightest stars (i.e., those with the most negative magnitudes) on the top. But remember that a star can appear bright either by really being intrinsically bright or alternatively by being very close to us. For the moment, we can get around this problem by plotting only stars that are at the same distance away from us. Later on we shall see what has to be done in order to determine the distances to particular stars.

How do we find a group of stars at the same distance? Luckily, there are clusters of stars in the sky (described in more detail in Chapter 4) that are really clusters of stars in space at approximately the same distance away from us. They are not stars that are, even though they are in the same direction, at different distances from us. We do not have to know

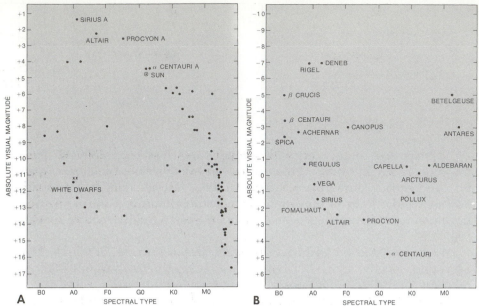

Figure 2–27 Hertzsprung-Russell diagrams (*A*) for the nearest stars in the sky and (*B*) for the brightest stars in the sky. The brightness scale is given in *absolute magnitude,* which is the magnitude the stars would have if they are all placed at the same standard distance. (This distance is 10 parsecs, equivalent to about 32.6 light years; parsecs will be discussed in the next section.) Because the effect of distance has been removed, the intrinsic properties of the stars can be compared directly on such a diagram.

Figure 2–28 A physical grouping of stars in space, the globular cluster NGC 5897.

what the distance is to plot a diagram of the type plotted by Hertzsprung; all we have to know is that the stars are really clustering in space (Fig. 2–28).

A plot of temperature versus brightness is known as a *Hertzsprung-Russell diagram,* or simply as an *H-R diagram.* Note that since H-R diagrams were sometimes originally plotted by spectral type, from O to M, the hottest stars are on the left side of the graph. Thus temperature increases from right to left. Also, since the brightest stars are on the top, magnitude decreases toward the top. In some sense, thus, both axes are plotted backwards from the way a reasonable person would choose to do it if there were no historical reasons for doing it otherwise.

When plotted on a Hertzsprung-Russell diagram, the stars mainly lie on a diagonal band from upper left to lower right (Fig. 2–29). Thus the hottest stars are brighter than the cooler stars. Most stars fall very close to this band, which is called the *main sequence.* Stars on the main sequence are called *dwarf stars,* or *dwarfs.* There is nothing strange about dwarfs: they are the normal kind of stars. The sun is a type G dwarf. Some dwarfs are quite large and bright; the word "dwarf" is used only in the sense that these stars are not a larger brighter kind of star that we will define below as giant.

Some stars lie above and to the right of the main sequence. That is, for a given spectral type the star is intrinsically brighter than a main sequence star. These stars are called *giants,* because their luminosities are large compared to dwarfs of their spectral type. Some stars, like Betelgeuse, are even brighter than normal giants, and are called *supergiants.* Since two stars of the same spectral type have the same temperature and, according to the Stefan-Boltzmann law, the same amount of emission from each area of their surfaces, the brighter star must be bigger than the fainter star of the same spectral type. Stars in a class of faint hot objects, called *white dwarfs,* are

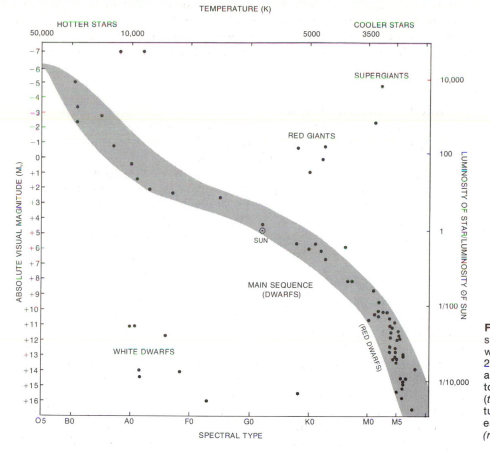

TEMPERATURE (K)

Figure 2–29 The Hertz-sprung-Russell diagram, with the stars of Figure 2–27 included. The spectral axis *(bottom)* is equivalent to the temperature axis *(top)*. The absolute magnitude axis *(left)* is equivalent to the luminosity axis *(right)*.

located below and to the left of the main sequence. They are smaller and fainter than main sequence stars of the same spectral type.

The use of the Hertzsprung-Russell diagram to link the spectrum of a star with its brightness is a very important tool for stellar astronomers. We shall return to further discussion of the H-R diagram after we discuss ways of measuring the distances to stars.

2.11 STELLAR DISTANCES

We can tell a lot by looking at a star, or by examining its radiation through a spectrograph. But such observations do not tell us directly how far away the star is. Since all the stars are but points of light in the sky even when observed through the biggest telescopes, we have no reference scale to give us their distances (Fig. 2–30).

The best way to find the distance to a star is to use the principle that is also used in the rangefinders in cameras. Sight toward the star from different locations, and see how the direction toward the star changes. You can see this effect by holding out your thumb at arm's length. Examine it first with one eye closed and then with the other eye closed. Your thumb seems to change in position as projected against a distant background; it appears to move across the background by a certain angle, about 5°. This is because your eyes are a few centimeters apart from each other, and when you look through each eye you are looking from a different point of view.

Figure 2–30 "Pardon me, I thought you were further away." Without clues to indicate distance, we cannot properly estimate size. (Reproduced by special permission of PLAYBOY Magazine; copyright © 1971 by Playboy.)

Box 2.2   Angular Measure

There are 360° to a full circle, and 90° to a right angle. Thus 1° (1 degree or, to prevent confusion with degrees of temperature, 1 degree of arc) is 1/360th of a full circle.

Each degree is divided into 60 minutes (60 minutes of arc, abbreviated 60 arc min). Each minute of arc, in turn, is divided into 60 seconds (60 seconds of arc, abbreviated to 60 arc sec).

Thus there are 60 arc sec in 1 arc min, and 3600 arc sec in 1°.

The moon is about ½° across, which is 30 arc min or 1800 arc sec.

---

1 astronomical unit (A.U.) is the average distance from the earth to the sun.

The recent development of automatic measuring engines has enabled trigonometric parallaxes to be measured for stars five times more distant than could previously be studied—trigonometric parallaxes can now be measured for stars as far as 300 light years from the sun and as faint as 16th magnitude.

Now hold your thumb up closer to your face, just a few centimeters away. Note that the angle your thumb seems to jump across the background as you look through first one eye and then through the other is greater than it was before. Just the same effect can be used for finding the distances to nearby stars. The nearer the star is to us, the farther it will appear to move across the background of distant stars. To maximize this effect, we want to observe from two places that are separated from each other as much as possible. For us on earth, that turns out to be the position of the earth at intervals of six months. In that six-month period, the earth moves to the opposite side of the sun from where it began. In principle, we observe the position of a star in the sky very precisely, with respect to the background stars, and six months later, when the earth has moved halfway around the sun, we repeat the observation. The straight line joining the points from which we observe is called the *baseline*. In this case we have observed with a baseline twice as long as the distance from the earth to the sun, 2 A.U. in length. We can think of differences in position in the sky as an angle whose vertex is at the earth. The angle that a star seems to move between two observations made from the ends of a baseline of 1 A.U. (half the maximum possible baseline) is called the *parallax* of the star (Fig. 2–31). From the parallax, we can use simple trigonometry to calculate the distance to the star. This is called the method of *trigonometric parallax*.

The basic limitation to the method of trigonometric parallax is that the farther away the star is, the smaller is its parallax. It turns out that this method can only be used for about eight thousand of the stars closest to us; most other stars have parallaxes too small for us to measure. Even Proxima Centauri, the star nearest to us, has a parallax of only 3/4 arc sec. This is the angle subtended (Fig. 2–32) by a dime at a distance of 2 km. Clearly we must find other methods to measure the distances to other stars.

Since parallax measures have such an important place in the history of astronomy, a unit of distance was defined that relates to these measurements. If we were outside the solar system, and looked back at the earth and the sun, we would see a certain angle between them as measured from our vantage point. From twice as far away as Pluto, for example, when the earth and the sun were separated by the maximum amount, they would be approximately 1° apart.

As we go farther and farther away from the solar system, 1 A.U. subtends a smaller and smaller angle. When we are approximately 60 times farther away, 1 A.U. will subtend only 1 arc minute, since 60 arc min = 1 arc degree. When we are approximately 60 times farther away again, 1 A.U. will subtend only 1 arc second, since 60 arc sec = 1 arc minute. Note that the angle subtended by the astronomical unit is a measure of the distance we have gone (Fig. 2–32).

When we are at the distance such that 1 A.U. subtends only 1 arc sec, we call that distance *1 parsec* (Fig. 2–33). A star that is 1 parsec from the sun has a **par**allax of one arc **sec**. 1 parsec is a long distance. Most astronomers tend to use parsecs instead of light years, the distance that light travels in a year (= 9.5×10^{12} km), when talking about stellar distances. It takes light 3.26 years to travel 1 parsec; thus 1 parsec = 3.26 light years. Note that parsecs and light years are **distances**, just like kilometers or miles, even though their names contain references to their definitions in terms of angles or time.

2.12 ABSOLUTE MAGNITUDES

We have thus far defined only the *apparent magnitudes* of stars, how bright the stars **appear** to us. However, if we chose a standard distance, we could consider how bright all stars would appear if they were at that standard distance. The standard distance that we choose is 10 parsecs. We define the *absolute magnitude* of a star to be the magnitude that the star would appear to have if it were at a distance of 10 parsecs. We normally write absolute magnitude with a capital *M*, and apparent magnitude with a small *m*.

Thus if a star happens to be exactly 10 parsecs away from us, its absolute magnitude is exactly the same as its apparent magnitude. If we were to take a star that is farther than 10 parsecs away from us, though, and moved it to be 10 parsecs away, then it would appear brighter to us than it does at its actual position. Since the star would be brighter, its absolute magnitude would be a lower number (for example, 2 rather than 6) than its apparent magnitude.

On the other hand, if we were to take a star that is closer to us than 10 parsecs, and moved it to 10 parsecs away, it would be farther away and therefore fainter. Its absolute magnitude would be higher (more positive, for example, 10 instead of 6) than its apparent magnitude.

Figure 3–31 The nearer stars seem to be slightly displaced with respect to the farther stars when viewed from different locations in the earth's orbit. We call the average separation of the earth and the sun, measured in linear units like kilometers or miles, 1 Astronomical Unit (1 A.U.). 1 A.U. = 150,000,000 km (93,000,000 miles). The maximum baseline possible, 2 A. U., corresponds to observations made six months apart.

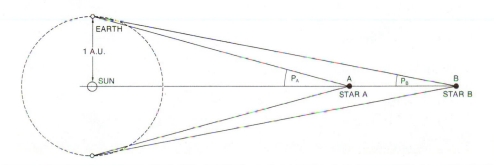

Figure 2–32 Star B is farther from the sun than star A and thus has a smaller parallax. The parallax angles are marked P_A and P_B. To *subtend* is to take up an angle; as seen from point A, 1 A.U. subtends an angle P_A.

1 A.U.

1 PARSEC

1 ARC SEC

Figure 2–33 A *parsec* is the distance at which 1 A.U. subtends an angle of 1 arc sec. One parsec is approximately equal to 3.26 light years. The drawing is not to scale. 1 arc sec is actually a tiny angle.

To assess just how much brighter or fainter the star would appear, we must realize that the intensity of light from a star follows the *inverse square law* (Fig. 2–34). That is, the intensity of a star varies inversely with the square of the distance of the star from us. (This law holds for all point sources of radiation, that is, sources that appear as points without length or breadth.) If we could move a star twice as far away from us, it would appear four times fainter. If we could move a star 9 times as far away from us, it would grow 81 times fainter. Of course, we are not physically moving stars (we would burn our shoulders while pushing), but merely considering how they would appear at different distances. In all this, we are assuming that there is no matter in space between the stars to absorb light. Unfortunately, this is not always a good assumption (see Chapter 20).

For many purposes, it is sufficient just to understand how the inverse square law affects brightness. Astronomers, however, normally speak in terms of magnitudes. Remember that a difference of one magnitude is a factor of approximately 2.5 in brightness. Thus, if a star is 2.5 times fainter (that is, gives off 2.5 times less light) than another star, it is 1 magnitude fainter. If it is 3 times fainter, it is slightly more than 1 magnitude fainter. If it is 6 times fainter, it is slightly less than 2 magnitudes fainter.

Example: A star is 20 parsecs away from us, and its apparent magnitude is +4. What is its absolute magnitude?

Answer: If the star were moved to the standard distance of 10 parsecs away, it would be twice as close and, therefore, by the inverse-square law, would appear four times brighter. Since 2.5 times is one magnitude, and $(2.5)^2 = 6.25$ is two magnitudes, it would be approximately 1½ magnitudes brighter. Since its apparent magnitude is 4, its absolute magnitude would be $4 - 1½ = 2½$.

Astronomers have a formula they can use to carry out this calculation, but the formula merely does numerically what we have just carried out

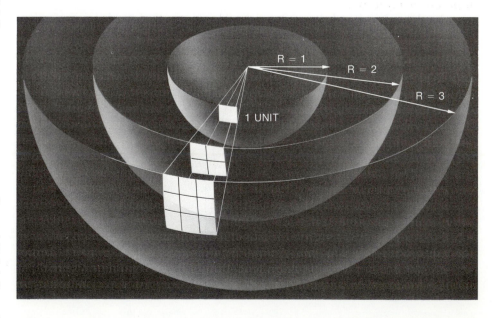

R = 1   R = 2   R = 3

1 UNIT

Figure 2–34 The inverse square law. Radiation passing through a sphere twice as far away as another sphere has spread out so that it covers $2^2 = 4$ times the area; n times farther away it covers n^2 times the area.

logically. The formula is $m - M = 5 \log_{10} r/10$, where m is the apparent magnitude, M is the absolute magnitude, and r is the distance in parsecs. Note that if $r = 10$, then $r/10 = 1$, $\log 1 = 0$, and $m = M$. If $r = 20$, as in our example above, $m - M = 5 \log 2 = 5 \times 0.3 = 1.5$ magnitudes. Be careful not to get carried away using the formula without understanding the point of the manipulations.

2.13 SPECTROSCOPIC PARALLAX

When we take a group of stars whose distances we can measure directly, by some method like that of trigonometric parallax, we can plot a Hertzsprung-Russell diagram for all these stars. We have thus established a "standard" Hertzsprung-Russell diagram, and from this standard diagram can read off the absolute magnitude that corresponds to a given spectral type.

We can apply the standard Hertzsprung-Russell diagram to find the distance to any star whose spectrum we can observe, if we know that the star is on the main sequence. We first find the absolute magnitude that corresponds to its spectral type. We must also know the apparent magnitude of the star, but we can get that easily and directly by simply observing it. Once we know both the apparent magnitude and the absolute magnitude, we have merely to figure out how far the star has to be from the standard distance of 10 parsecs to account for the difference $m - M$.

Example: We see a G2 star on the main sequence. Its apparent magnitude is $+8$. How far away is it?

Answer: From the H-R diagram, we see that $M = +5$. The star appears fainter than it would be if it were at 10 parsecs. It is approximately $(2.5)^3 = 6 \times 2.5 = 15$ times fainter. (More accuracy in carrying out this calculation would not be helpful.) By the inverse square law, this means that it is approximately 4 times farther away (exactly 4 times would be a factor of $4^2 = 16$ in brightness). Thus the star is 4 times 10 parsecs, or 40 parsecs, away.

We are measuring a distance, and not actually a parallax, but by analogy with the method of trigonometric parallax for finding distance, the method using the H-R diagram is called finding the *spectroscopic parallax.*

2.14 THE DOPPLER EFFECT

The Doppler effect is one of the most important tools that astronomers can use to understand the universe. This is because without having to measure the distance to an object they can use the Doppler effect to determine its *radial velocity,* its velocity in the direction toward or away from us (the radius), measured with respect to our own velocity. Thus if we were moving at the same velocity as a source of radiation or sound, it would have no net velocity with respect to us, and no Doppler shift would be observed. The Doppler effect in sound is familiar to most of us, and its analogue in electromagnetic radiation, including light, is very similar.

A   B   C

MOVING EMITTER

STATIONARY EMITTER

Figure 2–35 An object emits waves of radiation that can be represented by a sphere centered on the object and expanding. In the first 3 parts of the drawing, the emitter is moving in the direction of the arrow. In case A, we see that the wave emitted when the emitter was at point 1 is a sphere (labeled S_1) around point 1. In case B, some time later, sphere S_1 has continued to grow around point 1 (even though the emitter is no longer there), and we also see sphere S_2, which was emitted when the emitter had moved to point 2. Sphere S_2 is thus centered on point 2. In C, still later, yet a third wave has been emitted, S_3, this time centered on point 3, while spheres S_1 and S_2 have continued to expand.

Contrast the case at the extreme right, in which the emitter does not move, and so all waves are centered around the same point.

For the case of the moving emitter, observers (left) who are being approached by the emitting source see the waves compressed. The wavelengths are shorter, and so the spectrum is blueshifted. Observers (right) from whom the source is receding detect a redshift. We call this the Doppler effect.

2.14a *Explanation of the Doppler Effect*

Figure 2–36 The Doppler effect. In each pair of spectra, the position of the spectral line in the laboratory is shown on top and the position observed in the spectrum of the star is shown below it. Lines from approaching stars appear blueshifted, lines from receding stars appear redshifted, and lines from stars that are moving transverse to us are not shifted because at that particular moment the star has no component of velocity toward or away from us.

In sound, you may be familiar with the change of pitch of a train whistle, or the whine of a jet engine, as the train or plane first approaches you and then passes you and begins to recede. The object that is emitting the sound waves is approaching you at first. By the time it emits a second wave, it has moved closer to you than it had been when it emitted the first wave. Thus the waves pass more frequently than they would if the source were not moving. The wavelengths seem compressed, and the pitch is higher. When the emitting source passes you, the wavelengths become stretched, and the pitch of the sound is lower (Fig. 2–35).

With light waves, the physical mechanism is not as straightforward as is the stretching or compressing of the wavelengths of sound, but the effect is the same. As a body emitting light, or other electromagnetic radiation, approaches you, the wavelengths become slightly shorter than they would

NO SHIFT

REST WAVELENGTHS
STAR SPECTRUM

BLUE   RED

STAR A
SUN

STAR C MOVING TOWARD

STAR B MOVING AWAY

REST WAVELENGTHS
STAR SPECTRUM

BLUE   RED

BLUESHIFT

REST WAVELENGTHS
STAR SPECTRUM

BLUE   RED

REDSHIFT

Figure 2-37 Two photographs taken 11 months apart. In making this combined print, one of the negatives was shifted slightly. The proper motion of Barnard's star is clearly visible.

be if the body were at rest. Visible radiation is thus shifted slightly in the direction of the blue (it doesn't actually have to become blue, only be shifted in that direction). We say that the radiation is *blueshifted*. Conversely, when the emitting object is receding, the radiation is said to be *redshifted* (Fig. 2–36). We generalize these terms to types of radiation other than light, and say that radiation is blueshifted whenever it changes to shorter wavelengths, and redshifted whenever it changes to longer wavelengths.

2.15   STELLAR MOTIONS

On the whole, the network of stars in the sky is fixed. But radial velocities can be measured for all stars, and some of the stars are seen to move slightly across the sky with respect to the more distant stars. The actual velocity of a star in 3-dimensional space, with respect to the velocity of the sun, is called its *space velocity*.

We usually deal separately with the velocity of a star toward or away from us (the radial velocity), and the angular velocity across the sky, which is called the *proper motion*. These motions are perpendicular to each other.

Proper motion is ordinarily very small, and is accordingly very difficult to measure. Usually we must compare the positions of stars taken at intervals of decades to detect a sufficient amount of proper motion to allow a measurement of any accuracy to be made. Only a few hundred stars have proper motions as large as 1 arc sec per year; proper motions of other stars are sometimes measured in seconds of arc per century!

The star in the sky with the largest proper motion was discovered in 1916 by E. E. Barnard, and is known as Barnard's star (Fig. 2–37). It moves across the sky by 10¼ arc sec per year. Some call it "Barnard's runaway star" because of this proper motion, which, though the highest for any known star, is still minuscule. Barnard's star is seen to move so rapidly across the sky because it is relatively close to us. Only 1.8 parsecs away, it is the nearest star to us after the sun, Proxima Centauri and the Alpha Centauri double star. Since the moon appears half a degree across, Barnard's star moves the equivalent of the diameter of the moon across the sky in only 180 years.

SUMMARY AND OUTLINE OF CHAPTER 2

Aims: To learn fundamentals and basic definitions pertaining to stars and spectroscopy

Cause of twinkling

Magnitude: how to convert from magnitudes to factors of brightness (Section 2.1)

The spectrum: absorption lines, emission lines, the continuum (Sections 2.2 and 2.3)

Colors of stars (Section 2.4)

Radiation laws: Planck's law, Wien's displacement law, Stefan-Boltzmann law (Section 2.5)

Spectral types (Section 2.6)

Conversion of units, temperature scales (Section 2.6)

Definition of ion (Section 2.7)

Pattern of a hydrogen spectrum (Section 2.8)

Stellar spectra (Section 2.9)

The Hertzsprung-Russell diagram (Section 2.10)

Parallax, parsec (Section 2.11)

Inverse-square law, absolute magnitude (Section 2.12)

Spectroscopic parallax (Section 2.13)

Doppler effect, use in determining stellar motions, proper motion (Sections 2.14 and 2.15)

QUESTIONS

1. Venus can be as bright as magnitude -4. Betelgeuse is a first magnitude star (m = +1). How many times brighter is Venus at its brightest than Betelgeuse?

2. Star A has magnitude +12. Star B appears 10,000 times brighter. What is the magnitude of star B? Star C appears 10,000 times fainter than star A. What is its magnitude?

3. Star A has magnitude +10. The magnitude of star B is +8 and of star C is +3. How much brighter does star B appear than star A? How much brighter does C appear than A? How much brighter does C appear than B?

4. Why do the stars twinkle?

5. How much brighter is a -1 magnitude star than a +1 magnitude star?

6. What are the colors of the rainbow? What are the parts of the electromagnetic spectrum?

7. The sun's spectrum peaks at 5600 Å. Would the spectrum of a star whose temperature is twice that of the sun peak at a longer or a shorter wavelength? How much more energy than a square centimeter of the sun would a square centimeter of the star give off?

8. One black body peaks at 2000 Å. Another peaks at 10,000 Å. Which gives out more radiation at 2000 Å? Which gives out more radiation at 10,000 Å? What is the ratio of the total radiation given off by the two bodies?

9. Which contains more information, Wien's displacement law or Planck's law? Explain.

10. What is the ratio of energy output for a bit of surface of an average O star and a same-sized bit of surface of the sun?

11. Star A appears to have the same brightness through a red and through a blue filter. Star B appears brighter in the red than in the blue. Star C appears brighter in the blue than in the red. Rank these stars in order of increasing temperature.

12. What is the difference between the continuum and an absorption line? The continuum and an emission line? Draw a continuum with absorption lines. Can you draw absorption lines without a continuum? Can you draw emission lines without a continuum?

13. Are emission, absorption, or continuous spectra given off by (a) an ordinary incandescent electric bulb, (b) a fluorescent lamp, (c) a neon sign?

14. You are driving a car, with the speedometer showing that you are going 80 km/hr. The person next to you asks why you are only going 75 km/hr. Explain why you each saw different values on the speedometer.

15. Would the parallax of a nearby star be larger or smaller than the parallax of a more distant star? Explain.

16. A star has an observed parallax of 0.1 arc second. Another star has a parallax of 0.02 arc sec. Which star is farther away?

17. Estimate the farthest distance for which you can detect parallax by alternately blinking your eyes. To what angle does this correspond?

18. Must we wait a full 6 months to make the two parts of a parallax measurement? Explain.

19. What is the significance of the existence of the main sequence?

20. What factors determine spectral type? What instrument would you use to determine the spectral type of a star?

21. Why is the Balmer series the most commonly observed spectral series of atomic hydrogen?

22. Two stars have the same apparent magnitude and are the same spectral type. One is twice as far away as the other. What is the relative size of the two stars?

23. Two stars have the same absolute magnitude. One is ten times farther away than the other. What is the difference in apparent magnitudes?

24. A star has apparent magnitude of +4, and is 100 parsecs away from the sun. If it is a main sequence star, what is its spectral type? (Hint: refer to Fig. 2–29.)

25. A star is 20 parsecs from the sun and has apparent magnitude +2. What is its absolute magnitude?

26. A star has apparent magnitude +9 and absolute magnitude +4. How far away is it?

27. List spectral types of stars in order of strength, from strongest to weakest, of (a) hydrogen lines, (b) ionized calcium lines, (c) titanium oxide lines.

28. If a star is moving away from the earth at very high speed, will the star have a continuous spectrum that appears hotter or cooler than it would if the star were at rest? Explain.

29. If we measure the Doppler shift of a star's spectral lines, how can we calculate that star's velocity with respect to the local standard of rest? What other data do we need?

30. (a) What does the proper motion of Barnard's star indicate about its distance from us?
    (b) What would Barnard's star's proper motion be if it were twice as far away from us as it is?

3

Observing the Stars

If you forced a group of astronomers to choose one and only one instrument to be marooned with on a desert island, they would probably choose — a computer. The notion that astronomers spend most of their time at telescopes is far from the case in modern times. Still, telescopes have been and continue to be very important to the development of astronomy, and in this chapter we will discuss telescopes and other observational devices.

Even the question of how to define a telescope has no simple answer, for a "telescope" to observe gamma rays may be a package of layers of plastic flown into the atmosphere aboard a balloon, and a radio telescope may be a large number of small aerials strewn over acres of landscape. We will begin, nevertheless, by discussing telescopes of the traditional type, which observe the radiation in the visible part of the spectrum. They are important in and of themselves because of the many things we have learned over the years by studying visible radiation. Also, the types of telescopes most used by students and by amateur astronomers (some of whom are quite professional in their approach to the subject) are most likely to be optical. Further, the principles of focusing and detecting electromagnetic radiation that were originally developed through optical observations have widespread use throughout the spectrum.

3.1 THE PURPOSE OF A TELESCOPE

Contrary to popular belief, the most important purpose for which most optical telescopes are used is usually to gather light. (For the next few sections I shall drop the qualifying word "optical.") True, telescopes can be used to magnify as well, but for the most part astronomers are interested in observing fainter and fainter objects and so must collect more light to make these objects detectable. There are certain cases where magnification is important — such as for observations of the sun (which will be discussed in

The 4-meter Mayall telescope at the Kitt Peak National Observatory.

Chapter 5) or for observations of the planets—but stars appear as mere points of light no matter how much magnification is applied.

When we look at the sky with our naked eyes, there are several limitations that come into play in addition to that of spectral selectivity. One limitation is that we can see only the light that passes through an opening of a certain diameter—the pupils of our eyes. In the dark, our pupils dilate so that as much light as possible can enter, but the apertures are still only a few millimeters across. A second limitation is one of time. Our brains distinguish a new image several times a second, and so we are unable to store faint images for a long time to accumulate a brighter image. Astronomers overcome both these limitations with the combination of a telescope to gather light and a recording device, such as a photographic plate, to store the light. Other equipment may also be used, such as a spectrograph to analyze the spectral content of the light, or even simple filters.

A further advantage of a telescope over the eye, or of a larger telescope over a smaller telescope, is that of *resolution*, the ability to distinguish finer details in an image. For example, a telescope can distinguish the two components of a double star from each other in many cases where the unaided eye is unable to do so. If a telescope with a collecting area ten centimeters across can resolve double stars that are separated by a certain angle, 1 arc sec, a telescope twenty centimeters across, twice the diameter, can resolve stars that are separated by half that angle, ½ arc sec. In principle, *resolution is inversely proportional to the diameter of the telescope*, that is, to the diameter of the part of the telescope that is collecting the light.

For the larger telescopes the best resolution that can be achieved is limited to about 1 arc sec on an average good night by the turbulence in the earth's atmosphere rather than by the telescope size. Thus increasing the diameter of a telescope above a certain size no longer improves the resolution, even on the best of nights, though the advantages in terms of light-gathering power remain. A telescope is capable of considerably better resolution than the eye, which is limited to a resolution of 1 arc min.

We will begin by considering the telescopes themselves, and then go on to consider the equally important devices that are used in conjunction with the telescopes.

Actually, resolution depends not only on the aperture, as explained at right, but also on the wavelength of radiation being observed; the figures given in the text are for light of approximately 5000 Å. As the wavelength of radiation doubles, resolution is halved. For example, a telescope that can resolve two 5000 Å light sources that are 1 arc sec apart could only resolve two 10,000 Å sources if they were 2 arc secs apart. The limit of resolution for visible light, called Dawes' *limit, is approximately $2 \times 10^{-3} \lambda/d$ arc sec, when λ is in Å and d is in cm.*

3.2 REFRACTING TELESCOPES

Refraction is the bending of light (or other electromagnetic radiation) when the light passes from one medium (e.g., transparent object, air, interstellar space) into another. A lens uses the property of refraction to focus light, and a telescope that has a lens as its major element is called a *refracting telescope.*

A suitably curved piece of glass can be made so that all the light that travels through it is bent, bringing all the rays of each given wavelength to a focus. Such a curved piece of glass is called a lens. The lens in your eye does a similar thing to image objects of the world onto your retina. An eyeglass lens helps the lens in your eye accomplish this focusing task.

The distance of the image from the lens on the side away from the object being observed depends in part on the distance of the object from the front side of the lens. Astronomical objects are all so far away that they are, for the purpose of forming images, as though they were infinitely far away.

We say that they are "at infinity." The images of objects at infinity fall at a distance called the *focal length* behind the lens (Fig. 3–1).

The area of which an image is formed is called the *field of view*. A simple lens can only be designed to focus light from directly in front of it or from regions not too far to the side. Objects that are at angles far from the center of the field of view are not focused perfectly.

The technique of using a lens to focus faraway objects was, we think, developed in Holland in the first decade of the 17th century. It is not clear how Galileo heard of the process, but it is known that Galileo quickly bought or ground a lens and made a simple telescope that he demonstrated in 1610 to the Senate in Venice (Fig. 3–2). He put this small lens and another, smaller lens at opposite ends of a tube. The second lens, called the *eyepiece*, is used to examine and to magnify the image made by the first lens, called the *objective*. Galileo's telescope was only a few centimeters in diameter, but it magnified enough to impress the nobles of Venice, who had assembled to see the new invention.

Galileo turned his simple telescope on the heavens, and what he saw revolutionized not only astronomy but also much of seventeenth century thought. He discovered, for example, that the sun had spots—blemishes—on it, that Jupiter had satellites of its own, and that Venus had phases. We shall be discussing the fundamental importance of these discoveries later on.

Larger telescopes were made, and larger still. Not only did the telescopes magnify the image but they also collected light from a larger area than the area of the pupil of a person's eye, and thus allowed objects to be seen that are too faint for the naked, unaided eye.

There are many problems in making refracting telescopes. For one thing, lenses suffer from *chromatic aberration*, the effect whereby different colors are focused at different points (Fig. 3–3). (The speed of light in a medium—in glass, for example—depends on the wavelength of the light, so not all wavelengths can be brought to a focus at the same point. Ingenious methods of making "compound" lenses of different glasses that have slightly differing properties have succeeded in reducing this problem, but the residual chromatic aberration is a fundamental problem for the use of refracting telescopes.)

The largest refracting telescope in the world has a lens 1 meter (40 inches) across. It is at the Yerkes Observatory in Williams Bay, Wisconsin (Fig. 3–5).

With lenses that large, physical problems arise. It has been difficult to get a pure piece of glass sufficiently free of internal bubbles and sufficiently homogeneous throughout its volume to allow the construction of a lens larger than that at Yerkes. And a lens can be supported only from its rim when it is mounted in a telescope, since one cannot obstruct the aperture. Gravity then causes the lens to sag in the middle, and this effect changes as

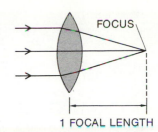

Figure 3–1 The focal length is the distance behind a lens to the point at which objects at infinity are focused.

Figure 3–2 A simple refracting telescope consists of an objective lens and an eyepiece.

VIOLET FOCUS   GREEN FOCUS   RED FOCUS

Figure 3–3 The focal length of a lens is different for different wavelengths, which leads to chromatic aberration.

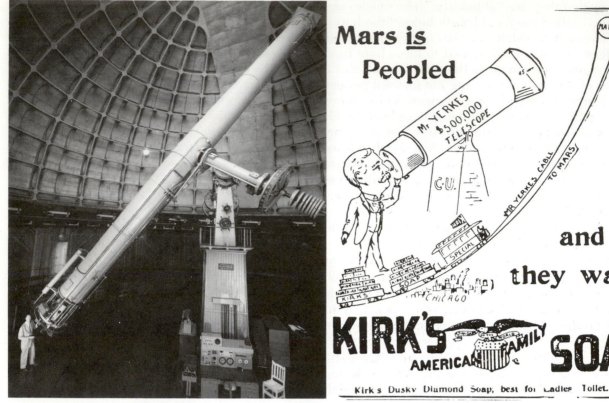

Figure 3–4

Figure 3–5

Figure 3–4 The 0.9-meter (36-inch) refractor of the Lick Observatory.

Figure 3–5 The opening of the 1-meter (40-inch) refractor of the Yerkes Observatory was the cause of much notice in the Chicago newspapers in 1893.

the telescope is pointed in different directions. One can try to make the lens more rigid by making it thicker, but this makes it heavier and even more difficult to get a sufficiently pure lens blank. The percentage of light that is transmitted through the glass diminishes too. The 1-meter telescope at Yerkes, for all of these reasons, is still the largest refracting telescope in the world.

3.3 REFLECTING TELESCOPES

Reflecting telescopes are based on the principle of reflection with which we are so familiar from ordinary household mirrors. A mirror reflects the

Figure 3–6 A spherical mirror focuses light that originates at its center of curvature back on itself, but suffers spherical aberration in that it does not perfectly focus light from infinity.

SPHERICAL MIRROR

SPHERICAL MIRROR

DIVERGING   DIVERGING SLIGHTLY   PARALLEL

Figure 3–7 Light rays from very distant objects are diverging so slightly by the time they reach us that we speak of *parallel light*.

light that hits it so that the light bounces off at the same angle at which it approached. A flat mirror gives an image the same size as the object being reflected, but funhouse mirrors, which are not flat, cause images to be distorted.

One can construct a mirror in a shape so that it reflects incoming light to a focus. Let us consider a spherical mirror, for example. If you were at the middle of a giant spherical mirror, in whatever direction you looked you would see your own image. A sphere images whatever is at its center back at its center (Fig. 3–6). If we were to use just a portion of a sphere, it would still image whatever was at the center of its curvature back to that same point.

But the stars are far away, and we must find a shape for a mirror that has the property of focusing light from celestial objects to a single point. The stars and planets are so far away that individual light rays emanating from there are diverging by such an imperceptible amount by the time they reach us on earth that they are practically parallel. We say that we are observing *parallel light* (Fig. 3–7).

A *parabola* is a two-dimensional curve that has the property of focusing parallel rays to a point. Telescope mirrors are actually *paraboloids,* which are the three-dimensional curves generated when parabolas are rotated around their axes of symmetry (Fig. 3–8). Over a small area, a paraboloid differs from a sphere only very slightly, and one can ordinarily make a parabolic mirror by first making a spherical mirror and then deepening the center slightly (Fig. 3–9).

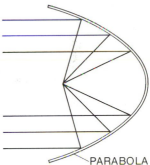

PARABOLA

Figure 3–8 A parabola focuses parallel light to a point.

The technique of making a small telescope mirror is not excessively arduous. Many amateur astronomers have made mirrors approximately 15 cm (6 inches) across without having had previous experience; it requires perseverance and about 50 hours of time. Making a large telescope mirror is another story, of course, and usually takes years.

The advantages of reflecting telescopes over refracting telescopes are several. The reflection of light is not color-dependent, so that chromatic aberration is not a problem for reflectors. One normally deposits a thin coat of a highly reflecting material on the front of a suitably ground and polished surface. Since light does not penetrate the surface, it does not matter too much what is inside the telescope mirror. The internal material does not have to be free of bubbles or striations; it only has to have a smooth surface, though it should be free of internal stress and strain and very stable whatever the temperature is. Further, one can have a network of supports all across the back of the telescope mirror, to prevent the mirror from sagging under the force of gravity. All these advantages allow reflecting telescopes to be made much larger than refracting telescopes. (At wavelengths shorter than those of visible light, glass does not pass radiation, so only reflecting telescopes can be used in these x-ray or shorter ultraviolet regions.)

A potential problem with small reflecting telescopes could be the fact

Figure 3–9 A parabolic mirror is usually made by deepening the center of a spherical mirror.

SPHERE   PARABOLOID

Figure 3–10A A Newtonian reflector.

NEWTONIAN

OBJECTIVE MIRROR

EYEPIECE

SECONDARY MIRROR

Figure 3–10B The replica of Newton's telescope in the visitor's gallery overlooking the 2.5-meter (98-inch) Isaac Newton telescope of the Royal Greenwich Observatory at Herstmonceux, England, showing the side visible only to the astronomers.

that one must gather the reflected light without impeding the incoming light. Clearly, if you put your head at the focal point of a small telescope mirror, the back of your head would block the incoming light and you would see nothing at all!

Isaac Newton got around this problem by putting a small diagonal mirror a short distance in front of the focal point to reflect the light so that the focus was outside the tube of the telescope. This type of apparatus, in use down to this day, is known as a *Newtonian telescope* (Fig. 3–10 A).

Two contemporaries of Newton invented alternative types of reflecting telescopes. G. Cassegrain, a French optician, and J. Gregory, a Scottish astronomer, invented types in which the light was reflected by the secondary mirror back toward the primary mirror and through a small hole in the center of the primary. These designs are called *Cassegrainian* (or *Cassegrain*) and *Gregorian telescopes* (Fig. 3–11).

Modern large telescopes also often have a *coudé* focus. In the coudé system, the light is bent with a series of mirrors until it travels down the axis of the telescope that is pointing at the north pole ("coudé" means "bent" or "elbowed" in French). The image never moves laterally, but only rotates. Thus the light at the coudé focus comes to the rear of the telescope in an unchanging direction. As a result, rather large spectrographs can be built into a room at the base of the telescope.

In all of these methods, a slight bit of the incoming light is blocked by the secondary mirror, and by the supports that suspend the secondary mirror in the middle of the tube. But the secondary mirror can be made small enough relative to the primary mirror that relatively little light is blocked.

Foci is the plural of focus.

The largest telescopes have several interchangeable secondary mirrors built in so that different foci (Newtonian, Cassegrainian, and so forth) can be used at different times. The very largest telescopes are so huge that the observer can even sit in a "cage" suspended at the end of the telescope at the *prime focus,* the position of the direct reflection of the light from the primary mirror (Fig. 3–13). The observer in the prime-focus cage blocks only a small fraction of the incoming light.

Modern reflectors began with the construction of the 0.9-meter (36-inch) telescope at the Lick Observatory on Mt. Hamilton in California and then, under the aegis of George Ellery Hale, of the 1.5-meter (60-inch) and 2.5-meter (100-inch) telescopes on Mt. Wilson in California. The 2.5-

Figure 3–11 A Cassegrainian (often called Cassegrain) telescope has a convex secondary mirror.

EYEPIECE   OBJECTIVE MIRROR

CASSEGRAINIAN

SECONDARY MIRROR

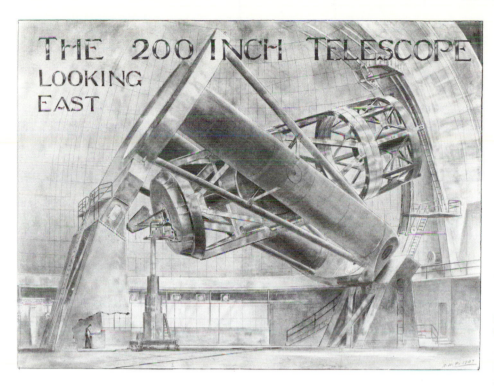

Figure 3–12 One of Russell W. Porter's set of drawings of the 5-meter (200-inch) telescope on Palomar Mountain.

Box 3.1 Conversion of Units

In order to know what to multiply and divide when converting from one system of units to another, some people find it convenient to multiply and divide the words representing the units as though they were numbers. They set up a chain of multiplication and division of known conversion factors so that the unit names they are trying to convert **from** cancel out. Then they do the same series of multiplications and divisions with the numbers themselves. For example, 1 inch = 2.54 cm. If we divide both sides of the equation by 2.54 cm, we have the following equation: $\frac{1 \text{ in}}{2.54 \text{ cm}} = 1$. Equivalently, we can divide by 1 in to get $\frac{2.54 \text{ cm}}{1 \text{ in}} = 1$. Now, we can always multiply any number by 1 without changing its value. Thus if we have, say, 200 inches and want to put this figure in units of centimeters, we can write 200 inches $\times \frac{2.54 \text{ cm}}{1 \text{ in}}$. The "inches" in the numerator and denominator cancel, and we have 508 cm as the converted figure. Similarly, $3 \text{ m} \times \frac{100 \text{ cm}}{\text{m}} \times \frac{1 \text{ in}}{2.54 \text{ cm}} = \frac{300}{2.54}$ in. This method works just as well with units of time (months, weeks, days, hours, minutes, seconds), units of volume, and so on.

Figure 3–13 The prime-focus cage of the 5-meter Palomar telescope.

meter telescope, with a mirror made of plate glass, began operation in 1917 and was the largest telescope in the world for 30 years.

One of the limitations of any telescope, reflecting or refracting, is the

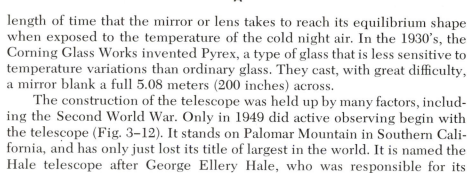

A

Figure 3–14 *(A)* The Soviet 6-meter (236-inch) telescope on Mount Pastukhov in the Caucasus. *(B)* The tube of the 6-meter telescope before installation in the dome.

B

length of time that the mirror or lens takes to reach its equilibrium shape when exposed to the temperature of the cold night air. In the 1930's, the Corning Glass Works invented Pyrex, a type of glass that is less sensitive to temperature variations than ordinary glass. They cast, with great difficulty, a mirror blank a full 5.08 meters (200 inches) across.

The construction of the telescope was held up by many factors, including the Second World War. Only in 1949 did active observing begin with the telescope (Fig. 3–12). It stands on Palomar Mountain in Southern California, and has only just lost its title of largest in the world. It is named the Hale telescope after George Ellery Hale, who was responsible for its construction.

This instrument has been known for years as the "200-inch telescope," but the scientific journals are now converting even this sacred name to metric units. We will round off 5.08 meters to 5 meters in this book. (It will not be amiss, I hope, to point out that the mirror is about 17 feet in diameter, the size of an ordinary room.)

Other telescopes have been built since, but with a single exception they have all been substantially smaller. The largest telescope in the world is a new 6-meter (236-inch) reflector in the Soviet Union, opened in 1976 on Mount Pastukhov in the Caucasus (Fig. 3–14).

Up to now, the largest astronomical telescope that has been sent into space has been the 0.9-meter (36-inch) telescope aboard the *Copernicus* satellite, the third Orbiting Astronomical Observatory launched by NASA. Currently the Space Telescope (ST) with a mirror 2.4 meters (95 inches) in diameter is on the drawing board for possible launch in 1983 to concentrate on visible and ultraviolet studies (Fig. 3–15). Free of the twinkling effects caused by turbulence in the earth's atmosphere, the ST would be able to observe finer details of objects in space than are visible with earth-based telescopes. Also, because it would not be limited by the brightness of the earth's atmosphere even above a remote mountain, the ST would be able to observe fainter objects than any of the ground-based equipment.

Figure 3–15 An artist's conception of the Space Telescope, with a cutaway showing the secondary mirror and its support.

3.4 OPTICAL OBSERVATORIES

Once upon a time, a hundred or so years ago, telescopes were put up wherever astronomers happened to be located. Thus all across the country and all around the world, on college campuses and near cities we find old observatories. Percival Lowell, who located an observatory in Arizona in 1894, and Hale, with his decision ten years later that the clear and dark skies of Southern California were an all-important consideration, were major forces in beginning the construction of observatories at remote sites chosen for their favorable observing characteristics rather than for their proximity to a home campus. We have already discussed the reflectors for which Hale was responsible: the 1.5-meter and 2.5-meter on Mount Wilson and the 5-meter on Palomar. Mount Wilson and Palomar are now jointly operated, and called the Hale Observatories, with offices in Pasadena. The second largest telescope in the country has, until recently, been the 3-meter (120-inch) reflector at the Lick Observatory in California.

The National Science Foundation has sponsored and built a United States National Observatory. Its telescopes are located on Kitt Peak, a sacred mountain of the Papago Indians, about 80 km (50 miles) southwest of Tucson, Arizona (Fig. 3–16). The headquarters of the Kitt Peak National Observatory are in Tucson, where the astronomers on the staff and the visiting astronomers work, except during the specific periods when they are observing with one of the telescopes.

A 2.1-meter (84-inch) telescope had been the workhorse of the Observatory; a 4-meter (158-inch) telescope (shown at the opening of this chapter and in Color Plate 1) was opened there in 1973. This latter telescope is named the Mayall telescope in honor of a former director of the Observatory.

Part of the sky is never visible from sites in the northern hemisphere, and many of the most interesting astronomical objects in the sky are only visible or best visible from the southern hemisphere. Therefore, considerable effort has gone into finding suitable southern hemisphere sites: tele-

Experience gained at the Lick Observatory, one of the earliest remote observatory sites, was an important influence on the later plans of Lowell and Hale.

Figure 3–16 An aerial view of Kitt Peak. The 4-meter telescope is at the right; the solar telescope is the sloping structure at the lower left. The white dome at extreme left is the 11-meter radio telescope of the National Radio Astronomy Observatory.

scopes in Peru, South Africa and, more recently, Australia have provided very important scientific data. In the last few years a special emphasis has been placed on constructing telescopes at southern hemisphere sites.

Many of the new sites are on coastal mountains west of the Andes range in Chile. The Kitt Peak staff has supervised the construction of the Cerro Tololo Inter-American Observatory there (Figs. 3–17 and 3–18), at which a 4-meter (158-inch) telescope has been constructed. The staff of the Hale Observatories has supervised, for the Carnegie Institution of Washington,

Figure 3–17 An aerial view of the Cerro Tololo Inter-American Observatory.

the construction of a 2.6-meter (101-inch) telescope on Cerro las Campanas, another Chilean peak. A consortium of European observatories and universities run the European Southern Observatory, where a 3.6-meter (142-inch) telescope is being erected on still another Chilean mountain top. It is quite common in the astronomical community to go off to Chile for a while to "observe."

Another important new site is in Hawaii. Already a 2.2-meter (88-inch) telescope has been installed on Mauna Kea, on the island of Hawaii. A Franco-Canadian-Hawaiian group is building a 3.6-meter (144-inch) telescope for a site there. This site is at an especially high altitude, 4145 m (13,600 ft), and the air above is relatively dry. Since water vapor blocks infrared radiation, these telescopes are especially suitable for infrared observations. A 3.2-meter (126-inch) telescope being built by NASA for Mauna Kea will concentrate on infrared work.

Australia and Great Britain remain at the forefront of research with an Anglo-Australian telescope of 3.9 meters (153 inches), which was opened at Siding Spring in eastern Australia in 1974.

The continued building of optical observatories, with the current unprecedented spate of large telescopes, shows the vitality of this type of astronomical research, even though new techniques involving space telescopes, radio astronomy, and other new technologies are now providing important data at an increasing rate.

This brief discussion mentions only optical observatories, and not the many telescopes that are now in existence on the ground and in space to observe other regions of the spectrum.

It is not useful to get too exact in specifying the diameters of telescopes. The outer 4 cm (almost 2 in) of the 4-m mirrors are beveled, thus limiting the usable diameter to 3.96 m (156 in). Further, bits of the outermost part of the Kitt Peak mirror do not match the desired shape as accurately as the rest of the mirror; the outermost 19 cm (8 in) are masked off, leaving a usable diameter of 3.81 m (150 in). The Siding Spring telescope, with 3.89 m (153 in) of usable aperture, is actually intermediate in size between the Kitt Peak and Cerro Tololo telescopes. We shall refer to them as "4-meter telescopes."

Figure 3–18 Two views of the Cerro Tololo Inter-American Observatory taken from the same location with a fish-eye lens. (A) The camera was held fixed, and star trails and the overexposed image of Venus show above the domes. (B) The camera tracked the stars; the domes are blurred but the Milky Way shows.

2.5-METER   5-METER
MIRROR   MIRROR

Figure 3–19 A telescope twice the diameter of another has four times the collecting area.

Figure 3–20 A model of the Multi-Mirror Telescope being installed on Mount Hopkins in Arizona.

3.5 LIGHT-GATHERING POWER AND MAGNIFICATION

The principal function of large telescopes is to gather light. The amount of light collected by a telescope mirror is proportional to the area of the mirror, πr^2. (It is often convenient to work with the diameter instead of the radius, since the diameter is the number usually mentioned. The formula for the area of a circle may, of course, also be written as $\pi d^2/4$.)

The ratio of the areas of two telescopes is thus the ratio of the square of their diameters $(d_1{}^2/d_2{}^2)$, which is usually more easily calculated as the square of the ratio of their diameters $(d_1/d_2)^2$. For example, the 200-inch telescope has an area 4 times greater than the area of the 100-inch telescope, $(200/100)^2$ (Fig. 3–19). In the metric system $(5/2.5)^2$ gives the same result. Thus during observations of equal duration the 200-inch would collect four times as much light as the 100-inch if all other things were equal, and for the same exposure time would record stars 1 or 2 magnitudes fainter.

Even though the light-gathering power increases rapidly with the size of the telescope, the cost goes up even more rapidly, perhaps even with the cube of the diameter. Thus the funds have simply not been available in the United States to build a larger telescope than the 5-meter (200-inch), or even a duplicate of the overworked 5-meter.

Alternative methods of getting more aperture are now under investigation. The Harvard-Smithsonian Center for Astrophysics and the University of Arizona are now developing a multi-mirror telescope (MMT) (Fig. 3–20), which has six 1.8-meter (72-inch) paraboloids linked together and aligned by lasers, in the hope that they can succeed in focusing all the light from these several mirrors at the same point. It is much cheaper to build several 1.8 meter telescopes than it is to build one larger telescope of 4.5 meters (176 inches), the equivalent total aperture. The MMT is mounted on Mount Hopkins, south of Tucson, Arizona.

A telescope's light-gathering power (and its ability to concentrate it in as small an area as possible) is really all that is important for the study of individual stars, since the stars are so far away that no details can be discerned no matter how much magnification is used. The main limitation on the smallest image size that can be detected is caused not in the telescope itself, but rather by turbulence in the earth's atmosphere. The limitations are connected with the twinkling effect of stars. The steadiness of the earth's atmosphere is called, technically, the *seeing*. We would say, when the atmosphere is steady, "The seeing is good tonight." "How was the seeing?" is a polite question to ask an astronomer about his most recent observing run. Another factor of importance is the *transparency*, how clear the sky is. It is quite possible for the transparency to be good but the seeing very bad, or for the transparency to be bad (a hazy sky, for example), but the seeing excellent.

For some purposes, including the study of the planets and the resolution of double stars (Section 4.1), the magnification provided by the telescope does play a role. Magnification can be important for the study of the sun, as well, and for extended objects (as distinguished from point objects) like the nebulae or galaxies.

Magnification is equal to the focal length of the telescope objective (the primary mirror or lens) divided by the focal length of the eyepiece. Thus the magnification of a telescope is only partly determined by the major part of the telescope; by simply substituting eyepieces of different focal lengths one can get different magnifications. However, there comes a point where even though using an eyepiece of short focal length will enlarge the image, it will not give you any more visible detail; when using a telescope one should not exceed the maximum useful magnification, which depends on such factors as the diameter of the objective and the quality of the seeing.

A telescope of one meter focal length will, then, give a magnification of 25 when used with a 40-millimeter eyepiece, an ordinary combination for amateur observing. If we substitute a 10-millimeter eyepiece, we will have a magnification of 100. We would say that we were observing with "100 power."

3.6 OTHER TYPES OF OPTICAL TELESCOPES

A parabolic reflector gives a good focus only for light that strikes the mirror parallel to the axis of the paraboloid. Light at an angle to the axis is not focused very well. The fact that such reflectors have only a very small field in good focus limits their usefulness in certain cases. Though modern computer methods of lens design have led to advances, the fields of view of even the new 4-meter telescopes at Kitt Peak and Cerro Tololo are only 52 arc min. Refractors focus light over a much wider field.

Bernhard Schmidt, working in Germany in 1930, invented a type of telescope that combines some of the best features of both refractors and reflectors. In a *Schmidt camera* (Fig. 3–21), the main optical element is a large mirror, but it is spherical instead of being paraboloidal. Before reaching the mirror, the light passes through a thin lens, called a *correcting plate*, that distorts the incoming light in just the way that is necessary to have the spherical mirror focus it over a wide field. The image falls at a location where the eye cannot be put, and thus the image is always recorded on film (Fig. 3–22). Accordingly, this device is often called a Schmidt **camera** instead of a Schmidt telescope.

One of the largest Schmidt cameras in the world is on Palomar Mountain, in a dome near that of the 5-meter telescope. The Schmidt camera has a correcting plate 1.2 meters (48 inches) across, and a spherical mirror 1.8 meters (72 inches) in diameter. Its field of view is about 7° across, much broader than the field of view, only about 2 minutes of arc across, of the 5-meter reflector.

Figure 3–21 Edwin P. Hubble (see Chapter 23) at the guide telescope of the 1.2-meter (48-inch) Schmidt camera on Palomar Mountain.

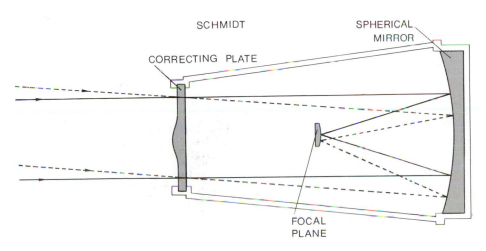

SCHMIDT

CORRECTING PLATE

SPHERICAL MIRROR

FOCAL PLANE

Figure 3–22 By having a non-spherical thin lens called a correcting plate, a Schmidt camera is able to focus a wide angle of sky onto a curved piece of film.

Figure 3–23 A 7° × 7° region of the sky in Orion taken with the 1.2-meter Schmidt camera as part of the National Geographic Society-Palomar Observatory Sky Survey. The faintest stars detectable are about 21st magnitude. The top photograph shows the image with a blue-sensitive plate, and the bottom photograph shows the image with a red filter and a red-sensitive plate. The print is negative, so brighter areas appear blacker. The three bright stars at the lower right are Orion's belt; they are B stars and are brighter in the blue than in the red. The bright region to their left radiates mostly H alpha, and so is brighter in the red than in the blue. The dust lanes, on the other hand, are more prominent in the blue. The Horsehead Nebula (also shown in Color Plate 42) is a region of dust obscuring part of the H alpha emission that runs south of this region. Near it is a small nebula that is brighter in the blue than in the red, which leads us to conclude that we are seeing reflected starlight rather than H alpha emission.

To the upper left is the brightest section of Barnard's ring, a ring of H alpha emitting gas that surrounds most of Orion. To the lower left center a large dust cloud causes the number of stars in this region to be less than in the upper right or lower left. This is best seen on the blue print. The three blotches at the upper left are artifacts called "ghost images" caused by internal reflections in the telescope of the three brightest stars.

The field of view of the 5-meter is so small that it would be hopeless to try to map the whole sky with it because it would take so long. But the 1.2-meter Schmidt has been used to map the entire sky that is visible from Palomar. The survey was carried out through both red and blue filters. Sample plates from this National Geographic Society/Palomar Observatory Sky Survey are shown in Figure 3–23.

The new 1-meter Schmidt camera at the European Southern Observatory and the British 1.2-meter Schmidt camera at Siding Spring, Australia, are now being used in a joint project to extend the survey to incorporate the one-quarter of the sky that cannot be seen from Palomar.

Many small telescopes now commercially available are of the Maksutov or Schmidt-Cassegrain designs, which are similar to that of a Schmidt camera in that they have both a spherical mirror and a correcting plate, but the light bounces back and forth inside until it exits through a hole in the main mirror. These designs allow a telescope to be made very compact and therefore portable.

3.7 SPECTROSCOPY

We have discussed the collection of quantities of light by the use of large mirrors or lenses, and the focusing of this light to a point or suitably small area. Often one wants to break up the light into a spectrum instead of merely photographing the area of sky at which the telescope is pointed. This allows us to study the spectral lines that display the excitation and ionization of atoms in astronomical sources (and thus determine the temperatures of those sources), to measure Doppler shifts, and so forth.

A prism breaks up light into its spectrum for the same reasons that light focused by a lens has chromatic aberration. A wavefront is refracted as it hits the side of the prism, but it is refracted by different amounts at different wavelengths.

When a beam of parallel light falls directly on a prism, the spectral lines that result are often indistinct because the spectrum from one place in the beam overlaps the spectrum from an adjacent spot in the beam. We can limit the blurring of spectral lines that results because of this by allowing only the light that passes through a long, thin opening called a *slit* to fall on the prism (Fig. 3–24). Then the spectral lines are sharp and relatively narrow.

Figure 3–24 The light entering the instrument from the left passes through a slit that defines a line of radiation, is made parallel and directed on the prism (or other dispersing element), and then is focused. The device is called a *spectroscope* if we observe with the eye, a *spectrograph* if we record the data on a photographic plate, and often a *spectrometer* if we record the data electronically.

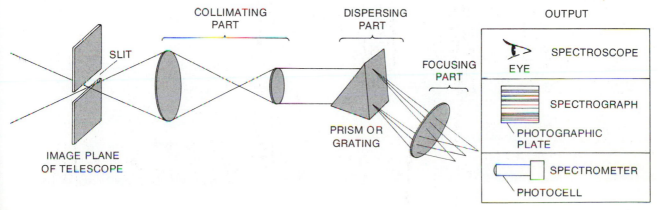

COLLIMATING PART

DISPERSING PART

OUTPUT

SLIT

FOCUSING PART

EYE

SPECTROSCOPE

SPECTROGRAPH

PHOTOGRAPHIC PLATE

PRISM OR GRATING

SPECTROMETER

PHOTOCELL

IMAGE PLANE OF TELESCOPE

Figure 3–25 The silicon vidicon spectrometer and data recording system of MIT, shown here, have been used in conjunction with the largest telescopes to study faint objects.

Actually, astronomical spectra are no longer usually made by means of prisms. Light may be broken up into a spectrum without need for a prism if lines are ruled on a surface very close together. And I do mean close together: one can have 10,000 lines ruled in a given centimeter (25,000 lines in a given inch)! Such a ruled surface is called a *diffraction grating*.

One can arrange a small telescope so as to examine the spectrum that comes off a prism or grating (Fig. 3–24). A device that makes a spectrum, usually including everything between the slit at one end and the viewing eyepiece at the other, is called a *spectroscope*. When the resultant spectrum is not viewed with the eye but is rather recorded either with a photographic plate or by some other means, the device is called a *spectrograph*. Sometimes when the resultant spectrum is measured with an electronic device, rather than recorded on film, the device is called a *spectrometer* (Fig. 3–25).

3.8 RECORDING THE DATA (*Optional*)

When the desired information emerges at the end of the telescope, whether it be an image of a field of stars or a spectrum, astronomers usually want a more permanent record than is afforded by merely observing the image and more accuracy than can be guaranteed in a sketch.

Traditionally, one puts a photographic film or "plate" (short for photographic plate, a layer of light-sensitive material on a sheet of glass) instead of an eyepiece at the observer's end of the telescope, so as to get a permanent record of the image or spectrum.

This use of photographic film also significantly increases our ability to detect signals from faint sources. The eye and brain can process information for only a fraction of a second at a time and do not function cumulatively. A photographic plate, however, can be left exposed to radiation for a long period (Fig. 3–26), sometimes for hours. Just as a long exposure with an ordinary camera can record objects that are only dimly lit, a long ex-

Figure 3–26 These three views of Orion (see also Fig. 8–2) show the effect of increased exposure time. The three stars representing the belt, shown in Fig. 3–23, appear at the center of the photograph. The Orion Nebula (see Color Plate 25) is below them. The red star Betelgeuse, at the upper left, does not appear very bright because these photographs were taken on blue-sensitive plates. The longest exposure shows that nebulosity covers almost the entire constellation.

A   B   C

posure with a telescope can record fainter objects than would a shorter exposure.

Many other methods of recording data have been developed—some of which are more sensitive or more accurate—and it is now being said that the day of film is past. The future seems to lie in electronic devices.

The photocell is a basic electronic device used by astronomers. Certain materials, when struck by light, give off an electric current. That current can be measured, perhaps by reading a meter or perhaps automatically, with the data recorded on punched cards or on magnetic tape. The readings can even be fed straight into a computer.

A modification of photocells has been developed whereby the electrons that become available inside the tube when light hits it can be directed to hit other special elements of the photocell. Each of the subsequent elements can be made to emit several electrons for each one hitting it. In this way, the original number of electrons can be multiplied, often by a substantial factor (perhaps 10^6), and the device is called a *photomultiplier*.

One advantage of a photocell or photomultiplier is that its response is "linear" up to some bright limit; that is, unlike film, when twice as much light hits it, it gives off twice as much current. One of the disadvantages of photocells and photomultipliers is that they measure only the total intensity of radiation falling on them from whatever direction; they do not form an image in the way that film does. So whether photocells and photomultipliers are useful to you depends on exactly what investigation you are undertaking. For measuring the intensity of light of a star, photomultipliers are fine. Such measurements of intensity are called *photometry*.

Photomultipliers are much more sensitive to faint signals than is photographic film, and also can be made to be sensitive in a wider region of the spectrum. Photometry has long been carried out with photocells and photomultipliers and can be done very precisely. In some observatories, the procedure is automated so that the signal level from the photomultiplier is monitored by a computer. When enough photons have been gathered to allow a measurement of a certain accuracy (say, to 1 per cent of the signal level), the computer ends the "exposure" and moves the telescope to the next object.

3.9 OBSERVING IN DIFFERENT SPECTRAL REGIONS (*Optional*)

The most familiar detector of electromagnetic radiation is the photographic plate. Basically, a photographic plate (or film) consists of a backing covered with an *emulsion*. The backing may be glass for stability or plastic for flexibility. The emulsion is the working layer. In it are grains of a silver compound. When these grains are struck by photons of radiation, they undergo a chemical change. Inspection of a photographic plate under high magnification shows the grainy structure of the image (Fig. 3–27).

The photons, as we have seen, have energies of hc/λ. Ultraviolet and x-ray photons have shorter wavelengths and thus greater energies than photons of visible light. Thus they also have enough energy to interact with the film grains. Photographic methods can be used, therefore, throughout the x-ray, ultraviolet, and visible parts of the spectrum.

Infrared photons have longer wavelengths and thus lower energies

Figure 3–27 This enlargement of the spectrum of a distant galaxy shows clumps of grain on the photographic plate.

Figure 3–28 A balloon launch carries aloft a 102-cm infrared telescope.

than visible photons. They do not have enough energy to interact with the silver grains on ordinary photographic plates. Some special types of film can be used at the very shortest infrared wavelengths, but for the most part astronomers have to employ other methods of detection in this region of the spectrum.

Another major limitation in the infrared is that there are very few windows of transparency in the earth's atmosphere. Most of the atmospheric absorption in the infrared is caused by water vapor, which is located at lower levels in our atmosphere than is the ozone that causes the absorption in the ultraviolet. Thus we do not have to go as high to observe infrared as we do to observe ultraviolet. It is sufficient to send up instruments attached to huge balloons (Fig. 3–28). The major balloon-launching site in the United States is in Palestine, Texas.

At even longer wavelengths, in the radio region of the spectrum, there is another window of transparency. Photocells do not respond to these photons of very low energy. We must use antenna techniques and then amplify the weak electrical signals generated by the antenna in much the same manner that we use to receive and amplify radio and television signals at home. The techniques of radio astronomy, now well established as one of the major branches of astronomy, will be discussed at length in Part IV.

At the shorter end of the spectrum, gamma rays, x-rays, and ultraviolet light do not come through the earth's atmosphere. Ozone, a molecule of three atoms of oxygen (O_3), is located in a broad layer between about 20 and 40 kilometers in altitude, and prevents all the radiation at wavelengths less than approximately 3000 Å from penetrating. Fears that the earth's ozone layer is in danger will be discussed in Section 13.4.

To get above the ozone layer, astronomers have launched telescopes and other detectors in rockets and in orbiting satellites. Often photocells are used in the orbiting spacecraft, as the signals can then be radioed back without the need to develop film.

In the short wavelength regions, we cannot merely use parabolic mirrors to image the incident radiation, as the x-rays will pass right through the mirror! Fortunately, x-rays can still be bounced off a surface if they strike the surface at a very low angle. This is called *grazing incidence* (Fig. 3–29). The principle is similar to that of skipping stones across the water. If you throw a stone straight down at a lake surface, the stone will sink immediately. But if you throw a stone out at the surface some distance in front of you, the stone could bounce up and skip along a few times. By carefully choosing a variety of curved surfaces that suitably allow x-radiation to "skip" along, astronomers can now make telescopes that actually make x-ray images. But the appearances of the telescopes are very different from those of optical telescopes (Fig. 3–30).

In order to form such high-energy photons, processes must be going on out in space that involve energies very much higher than most ordinary processes that go on at the surface layers of stars. The study of the processes that bring photons or particles of matter to high energies is called *high-energy astrophysics*. The astronomical community has placed a priority on work in this field for the next few years, and the next emphasis in space observations will be in x-ray and γ-ray spectral regions that give us the most information about these processes. NASA has launched the first of three spacecraft, HEAO's (for *High-Energy Astronomy Observatories*) to observe in the x-ray and γ-ray spectral regions. Additional HEAO's are planned for 1978 and 1979.

Figure 3–29 Grazing incidence.

3.10 A NIGHT AT PALOMAR *(Optional)*

In some sense, the opportunity to observe with what has been the largest optical telescope in the world, the 5-meter telescope on Palomar Mountain, is the ultimate reward for years of study and research in astronomy.

Before using the 5-meter Hale telescope, you must have qualified by acquiring experience at some other smaller telescope, probably a large telescope in its own right such as the 2.5-meter on Mount Wilson. You must then prepare a detailed proposal to a selection committee specifying what it is you want to observe and why you need the 5-meter telescope in particular to observe it. There may be 320 clear nights a year on Palomar Mountain, but many more nights a year would be needed if all the requests were to be granted.

Shortly before your observing time comes, there is a variety of last-minute preparations to make. You might have complicated new equipment to test out, or you might be using one of the standard spectroscopic, photoelectric, or photographic systems permanently installed at the telescope.

Even if you are using the standard equipment, you might have special filters to use. You will want to have prepared a list of the objects that you plan to observe, with an ample number of objects in a second category of priority in case everything goes fantastically well.

Finally, the big day comes and you set off for Palomar. If you are on the staff of the Hale Observatories, your office is in Pasadena, California, and Palomar is about a 2½ hour drive southward and slightly east. If you are at an institution farther away, you would first have to fly to Los Angeles. In any case, you have to have your suitcase packed and be prepared to spend a few days and nights away from home.

Figure 3–30 *(A)* The cylindrical mirror of HEAO-B does not resemble mirrors for optical telescopes. This is the outermost of four concentric mirrors. Skylab astronaut Owen Garriott is on the right in this view. *(B)* Caltech scientist Gordon Garmire looks through the mirror, polished for grazing incidence. The mirror has parabolic and hyperbolic sections, and consists of an aluminum structure coated with nickel. It has been launched in a rocket.

A

B

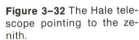

Figure 3–31 A moonlight view of the dome of the 5-meter Hale telescope on Palomar Mountain in California.

Figure 3–32 The Hale telescope pointing to the zenith.

As you drive into the mountains from the desert, the foliage changes. By the time you have finished the steep last bit of the climb up Palomar, you find yourself in a pine forest. By noontime of the day you are to start observing, you check into the residence hall, which is called "the Monastery."

Meals are served at the Monastery. The noon meal is for the new arrivals and for the daytime workers; the astronomers who worked the preceding night are usually asleep. After lunch, you set off for the telescope.

The 5-meter telescope stands in a huge domed building that is 40 meters (135 ft) high, the equivalent of a 13-story building (Fig. 3–31). Let us say that you are doing photographic spectroscopy. The first thing that you would do is to go to the plate vault, a large walk-in refrigerator where special stocks of photographic emulsions are kept. The emulsions are specially chosen for their astronomical characteristics, and are not the ones that you buy in your corner camera store. In addition to preparing your plates, you would also go to the darkroom and mix the photographic chemicals you would need for developing them after they have been exposed.

Then you would make certain that the spectrograph you plan to use is in place. There are so many interchangeable pieces of equipment available for use in studying the light reflected by the 5-meter mirror that you usually have to set up your own equipment. The rest of the afternoon might well be spent in carefully focusing the spectrograph.

It is finally dinner time, just before dark, and you head back to the Monastery. All the daytime personnel have left the mountain, and it is mostly the observers at this meal.

In the twilight you walk back to the telescope. A night assistant is assigned to the telescope at all times and is responsible for all the mechanical aspects of the telescope. The night assistants know the telescope thoroughly, and also know when a strange noise might mean that the telescope must be shut down for repairs, or when high wind or blowing dust mean that the telescope dome must be closed for the safety of the mirror.

The telescope is very well lubricated and very well balanced, so even though it weighs over 500 tons it turns very smoothly and needs only a moderate sized motor. But refraction effects in the earth's atmosphere may cause the light from stars to vary in position at a slightly different rate from the normal average, and you must continually check that the telescope is pointing in absolutely the right direction. This is called guiding. Especially in direct photographic work, where you are making an image, observers who are the most careful with their guiding get the best results.

After an hour or so, your exposure is finished. You cover the slit, and close the film holder. Then you replace the exposed plate with an unexposed plate, and prepare to take another exposure, perhaps of another object. Then you will go to the darkroom and develop your plate, gently rocking it back and forth in the developer for several minutes. The night assistant might be able to help you by guiding on the second star while you are developing the plate of the spectrum of the first star. It is important to develop your first plate right away to make sure that the system is working and that there are no blockages in the light path.

In the course of the night you will probably want to go up to the floor of the dome for a short while just to appreciate the stars and the marvelous monster of an instrument that you are controlling to gather photons from afar.

All too soon the night is over. The dawn is visible in the east, and you must shut down or else the sky light will fog your plate. After developing your last plate (or perhaps leaving it to do at the beginning of the next night), you trudge home to the Monastery to bed.

After a few days on the mountain, you may have a dozen or more plates (Fig. 3-33). Your work has just begun. It may take months or years to analyze the data completely, often using measuring devices and computers at your own institution. The thrill of having yourself captured those few photons from a distant star remains.

Figure 3-33 The collecting area of the Hale telescope is large, but the light is focused down to a small area on a spectrographic plate. The spectrum shown is that of the variable star R Coronae Borealis, and was taken by the author. Most of the spectrum visible is that of an iron arc operated in the telescope dome to provide known wavelengths for calibration purposes; the spectrum of the star is the few emission lines between the left and right halves of the calibration spectrum (the two vertical dark streaks).

3.11 COORDINATE SYSTEMS

Astronomers and geographers share the need to set up systems of coordinates—*coordinate systems*—to designate the positions of places in the sky or on the earth.

The geographers' system is familiar to most of us: longitude and latitude. Lines (actually half-circles) of longitude called *meridians* run from the north pole to the south pole. The zero circle of longitude has been adopted, by international convention, to run through the former site of the Royal Greenwich Observatory in England. We measure longitude by the number of degrees east or west an object is from the meridian that passes through Greenwich.

Latitudes are defined by parallel circles that run around the earth, all parallel to the equator. Zero degrees of latitude corresponds to the equator; ninety degrees of latitude corresponds to the poles (90° of north latitude for the north pole and 90° of south latitude for the south pole).

The astronomers' system corresponds exactly to the geographers' sys-

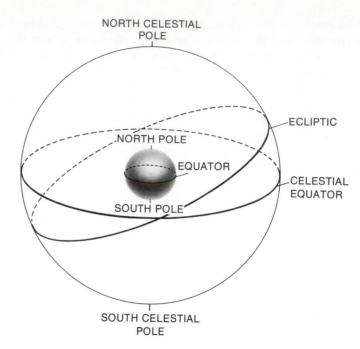

Figure 3–34 The celestial equator is the projection of the earth's equator onto the sky, and the ecliptic is the sun's yearly path through the stars. The vernal equinox is one of the intersections of the ecliptic and the celestial equator. From a given location at the latitude of the United States, the stars nearest the north celestial pole never set and the stars nearest the south celestial pole never rise above the horizon.

tem, except that the astronomers use the names *right ascension* for celestial longitude and *declination* for celestial latitude. They are measured with respect to a *celestial equator,* which is the extension of the earth's equator into space, and *celestial poles,* which are on the extensions of the earth's axis of spin into space and thus lie above the earth's poles (Fig. 3–34).

Right ascension and declination form a coordinate system fixed to the stars. To observers on earth, the stars appear to revolve every 24 hours. The coordinate system thus appears to revolve at the same rate. Actually, of course, the earth is rotating and the stars and celestial coordinate system remain fixed.

Astronomers set up a timekeeping system called *sidereal time* (sidereal means "by the stars"). The sidereal time is the right ascension of any star on the meridian, such as one that is passing directly overhead. Thus each location on the earth has a different sidereal time at each instant. Astronomers find this system convenient because they can consult clocks that are set to run on sidereal time. Thus by merely knowing the sidereal time, they can tell whether a star is favorably placed for observing.

Sidereal time and solar time are not exactly the same. They are both caused by the rotation of the earth on its axis. A *sidereal day* is the length of time that a given star in the sky takes to return to the same position in the sky. A *solar day* is the length of time that the sun takes to return to the same position in the sky. Since the earth is revolving around the sun every year, by the time a day has gone by, the earth has moved 1/365 of the way around the sun. Thus after the stars have returned to their same positions in the sky, it takes 1/365 of approximately 24 hours = 4 minutes for the sun to return to its position. A solar day is thus approximately 4 minutes (actually 3 minutes 56 seconds of time) longer than a sidereal day.

Every observatory has both sidereal clocks, for the astronomers to tell when to observe their stars, and solar clocks, for the astronomers to gauge when sunrise will come and to know when to go to dinner. A solar clock and a sidereal clock show the same time (on a 24-hour system) on only one day each year. The next day the sidereal clock is 4 minutes ahead, the sec-

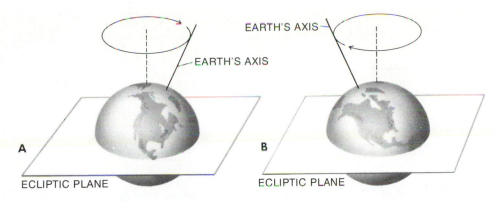

EARTH'S AXIS

EARTH'S AXIS

A

ECLIPTIC PLANE

B

ECLIPTIC PLANE

Figure 3–35 The earth's axis precesses with a period of 26,000 years.

ond day afterward it is 8 minutes ahead, and so on. Six months later the two clocks differ by 12 hours and the stars that were formerly at their highest at midnight are then at their highest at noon. As a result, they may not be visible at all at that season.

The coordinate in the sky that is perpendicular to the celestial equator marks the declination. Declination, latitude in the sky, is measured in degrees north or south of the celestial equator in exactly the same manner that latitude on the earth is measured with respect to the terrestrial equator.

Actually, there are minor effects that cause the celestial coordinates of the stars to change slightly. One such effect is *precession*, which causes a slight drift in the coordinate system with a 26,000 year period (Fig. 3–35). Precession takes place because the earth's axis doesn't always point exactly at the same spot in the sky; it rather traces out a small circle and takes approximately 26,000 years to return to the same orientation. As a result of precession, one has to make small corrections in any catalogue list of celestial positions to update them to the present time. Precession is a small effect that must be taken into account only for observing with large telescopes—changes in celestial coordinates may be a few minutes in a few years. Don't worry, Polaris will be the North Star again in about 26,000 years, but in between—in, say, A.D. 14,980—the North Star will be Vega. An analogous change on earth, with which some readers may be familiar, is that of compass corrections that must be made because of the drift of the earth's magnetic poles.

Another minor effect that causes changes in the right ascensions and declinations of celestial objects is proper motion (see Section 2.15), but this effect is so minuscule that even when astronomers work at it, the proper motion is difficult to measure.

Note that although the stars are fixed in their positions in the sky, the sun's position goes through the whole range of right ascension each year. The path of the sun in the sky with respect to the stars is called the *ecliptic* (Fig. 3–34). The earth's axis is inclined by $23\frac{1}{2}°$ with respect to the perpendicular to the plane of its orbit (Fig. 3–36). Because of this, the ecliptic is inclined by $23\frac{1}{2}°$ with respect to the celestial equator. The ecliptic and the celestial equator cross at two points. The sun crosses one of those points, called the *vernal equinox*, on the first day of spring. The sun crosses the other intersection, the *autumnal equinox*, on the first day of autumn. In the northern hemisphere, these occur on approximately March 21st and September 21st, respectively. On these days, the sun's declination is 0°; the sun's declination varies over the year from $+23\frac{1}{2}°$ to $-23\frac{1}{2}°$.

Nox is Latin for night.

ZENITH

N   EARTH   S

HORIZON

Figure 3–36 From the equator, the stars rise straight up, pass right across the sky, and set straight down.

ZENITH

EARTH

HORIZON

Figure 3–37 From the pole, the stars move around the sky in circles parallel to the horizon, never rising or setting.

The moon goes around the earth once each month, and thus the moon's right ascension changes through the entire range of right ascension once each month. Since the moon's orbit is inclined to the celestial equator, the moon's declination also varies. The planets' motions in the sky are less easy to categorize, but they also change their right ascension and declination from day to day. Tables of the daily positions of the sun, moon, and planets are published each year by the U.S. Naval Observatory in Washington, D.C., in a book entitled *The American Ephemeris and Nautical Almanac*.

3.12   OBSERVING THE STARS

At the latitudes of the United States, which range from +25° for the tip of the Florida Keys to +49° for the Canadian border, or down to +19° in Hawaii and up to +67° in Alaska, the stars rise and set at angles to the horizon other than 0° or 90°. In order to understand the situation, it is best first to visualize simpler cases.

If we were standing on the equator, the stars would rise perpendicularly to the horizon (Fig. 3–36). The north celestial pole would lie exactly on the horizon in the north, and the south celestial pole would lie exactly on the horizon in the south. Each star would rise somewhere on the eastern half of the horizon; each would remain "up" for twelve hours, and then would set. We would be able to see all stars, no matter what their declinations, for twelve hours a day. The sun, no matter what its declination, would also rise and set twelve hours apart, so the day and night would each last twelve hours.

If, on the other hand, we were standing on the north pole, the north celestial pole would be directly overhead, and the celestial equator would be on the horizon (Fig. 3–37). All the stars would move around the sky in circles parallel to the horizon. Since the celestial equator is on the horizon, we could see only the stars with northern declinations. The stars with southern declinations would never be visible.

Whenever the sun had a northern declination, it would be above the horizon and we would have daytime. It would move in a circle all around us and essentially parallel to the horizon. From day to day it would appear

Figure 3–38 In this series taken in June from northern Norway, above the Arctic Circle, one photograph was taken each hour for an entire day. The sun never set, a phenomenon known as the *midnight sun*. Since the site was not at the north pole, the sun and stars move somewhat higher and lower in the sky in the course of a day.

slightly higher in the sky for 3 months, and then move gradually lower. The date when it is highest in the sky is called the *summer solstice;* it occurs approximately on June 21st each year, and on that date the sun is 23½ degrees above the horizon because its declination is +23½ degrees. The six months of the year when the sun never sets are known as the time of the *midnight sun* (Fig. 3–38). The sun crosses the celestial equator and begins to have northern declination at the vernal equinox. The sun crosses the celestial equator in the other direction six months later at the autumnal equinox.

Let us consider in this paragraph a latitude between the equator and the north pole, say 40° north latitude. Then the stars seem to rise out of the

Figure 3–39 Star trails form circles around the north celestial pole; the dome of the 2.6-meter telescope of the McDonald Observatory in Texas is in the foreground.

Figure 3–40 Near the celestial equator, the star circles are so large that they appear almost straight in this view past the Yerkes Observatory in Wisconsin.

Figure 3–39

Figure 3–40

SUMMER IN NORTHERN HEMISPHERE
WINTER IN SOUTHERN HEMISPHERE

WINTER IN NORTHERN HEMISPHERE
SUMMER IN SOUTHERN HEMISPHERE

Figure 3–41 The seasons occur because the earth's axis is tipped with respect to the plane of its revolution around the sun. When the northern hemisphere is tilted toward the sun, it has its summertime; at the same time, the southern hemisphere is having its winter.

Figure 3–42 For each time of the year, we can see the constellations that are in the direction away from the sun. Because the earth rotates much more rapidly than it revolves around the sun (24 hours compared with 365 days), all observers on the earth see the same constellations in a given season, even though it is daytime for some while it is nighttime for others.

horizon at oblique angles. The north celestial pole (with the north star nearby it) is always visible in the northern sky, and is at an *altitude* of 40° above the horizon. The star Polaris, a 2nd magnitude star, happens to be located within 1° of the north celestial pole, and so is called the *pole star*.

When the sun is at the summer solstice, it is at its greatest northern declination and is above the horizon of northern hemisphere observers for the longest time each day. Thus daytimes in the summer are longer than daytimes in the winter, when the sun is at its lowest declinations. In the winter, the sun not only is above the horizon for a shorter period each day but also never rises very high in the sky. The time of its lowest declination is called the *winter solstice*. Winter in the northern hemisphere corresponds to summer in the southern hemisphere.

The *seasons* (Fig. 3–41), thus, are caused by the variation of declination of the sun, which, in turn, is caused by the fact that the earth's axis of spin is tipped by 23½° with respect to the perpendicular to the plane of the earth's orbit around the sun. This cycle, of course, repeats once each year.

When an astronomer wants to know if a star is favorably placed for observing, he must know both its right ascension and declination. By comparing the right ascension with the sidereal time to compute the hour angle, he can tell if it is the best time of year at which to observe (Fig. 3–42). But

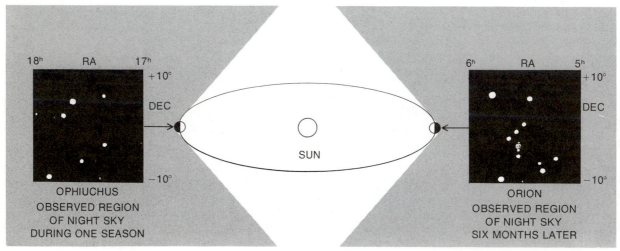

OPHIUCHUS
OBSERVED REGION
OF NIGHT SKY
DURING ONE SEASON

SUN

ORION
OBSERVED REGION
OF NIGHT SKY
SIX MONTHS LATER

© 1970, United Features Syndicate, Inc.

© 1970, United Features Syndicate, Inc.

Figure 3–43 Cartoons by Charles Schulz.

he must also know the star's declination to know how long it will be above the horizon each day.

These concepts involve spherical geometry, which many individuals find difficult to visualize. But a little experience with a telescope can make right ascension and declination seem very easy.

Telescopes are often mounted at an angle such that one axis points directly at the north celestial pole, i.e., at a position in the sky very close to that of Polaris. (In the southern hemisphere, the axis would point at the south celestial pole.) This axis is called the *polar axis*. Thus since all stars move across the sky in circles centered nearly at Polaris, the telescope must merely turn about that axis to keep up with the stellar motions. The other axis of the telescope is used to point the telescope in declination.

Since motion around only one axis is necessary to track the stars, one need have only a small motor set to rotate once every 24 sidereal hours. This motor is linked to turn the polar axis in the direction opposite to the rotation of the earth. The principle is the same for a small telescope in your backyard as for the 5-meter telescope at Palomar. Arrangements of this type are called *equatorial mounts*, since one axis rotates perpendicularly to the celestial equator.

The alternative to this system is to mount a telescope such that one axis rotates parallel to the horizon. This motion is called *azimuth*. A second, perpendicular, axis points the telescope up and down in *altitude* above the horizon. To track a star in this system, continual adjustments would have to be made in both axes, which was formerly very inconvenient. However, the availability of inexpensive minicomputers to make the necessary calculations of spherical geometry has allowed the mounting of large new telescopes in this system, which is called *alt-azimuth*. A large alt-azimuth mount is often less expensive to construct than an equatorial mount of the same size.

3.13 TIME AND THE INTERNATIONAL DATE LINE (*Optional*)

Every city and town on earth used to have its own time system, based on the sun, until widespread railroad travel made this inconvenient. In

Figure 3–44 Although in principle the earth is neatly divided into 24 time zones, in practice, political and geographic boundaries have made the system much less regular. The existence of daylight-saving time in some places and not in others further confuses the time zone system. Some countries even have double daylight-saving time, adding 2 hours. Other countries, shown with cross hatching, have time zones that differ by one-half hour from a neighboring zone. At the international date line, not only the hour but also the date changes. In the U.S., most states have daylight-saving time for six months a year—from the last Sunday in April until the last Sunday in October. But Arizona, Hawaii, and parts of Indiana have standard time year round.

1884, an international conference agreed on a series of longitudinal time zones. Now all localities in the same zone have a *standard time* (Fig. 3–44). Since there are twenty-four hours in a day, the 360° of longitude around the earth are divided into 24 standard time zones, each 15° wide. Because the time is the same throughout each zone, the sun is not directly overhead at noon at each point in a given zone, but in principle is less than about half an hour off.

Standard time is based on a *mean solar day*, the average length of a solar day. The length of a particular day can be slightly different (up to half a minute) from the mean solar day because the earth travels at varying speeds in its elliptical orbit around the sun (as discussed in Section 10.3) and because its axis of rotation is inclined with respect to the plane of its orbit (as discussed in Section 3.11).

As the sun seems to move in the sky from east to west, the time in any one place gets later. We can visualize noon, and each hour, moving around the world from east to west, minute by minute. We would get a particular time back 24 hours later, but if the hours circled the world continuously the date would not have changed. So we specify a line of longitude and have the date change there. We call it the *international date line*. England won for Greenwich, then the site of the Royal Greenwich Observatory, the distinction of having the basic line of longitude, zero degrees. Realizing that the international date line would disrupt the calendars of those who crossed it, that line was thus put as far away from the populated areas of Europe as possible: along the 180° longitude line. The line passes from north to south through the Pacific Ocean and actually bends slightly to avoid cutting through groups of islands, thus providing them with the same date as their nearest neighbor, as shown in Figure 3–44.

In the summer, in order to make the daylight last into later hours, many countries have adopted *daylight-saving time* (D.S.T.). Clocks are set ahead 1 hour on a certain date in the spring. Thus if darkness falls at 6 pm E.S.T., that time will be called 7 pm E.D.T., and most people will have an extra hour of daylight after work. In most places, that hour will be taken away in the fall, though some places have adopted daylight-saving time all year. The phrase to remember to help you set your clocks is "fall back, spring ahead." Of course, daylight-saving time is just a bookkeeping change in how we name the hours, and doesn't result from any astronomical changes.

Some countries have adopted daylight-saving time all year long, in order to provide more light in the evening hours. France now adds this hour all year and adds a second hour—double daylight-saving time—in the summer.

3.14 CALENDARS *(Optional)*

The period of time that the earth takes to revolve once around the sun is called, of course, a *year*. This period is about 365¼ mean solar days. A *sidereal year* is the interval of time that it takes the sun to return to a given position with respect to the stars. A *solar year*, or *tropical year*, is the period of time that it takes the sun to return to the vernal equinox, the point where the ecliptic crosses the celestial equator. (A sidereal year differs from a solar year by 20 minutes 24 seconds because of the earth's precession, discussed in Section 3.11.) Solar years are growing shorter by 0.005 second per year (half a second per century).

The Roman calendars had, at different times, different numbers of days in a year, and so the dates rapidly drifted out of synchronization with the seasons (which follow tropical years). Julius Caesar decreed that 46 B.C. would be a 445 day year in order to catch up, and defined a calendar, the *Julian calendar*, that would be more accurate. This calendar had years that were normally 365 days in length, with an extra day inserted every fourth year in order to bring the average year to 365¼ days in length. The fourth years were, and are, called *leap years*.

The Julian calendar was much more accurate than its predecessors, but the actual solar year 1980 will be 365 days 5 hours 48 minutes 45.6 seconds long, some 11 minutes 14 seconds shorter than 365¼ days (365 days 6 hours). By 1582, the calendar was about 10 days out of phase with the date at which Easter had occurred at the time of a religious council 1250 years earlier, and Pope Gregory XIII issued a bull—a proclamation—to correct the situation. He dropped 10 days from 1582.

In the *Gregorian calendar*, years that are evenly divisible by four are leap years, except that three out of every four century years, the ones not divisible evenly by 400, have only 365 days along with all the other years. Thus 1600 was a leap year, 1700, 1800, and 1900 were not, and 2000 will again be a leap year. Although many countries adopted the Gregorian calendar as soon as it was promulgated, Great Britain (and its American colonies) did not adopt it until 1752, when 11 days were thus skipped. As a result, we celebrate George Washington's birthday on February 22nd, even though he was born on February 11. (Actually, since the year used to begin in March instead of January, some months had to be transferred from one year to another. Washington was born in 1731, but we refer to his date of birth as February 22nd, 1732.) The Gregorian calendar is the one in current use. It will be over 3000 years before this calendar is as much as one day out of step with the seasons.

The name of the fifth month, formerly Quintilis, was changed to honor Julius Caesar; in English we call it July. The year then began in March; the last four months of the year still bear names from this system of numbering. Augustus Caesar, who carried out subsequent calendar reforms, renamed August after himself. Also he transferred a day from February in order to make August last as long as July.

Many citizens of that time objected to the supposed loss of the time from their lives and to the commercial complications. Does one pay a full month's rent for the month from which the days were omitted, for example? "Give us back our fortnight," they cried.

SUMMARY AND OUTLINE OF CHAPTER 3

Aims: To understand the different types of observing techniques and instruments used in astronomy

Optical telescopes (Sections 3.1 to 3.6)
    Resolution and twinkling of stars
    Refracting and reflecting telescopes, Schmidt cameras
    Optical observatories and their distribution across the world
    Light-gathering power and magnification
Spectroscopy and photometry (Section 3.8)
Types of astronomical detectors for different types of radiation (Section 3.9)
Coordinate systems: right ascension and declination, sidereal time, precession (Sections 3.11 and 3.12)
Standard time and the international date line (Section 3.13)
Calendars (Section 3.14)

QUESTIONS

1. What advantages does a reflecting telescope have over a refracting telescope?

2. What limits the resolving power of the 5-meter telescope?

3. List the important criteria in choosing a site for an optical observatory meant to study stars and galaxies.

4. Two reflecting telescopes have primary mirrors 2 m and 4 m in diameter. What is the difference in magnitude between the faintest stars you can observe with each telescope in a given observing time?

5. Why can't we use ordinary photographic plates to record infrared images?

6. Why must we observe ultraviolet radiation using rockets or satellites, while balloons are sufficient for infrared observations?

7. Why might some stars appear double in blue light though they could not be resolved in red light?

8. For each of the following, identify whether it is a characteristic of a reflecting telescope, a refracting telescope, both, or neither. Give any limitations on the applicability of your answer.

   (a) Free of chromatic aberration.

   (b) Has more severe spherical aberration.

   (c) Requires aluminizing.

   (d) Can be used for photography.

   (e) Has an objective that must be supported only by its rim.

   (f) Can be made in larger sizes.

   (g) Has a coudé focus.

   (h) Has a prime focus at which a person can work without blocking the incoming light.

9. Why can radio astronomers observe during the day, while optical astronomers are (for the most part) limited to nighttime observing?

10. What are the advantages of the Multi-Mirror Telescope? How will it compare with the 5-meter telescope for studying faint objects?

11. What are the similarities and differences between making radio observations and using a reflector for optical observations. Compare the radiation path, the detection of signals, and limiting factors.

12. Why is it sometimes better to use a small telescope in orbit around the earth than it is to use a large telescope on a mountaintop?

13. Between the vernal equinox, March 21st, and the autumnal equinox, 6 months later,

   (a) by how much does the right ascension of the sun change;

   (b) by how much does the declination of the sun change;

   (c) by how much does the right ascension of Sirius change;

(d) by how much does the declination of Sirius change?

14. (a) How does the declination of the sun vary over the year?

    (b) Does its right ascension increase or decrease from day to day? Justify your answer.

15. We normally express longitude on the earth in degrees. Why would it make sense to express longitude in units of time?

16. If a planet always keeps the same side toward the sun, how many sidereal days are there in a year on that planet?

17. What is the advantage of an equatorial mount? Why did large telescopes used to be made with equatorial mounts while they are now sometimes made with alt-azimuth mounts?

Variable Stars and
Stellar Groupings

We often think of stars as individual objects that shine steadily, but many stars vary in brightness and most stars actually have companions close by. Also, many stars appear as members of groupings called associations or clusters. By studying the effects that the stars have on each other, or by studying the nature of all the stars in a stellar cluster, we can derive much information that we could not discover by studying the stars one at a time. This information even leads us to a general understanding of the life history of the stars.

4.1 BINARY STARS

Most of the objects in the sky that we see as single "stars" really contain two or more component stars. The easiest such object to see with the naked eye is "one" of the stars in the handle of the Big Dipper. The brighter star is called Mizar and the companion is called Alcor. The ability to see with the naked eye that the system is, in fact, a double star is one of the standard tests of a person's visual acuity.

Through a telescope of any size, it is easy to see that the system is a multiple star. In fact, one can even see that one of the two major components, Mizar, is itself a double star (Fig. 4–1).

Sometimes a star appears double merely because two stars that are located at different distances from the sun appear in the same line of sight. Such systems are called *optical doubles*, and will not concern us here. We are more interested in stars that are physically associated with each other.

For actual doubles, the easiest way to tell that there are two instead of one is by looking through a telescope of sufficiently large aperture. Stars that appear double when observed directly are called *visual binaries*. The resolution of small telescopes may not be sufficient to allow the com-

Albireo (β Cygni) contains a B star and a K star, which make a particularly beautiful pair because of their different colors.

M13, a globular cluster in Hercules.

Figure 4–1 Alcor and Mizar provide examples of both visual and spectroscopic binaries.

ponents of a double star to be "separated" from each other, and larger telescopes can thus distinguish more double or multiple stars. When the stars have different colors, they form particularly beautiful objects to observe in even small telescopes. Five or ten per cent of the stars in the sky are visual binaries.

We are not always fortunate enough, however, to have the two components of a double star appear far enough apart from each other to be observed visually. Sometimes a star appears as a single object through a telescope, but one can see that the spectrum of that "object" actually consists of the overlapping spectra of at least two objects. If we can detect the presence of two spectra of different types—of, say, one hot star and one cool star—then we say that the object has a *composite spectrum.* If, on the other hand, over a period of time we can actually tell that the stars are revolving around each other by the fact that the spectra are changing in wavelength with respect to each other because of the Doppler shift, then we call the "object" a *spectroscopic binary* (Fig. 4–2). As the objects orbit each other, unless we are looking straight down on the orbit from above, they each spend half of each orbit at velocities approaching us and the other half at velocities receding from us, relative to their average space motions. The velocity variations in spectroscopic binaries are periodic and, of course, the spectrum of each component varies separately in wavelength. Note that even if the spectrum of one of the stars is too faint to be seen, we can still

Figure 4–2 Two spectra of Mizar (ζ Ursae Majoris) taken 2 days apart show that it is a spectroscopic binary. The lines of both stars are superimposed in the upper stellar absorption spectrum, but are separated in the lower spectrum by 2 Å, which corresponds to a relative velocity of 140 km/sec. Emission lines from a laboratory source are shown at the extreme top and extreme bottom to provide a comparison with a source at rest.

4415.1 Å 4481 Å 4526.6 Å

Figure 4–3 Two spectra of α Geminorum taken at different times show a Doppler shift. Thus the star is a spectroscopic binary, even though lines from only one of the components can be seen. The comparison spectrum of a laboratory source appears at the top and bottom. Note that emission lines from this laboratory source are in the same horizontal positions at extreme top and bottom, while the absorption lines of the stellar spectra are shifted laterally (that is, in wavelength) with respect to each other.

tell if the star is a spectroscopic binary from the variations in radial velocity of the visible component (Fig. 4–3). Two-thirds of all solar-type stars have stellar companions; the other third may have less massive companions (see Section 19.2). Though the presence of companions of stars of other spectral types has not been studied in such detail, we can reasonably conclude that few stars are single.

Sometimes the stars pass in front of each other, as seen from our viewpoint on the earth. The "double star" then changes in brightness periodically, as one star cuts off the light from the other. Such a pair of stars is called an *eclipsing binary* (Fig. 4–4). The easiest to observe is Algol, β Persei (beta of Perseus), in which the eclipses take place every 69 hours, dropping the total brightness of the system from magnitude 2.3 to 3.5. (A third star is present in the Algol system. It orbits the other two every 1.87 years, and does not participate in the eclipses.)

A double star may fall in more than one of these categories. For example, two stars that revolve around each other in an orbit that makes them eclipse would compose a spectroscopic binary too. Only nearby binary stars are close enough to be seen as visual binaries, but we can detect spectroscopic binaries or binaries with composite spectra at much greater distances, even in other galaxies.

Note that the way that a binary star appears to us depends on the orientation of the two stars not only with respect to each other, but also with respect to the earth. If we are looking down at the plane of their mutual orbit (Fig. 4–5) then we might see a composite spectrum; under the most favorable conditions the star might be seen as a visual binary. But in this orientation we would never be able to see the stars eclipse. We would also not be able to see the Doppler shifts typical of spectroscopic binaries, since only radial velocities contribute to the Doppler shift.

It is possible that a star could be a "double," but still not be detectable by any of the above methods. There are cases where the existence of a double star shows up only as a deviation from a straight line in the proper motion of the "star" across the sky. Such stars are called *astrometric binaries* (Fig. 4–6). We will say more of this later on, as this identification technique

Figure 4–4 The shape of the light curve of an eclipsing binary depends on the sizes of the components and the angle from which we view them. At lower left, we see the light curve that would result for star B orbiting star A, as pictured at upper left. When star B is in the positions shown with subscripts at top, the regions of the light curve marked with the same subscripts result. At right, we see the appearance of the orbit and the light curve for star D orbiting star C, with the orbit inclined at a greater angle than at left. From earth, we observe only the light curves, and use them to determine what the binary system is really like, including the inclination of the orbit and the sizes of the objects.

Figure 4–5 The appearance and the Doppler shift of the spectrum of a binary star depend on the angle from which we view the star.

Figure 4–6 Sirius A and B, an astrometric binary. From studying the motion of Sirius A (often called, simply, Sirius) astronomers deduced the presence of Sirius B before it was seen directly. Sirius B's orbit is larger because Sirius A is a more massive star.

has been important in the discovery of white dwarfs (see Section 7.4) and of very cool ordinary dwarfs, and is important for the discussion of the chances of finding extraterrestrial life (see Section 19.2).

Many of the celestial sources of x-rays that have been observed in recent years from orbiting telescopes turn out to be binary systems. Matter from one member of the pair falls upon the other member, heats up, and radiates x-rays. These binary x-ray sources will be discussed in Sections 8.10 and 9.6.

Though one usually speaks of double stars, there is actually nothing to prevent stars from existing in threes, or fours, or with even more components. For example, Mizar's two visual components, called Mizar A and Mizar B, have each been found to be spectroscopic binaries. Thus Mizar is a quadruple star.

4.2   STELLAR MASSES

The study of binary stars is of fundamental importance in astronomy because it allows us to determine stellar masses. If we can determine the orbits of the stars around each other, we can calculate theoretically the

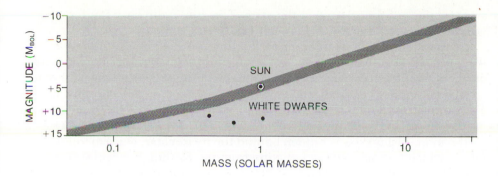

Figure 4-7 The mass-luminosity relation, measured from binary stars. Most of the stars fit within the shaded band. The relation is defined by straight lines of different slopes for stars brighter than and fainter than magnitude +7.5; we do not know if there is a real difference between the brighter and fainter stars that causes this or whether it is an effect introduced when the data are reduced, at which stage we try to take account of the energy that the stars radiate outside the visible part of the spectrum. Note that white dwarfs, which are relatively faint for their masses, lie below the mass-luminosity relation. Red giants lie above it. Thus we must realize that the mass-luminosity relation holds only for main-sequence stars.

masses of the stars necessary to produce the gravitational effects that lead to those orbits. For a star that is a visual binary with a sufficiently short period (only 20 or even 100 years, for example), we are able to determine the masses of both of the components. If a star is a spectroscopic binary, even one with spectral lines of both stars present, we can find only lower limits for the masses—that is, we can say that the masses must be larger than certain values. This limitation occurs because we cannot usually tell the inclination of the plane of the orbit of the stars around each other.

So we do not always know as much as we would like about the masses of the stars in binary systems. Of course, we are better off than we are for stars outside binary systems, for we cannot directly measure their masses at all! From studies of the several dozen binaries for which we can accurately tell the masses, astronomers have graphed the brightnesses of the stars against the masses. ("Against" in this context means that the brightnesses are plotted on one axis and the masses on the other.) Stars lie on a narrow band, the *mass-luminosity relation* (Fig. 4-7), which shows that the masses and luminosities (measured from the brightnesses) of stars are closely related. The mass-luminosity relation in this form is valid only for stars on the main sequence, that is, for normal dwarfs. The more massive a star is, the brighter it is. The mass, in fact, is the prime characteristic that determines where on the main sequence a star will settle down to live its lifetime.

About 90 per cent of stars, those on the main sequence, follow the mass-luminosity relation. The other stars, such as giants or white dwarfs, are special cases that we will discuss later on.

There are many more stars of low mass than there are of high mass. In fact, the number of stars of a given mass in any sufficiently large volume of space increases as we consider stars of lower and lower mass. Thus most of the mass in a cluster of stars, or in a galaxy, comes from stars at the lower end of the H-R diagram. Even though the many faint stars contain most of the total mass in a group of stars, most of the light that we receive from a group of stars, on the contrary, comes from the few very brightest stars. A single O star can outshine a thousand K main-sequence stars. Thus the majority of the stars in a star system together give off a minor fraction of the light; the few brightest stars in that system give off most of the radiation.

4.3 STELLAR SIZES

Stars appear as points to the naked eye and as small, fuzzy disks through large telescopes. The blurring of the stellar images into disks results from distortion by the earth's atmosphere, and hides the actual size and structure of the surface of the stars. The sun is the only star for which we can measure its angular diameter easily and directly.

Astronomers use an indirect method to find the sizes of most stars. If we know the absolute magnitude of a star (from some type of parallax measurement) and the temperature of the surface of the star (from measuring its spectrum), then we can tell the amount of surface area the star must have. This, of course, depends on the radius.

One type of direct measurement works only for eclipsing binary stars. As the more distant star is hidden behind the nearer star, one can follow the rate at which the intensity of radiation from the further star declines. From this information together with Doppler measurements of the velocities of the components, one can calculate the sizes of the stars. One can sometimes even tell how the brightness varies across a star's disk.

It is much more difficult to measure the size of a single star directly. One way of measuring the diameter is by *lunar occultation.* As the moon moves with respect to the star background, it occults (hides) stars. By studying the light from a star in the fraction of a second it takes for the moon to completely block it, we can deduce the size of the stellar disk. Only a handful of stars have had their diameters measured by this technique up to the present time.

Other methods for measuring the diameters of single stars use a principle called *interferometry,* which will be discussed in the context of radio observations in Chapter 23. Even though the starlight passes through the earth's atmosphere without being occulted by either the moon or by a companion star, the method overcomes the problems of blurring by the earth's atmosphere.

The first such measurements were carried out at Mt. Wilson 50 years ago; only 7 of the stars that subtended the largest angles in the sky had their diameters measured. These were the largest of the closest stars, that is, nearby red giants and supergiants. A modern variation of this technique has been worked out at the University of Maryland.

Over the last ten years, a new procedure called *speckle interferometry* has been developed by a team of scientists at the Paris Observatory in France and at the State University of New York at Stony Brook. At any one instant, the image of a star through a large telescope looks speckled because different parts of the image are affected by different small turbulent areas in the earth's atmosphere. A long exposure blurs these speckles, and gives us the fuzzy disk normally photographed. The speckle interferometry technique involves taking photographs of the speckle pattern with very short exposures—on the order of 1/100 second—or using electronic detection devices and then using mathematical techniques and computer assistance to deduce the properties of the starlight that entered the telescope. Thus, in this manner, from many short exposures of a stellar image, scientists can measure the diameter and the limb darkening. The Kitt Peak National Observatory is an especially active center for speckle interferometry. All the above interferometric methods work best for large, cool stars. For the last twenty years, R. Hanbury Brown and his associates in Australia have used another type of stellar interferometer (Fig. 4–8) that works best for hot, bright stars. They have determined the diameters of three dozen stars whose diameters cannot be measured with the other techniques.

Only a few dozen stellar diameters have been measured directly, and they confirm the indirect measurements (Fig. 4–9). Direct and indirect measurements show that the diameters of main-sequence stars decrease as

The surface of a star is darker at its limb (edge) than at the center of the disk. You can see this phenomenon for the sun on page 94.

Figure 4–8 Hanbury Brown's intensity interferometer in Australia, used to measure the diameters of hot stars. The two sets of mirrors can be placed up to 188 meters apart from each other.

we go from hotter to cooler; that is, O dwarfs are relatively large and F dwarfs are relatively small. (Recall that "dwarf" simply means "main-sequence star.") Indirect measurements alone show that this trend continues for K and M stars, which are smaller still. As for stars that are not on the main sequence, red giants are indeed giant in size as well as in brightness.

Interferometric techniques have been used not only to measure the diameters of individual stars but also to resolve the components of close binary stars. The speckle interferometry technique, for example, has already resolved the components of dozens of close binary stars. Several of the systems had not been previously known to be binary.

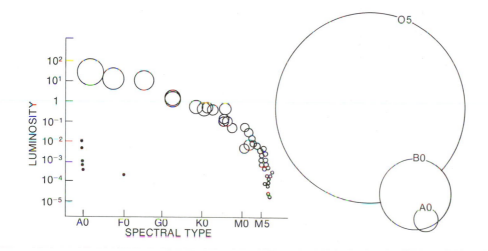

Figure 4–9 The H-R diagram for stars within 17 light years of the sun. The open circles indicate approximately the relative diameters of the main-sequence stars; the white dwarfs are shown as dots. In addition, relative sizes of O, B, and A stars are shown at right.

4.4 VARIABLE STARS

Some stars can be seen to vary in brightness with respect to time. One basic parameter that characterizes the variation of a variable star is the *period*. The period is the time it takes for a star to go through its entire cycle of variation and to return to the original degree of brightness.

One way for a star to vary in brightness, as we have seen, is for it to be an eclipsing binary. Sometimes we get the light simultaneously from both members of the binary, and sometimes the light from one of the members is at least partially blocked by the other member. But it is possible for individual stars to vary in brightness all by themselves. Thousands of such stars are known in the sky. The periods of the variations can range from seconds for some types of stars to years for others. Some types of variable stars, as we shall see, can be used to provide vital information about the distance scale.

Many variable stars are studied on photographic plates taken at professional observatories. In the early part of this century, visual and photographic study of variable stars was an important activity for astronomers, and major discoveries followed. Nowadays, not many professionals choose to spend their time studying variable stars, yet it takes a lot of hard work over an extended period of time to keep track of what the variables are doing. An international amateur group called the American Association of Variable Star Observers (AAVSO), with its headquarters in Cambridge, Massachusetts, is now the center of much of the observational activity on variable stars. The headquarters staff averages the measurements of all the observers of a given star on a given day, and plots the variations of the magnitude of the observed stars with time. Such plots of magnitude versus time are called *light curves*.

Finding the distance scale in the universe, the distances to astronomical objects in linear units like km or A.U., is one of the basic problems that astronomers attack.

There is a varied menagerie of variable stars corresponding to different physical conditions in the stars. We will limit our discussion to two types of variables that have provided important information about the scale of distance in the universe.

4.4a   Cepheid Variables

In Greek mythology, Cepheus was a king of Ethiopia, Cassiopeia's husband, and Andromeda's father.

The most important type of variable star in astronomy is the *Cepheid variable* (cef′e-id). The prototype is δ Cephei (delta of Cepheus). Cepheid variables have very regular periods that, for individual Cepheids (as they are called), can be from 1 to 100 days.

Cepheids are relatively rare stars in our galaxy; only about 700 are known. Some have been detected in other galaxies. δ Cephei itself, as an example from our own galaxy, varies between apparent magnitudes 3.6 and 4.3 with a period of 5.4 days (Fig. 4–10). Another Cepheid in our galaxy is

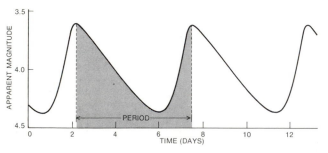

Figure 4–10   The light curve of δ Cephei, the prototype of the class of Cepheid variables.

Polaris, the north star, which is not as obviously varying since the maximum change in brightness it undergoes is only 0.1 magnitude; it varies between apparent magnitudes 2.5 and 2.6 with a period of about 4 days. (Polaris is actually a double star; a 9th magnitude companion is also part of the system.)

Cepheids are important because a relation has been found that links the periods of their light changes, which are simple to measure, with their absolute magnitudes. For example, if we measure that the period of a Cepheid is 10 days, we need only look at Figure 4–11 to see that the star is of absolute magnitude −3. From the absolute magnitudes, as we shall review later in this section, one can find their distances from the sun by using reasoning similar to that used in our earlier discussion of absolute magnitudes of stars.

Figure 4–11 The period-luminosity relation for Cepheid variables.

The story started 70 years ago with Henrietta Leavitt (Fig. 4–12) who, at the Harvard College Observatory, was studying the light curves of variable stars in the southern sky. In particular, she was studying the variables in the Large and Small Magellanic Clouds (Fig. 4–13), two hazy areas in the sky that were discovered by the crew of Magellan's expedition around the world when they sailed far south. Whatever the Magellanic Clouds were—we now know them to be galaxies, but Leavitt did not know this— she knew that all the stars in each cloud were at approximately the same distance away from the earth. Thus even though she could only plot the apparent magnitudes, the relation of the absolute magnitudes to each other was exactly the same as the relation of the apparent magnitudes.

By 1912, Leavitt had established the light curves and determined the periods for two dozen stars in the small Magellanic Cloud. She plotted the magnitude of the stars (actually a median brightness, a value between the maximum and minimum brightness) against the period. She realized that there was a fairly strict relation between the two quantities, and that the Cepheids with longer periods were brighter than the Cepheids with shorter periods. By simply measuring the periods, she could determine the magnitude of one star relative to another; each period uniquely corresponds to a magnitude.

Figure 4–12 Henrietta S. Leavitt, who worked out the period-luminosity relation, at her desk at the Harvard College Observatory.

Henrietta Leavitt could measure only **apparent** magnitude but did not know the distance to the Magellanic Clouds, so she did not know the **absolute** magnitude (intrinsic brightness) of the Cepheids. To find the distance to the Magellanic Clouds, we first had to be able to find the distance to *any* Cepheid—even one not in the Magellanic Clouds—to tell its absolute magnitude. A Cepheid of the same period but located in the Magellanic Clouds would presumably have the same absolute magnitude.

To find the absolute magnitude of a Cepheid it would seem easiest to start with the one nearest to us. Unfortunately, not a single Cepheid is close enough to the sun to allow its distance to be determined by the method of trigonometric parallax. More complex, statistical methods had to be used to study the relationships between stellar motions and distances. This gave the distance to a nearby Cepheid. Once we have the distance to and thus the absolute magnitude of a nearby Cepheid, all stars of that same period in the Magellanic Clouds would have the same absolute magnitude as the nearby one. From that point on it is easy to tell the absolute magnitude of Cepheids of **any** period in the Magellanic Clouds. After all, if a Cepheid has **apparent** magnitude that is, say, 2 magnitudes brighter than the apparent magnitude of our Cepheid of known intrinsic brightness, then its **absolute** magnitude is also 2 magnitudes brighter. After this process, we

Figure 4–13 From the southern hemisphere, the Magellanic Clouds are high in the sky. In this view from Perth, Australia, the Large Magellanic Cloud is about 20° above the horizon and the Small Magellanic Cloud is twice as high. They are not quite this obvious to the naked eye. Canopus, the second brightest star in the sky, appears at the lower left.

have the period-luminosity relation in a more useful form—period vs. absolute magnitude. We call this process "the calibration" of the period-luminosity relation.

Thus Cepheids can be employed as indicators of distance, using the same method that we have already discussed (Section 2.11) to determine the distance: (1) the period-luminosity relation gives us the absolute magnitude of the Cepheid; (2) we then calculate how far a star of that absolute magnitude would have to be moved from the standard distance of 10 parsecs to appear as a star of the apparent magnitude that we observe.

When the calibration of the period-luminosity relation was worked out quantitatively by the American astronomer Harlow Shapley in 1917, the distance to the Magellanic Clouds could be calculated. They were very far away, a distance that we now know means that they are not even in our galaxy! Instead, they are galaxies by themselves, two small irregular galaxies that are companions of our own, larger, galaxy.

Cepheids can also be seen not only in our own galaxy and in the Magellanic clouds but also in more distant galaxies. They are bright stars (giants and supergiants, in fact) and so can be seen at quite a distance. The use of Cepheids is the prime method of establishing the distance to all the nearer galaxies.

4.4b RR Lyrae Variables

Many stars are known to have short regular periods, less than one day in duration. Certain of these stars, no matter what their period, have the same average absolute magnitude. Again, as with Cepheids, none of these short-period variables is within range to obtain a trigonometric parallax, and the calibration of the absolute magnitude scale had to be accomplished by their membership in clusters.

Such stars are called *RR Lyrae stars* after the prototype of the class (Fig. 4–14). Since many of these stars appear in globular clusters (which will be described in Section 4.5), RR Lyrae stars are also called *cluster variables.*

Once we detect an RR Lyrae star, we immediately know its absolute magnitude, since all the absolute magnitudes are the same (about 0.5). Just as before, we measure the star's apparent magnitude and can thus easily calculate its distance and the distance to the cluster.

Figure 4–14 The light curve of RR Lyrae, the prototype of the class known as RR Lyrae variables or cluster variables.

4.5 CLUSTERS AND STELLAR POPULATIONS

Even aside from the hazy band of the Milky Way, the distribution of stars in our sky is not uniform. There are certain areas where the number of stars is very much higher than the number in adjacent areas. Such sections of the sky are called *star clusters* (Color Plates 4 and 5).

One type of star cluster appears as just an increase in the number of stars in that limited area of sky. Such clusters are called *open clusters,* or *galactic clusters* (Fig. 4–15). The most familiar example of a galactic cluster is the Pleiades, a group of stars visible in the winter sky. The eye sees at least six stars very close together. With binoculars or the smallest telescopes, dozens more can be seen. A larger telescope reveals hundreds of stars. Another galactic cluster is called the Hyades, which form the "v" that outlines the face of Taurus, the bull, a constellation best visible in the winter sky (Fig. 4–16). More than a thousand such clusters are known, most of them too faint to be seen except with telescopes. (The distances to many of these clusters can be determined, as discussed in Section 4.5a.) It has been

The Pleiades are often known as the Seven Sisters, *after the seven daughters of Atlas who were pursued by Orion and who were given refuge in the sky. That one is missing—the Lost Pleiad—has long been noticed. Of course, a seventh star is present (and hundreds of others as well), although too faint to be plainly seen with the naked eye. The Pleiades seem to be riding on the back of Taurus.*

Figure 4–15 The galactic cluster M67 in the constellation Cancer, the crab.

Figure 4–16 The Hyades are a galactic cluster that outlines the face of Taurus, the bull, pictured here in a plate from the atlas of Hevelius published in 1690. The star map is reversed left to right compared to the view we see in the sky, because it is the view that one would see if one viewed a star globe from outside.

found that all the stars in a galactic cluster are packed into a volume not more than 10 parsecs across.

Stars in galactic clusters seem to be representative of stars in the spiral arms of our galaxy and of other galaxies. (Spiral structure is described in Section 20.2.) When the spectra of stars in galactic clusters are analyzed to find the relative abundances of the chemical elements in their atmospheres, we find that over 90 per cent of the atoms are hydrogen, most of the rest are helium, and less than 1 per cent are elements heavier than helium.* This is similar to the composition of the sun (which is discussed in Section 5.2b and Table 5–1).

The second major type of star cluster appears in a small telescope as a small, hazy area in the sky. Observation with larger telescopes can distinguish individual stars, and reveals that these clusters are really composed of many thousands of stars packed together in a very limited space. The clusters take spherical forms, and are known as *globular clusters* (Fig. 4–17).

Globular clusters can contain 10,000 to one million stars, in contrast with the 20 to several hundred stars in a galactic cluster. A globular cluster can fill a volume up to 100 parsecs across. A globular cluster has fewer stars at its periphery and more closely packed stars toward the center. One can find the distance to many of these clusters by observing the RR Lyrae stars, which are found in many of them.

Globular clusters seem to avoid the plane of our galaxy and are typically

---

*The structure of the nuclei of the elements is discussed in Section 6.3. For the present, we need only know that hydrogen, whose dominant form contains only one nuclear particle, is the least massive element, and helium, whose dominant form contains 4 nuclear particles, is the second least massive element.

Figure 4–17 The globular cluster M3 in the constellation Canes Venatici, the hunting dogs.

found only in regions above and below the galactic plane. We say that they are in the galactic *halo*. The abundances of the elements heavier than helium in globular clusters are much lower, by a factor of 10 or more, than their abundances in the sun.

4.5a H-R Diagrams for Galactic (Open) Clusters

The Hertzsprung-Russell diagram for several galactic clusters is shown in Figure 4–18. For the purposes of this discussion it is most important to note that the horizontal axis is a measure of temperature and the vertical axis is a measure of brightness.

The H-R diagrams for different galactic clusters appear to agree over part of the main sequence but diverge at the upper left. These stars at the upper left are the more massive ones. They are more luminous and use up their nuclear fuel at a faster rate than the cooler, more numerous, ordinary stars like the sun. The massive stars are on the main sequence in the stable, middle-age part of their lifetimes. Later in their lives they become larger and thus brighter, since they then have more surface area to radiate light. At that time they move upward on the H-R diagram from the main sequence. Their outer layers may also, after a time, become cooler, and the star will then move toward the right on the H-R diagram.

In Fig. 4–18, we see graphs from several clusters overlaid.

Figure 4–18 This Hertzsprung-Russell diagram of several galactic clusters shows that their fainter stars are on the main sequence, while the brighter stars are above and to the right of the main sequence. Almost all stars in the younger clusters, like h and χ Persei, follow the main sequence, while the hotter and more massive members of older clusters, like M67, have had time to evolve toward the red giant region. By observing the point where the cluster turns off the main sequence, we deduce the length of time since its stars were formed, i.e., the age of the cluster. The presence of the *Hertzsprung gap* (shaded), a region in which few stars are found, indicates that stars evolve rapidly through this part of the diagram.

Part of the H-R diagram for the globular cluster M3 (hollow bar) is graphed for comparison.

When we examine the H-R diagram of a cluster, we are observing a group of stars all of which were presumably formed at the same time in the same region of space and so have similar chemical compositions. We presume that all the stars "fell" on the main sequence during most of their lives (that is, the points representing their magnitudes and temperatures were on the main sequence). Points representing the fainter stars usually fall on the main sequence of the H-R diagram. But for some galactic clusters (Fig. 4–18) the points representing the brighter, more massive stars are to the upper right of the main sequence. We interpret this to mean that these stars have passed through the phase of their lives in which they fell on the main sequence, and have "moved off" the main sequence as they aged and evolved. Thus we deduce that a given cluster is older than the main-sequence lifetimes of those stars in that cluster that have moved off the main sequence. On the other hand, the fact that stars of a certain temperature and absolute magnitude, and thus a certain mass, are still on the main sequence indicates that the cluster is younger than their main-sequence lifetimes. The age of the cluster is between these two values.

If we measure apparent magnitudes for a large number of stars in a cluster, we can deduce the absolute magnitude by the following method. We already know that all the stars in the cluster are at the same distance. Since we know that the lower part of the main sequence will agree for all galactic clusters, we can take an H-R diagram measured in apparent magnitudes and merely slide it up and down along lines of constant color until we see the best agreement with respect to a standard temperature-magnitude diagram. Then we have found the number of magnitudes by which the absolute

Figure 4–19 The double cluster in Perseus, h and χ Persei, a pair of galactic clusters that are readily visible in a small telescope and close enough together that they appear in the same field of view. Perseus is a northern constellation that is most prominent in the winter sky. In Greek mythology, Perseus slew the Gorgon Medusa and saved Andromeda from a sea monster.

and apparent magnitude scales differ and thus *from the H-R diagram we can easily calculate the distance to the cluster.*

For example, let us consider just one point on the main sequence. A-type stars are known to have absolute magnitude of about +2. If we see an A star of apparent magnitude +7, we know that the cluster is farther enough away than 10 parsecs that its stars appear 5 magnitudes = 100 times fainter. Because of the inverse square law of brightness, we know that the star and its cluster are 10 times further away than 10 parsecs, which is 100 parsecs (see Section 2.12). Actually, this procedure is more precise when we *fit* (superimpose) a whole segment of a main sequence, or some other part of an H-R diagram. The procedure is thus called *fitting*, as in "fitting the main sequence."

For a single star, study of its spectrum can tell us its spectral type and its luminosity class. Then the H-R diagram gives us its absolute magnitude. By comparison with the apparent magnitude, we find its distance; this is called the method of *spectroscopic parallax.*

Figure 4–18 shows that stars in the double cluster in Perseus, h and χ Persei, follow the main sequence for most of its length. The line of points representing the stars turns off the main sequence only at its upper left. Thus h and χ Persei are very young clusters. They may have formed only 10^7 years ago, a short time scale from an astronomical point of view. The Pleiades, which turns off closer to spectral type A than to B, is still a young cluster but is somewhat older than the double cluster in Perseus. It may be 10^8 years old. The Hyades, which turns off the main sequence still further down, may be much older, possibly 10^9 years.

4.5b H-R Diagrams for Globular Clusters

The H-R diagram for a globular cluster is shown in Figure 4–20. H-R diagrams for all globular clusters have similar forms.

These H-R diagrams have only short main sequences, so evidently the globular clusters must be very old. Some stars even appear in a *horizontal branch* at approximately absolute magnitude zero. This horizontal branch goes between the left side of the diagram and the stars on the right side of the diagram that have long since turned off the main sequence. In the horizontal branch there is a gap where stars of constant brightness are not found. Only the RR Lyrae stars fall in this gap.

Figure 4–20 The Hertz-sprung-Russell diagram for the globular cluster M3. There is a gap in the horizontal branch where no stars of constant brightness are found. The RR Lyrae variables fall here, some 200 for this cluster. We can observe only the members of the cluster that are brighter than a certain limit that depends on the size of our telescope and the darkness of the night sky. Thus no stars appear in the shaded region at the bottom.

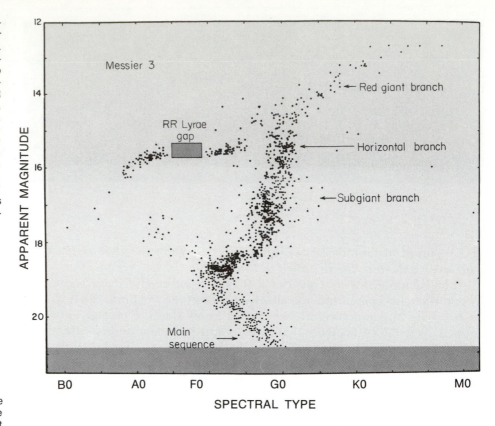

Figure 4–21 The circle shows the position of the source of x-ray pulses that are being interpreted as indicating the presence of a massive black hole. When this circle, whose size indicates the uncertainty of the position measurement from the SAS-3 satellite, is drawn on a photograph of the sky, globular cluster NGC 6624 is included in it.

4.5c Of Globular Clusters, X-Ray Bursts, and Black Holes (Optional)

For the last few years, since scientists have been able to place telescopes in orbit, they have been able to observe x-rays from celestial sources. (Chapter 22 contains a discussion of x-ray sources in our galaxy.) Some of

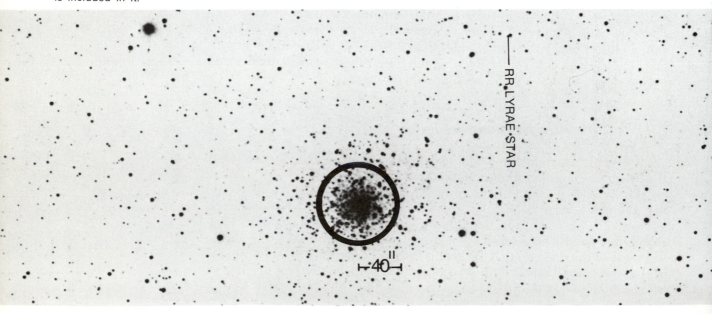

the x-ray sources are associated with globular clusters; when one measures the position in the sky of one of these x-ray sources and looks at a visual image of the region, a globular cluster is right there.

The strength of the x-radiation varies strongly. In one of the globular clusters, NGC 6624 (Fig. 4–21) (the 6624th object in the New General Catalogue, a standard list of visible objects in the sky compiled a hundred years ago), the x-radiation sometimes changes by a factor of 5 from one minute to the next. In 1975, telescopes aboard the Astronomical Netherlands Satellite and other satellites recorded an even stranger variation (Fig. 4–22): intense bursts of x-rays from this globular cluster, each representing an increase in intensity of 20 to 30 times in half a second, followed by a decay in the intensity for the next 8 seconds (Fig. 4–23). The bursts occur about every 4½ hours; the interval between them varies by about 15 minutes. In each burst the cluster emits about a million times the energy radiated by the sun in the same time. At least 5 globular clusters are known to give off such bursts. Similar strange bursts also come from other directions in the sky, and it is possible that there are globular clusters there too, perhaps hidden by interstellar dust.

The most likely current interpretation of these bursts involves a very strange notion that had been advanced on theoretical grounds even prior to the discovery of bursts: a very massive black hole may exist at the center of many globular clusters. (We discuss at length the properties of black holes in Chapter 9, but suffice it to say here that they are places in space where matter has collapsed in on itself to such an extent that the gravity has increased to a point where even electromagnetic radiation is prevented from escaping. Matter in the process of falling into a black hole can become very hot—after it enters the black hole itself the matter is heard from no more—and x-rays can result.)

Some theoreticians predict that the observed kinds of x-rays—with rapid variations plus specific sharp bursts—would result if a black hole containing about 100 times the mass of the sun existed in the center of a globular cluster such as NGC 6624. Even the fact that the sharp bursts include rises in intensity over a time scale of half a second is predicted by the theoretical calculations. Strange as it sounds, the theory based on the existence of a black hole in the globular cluster is one of the most reasonable theories yet presented for the observed effects, though other possible explanations are also being considered.

Some two dozen sources of these x-ray bursts are now known, and they are being called "bursters." Since most of them are not associated with globular clusters, this particular model based on black holes can't explain all the bursters. But black holes of one size or another may still explain many or most of them. Bursters will be discussed in Section 22.2.

The deduction that black holes may well exist in globular clusters is an interesting merger of modern and traditional astronomy. For example, to know how much energy is being emitted in the form of x-rays by the globular cluster, we must know how far away the cluster is. The distance to NGC 6624 is found by fitting the horizontal branch of its Hertzsprung-Russell diagram to a standard H-R diagram. The value comes out to be 8 kiloparsecs, most of the way between the sun and the center of our galaxy.

Globular clusters, far from being uninteresting relics of old stars, turn out to be fascinating places that are earning increased attention.

Figure 4–22 The SAS-1 satellite, one of NASA's series of Small Astronomy Satellites. SAS-3 has been observing the strange x-ray bursts.

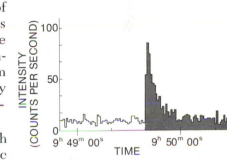

Figure 4–23 An x-ray burst (shaded) from the globular cluster NGC 6624 observed on September 28, 1975. The burst lasted about 10 seconds.

SUMMARY AND OUTLINE OF CHAPTER 4

Aims: To discuss multiple stars, variable stars, and stellar clusters, and to draw conclusions about the distances to stars and about the evolution of stars from their study

Binary stars (Section 4.1)
    Optical doubles
    Visual binaries
    Binaries with composite spectra
    Spectroscopic binaries
    Eclipsing binaries
    Astrometric binaries
Determination of stellar masses (Section 4.2)
    Mass-luminosity relation
Determination of stellar sizes (Section 4.3)
    Indirect: calculate from absolute magnitude or measure in eclipsing binary system
    Direct: interferometry
Variable stars (Section 4.4)
    Cepheid variables (Section 4.4a)
        Period-luminosity relation
        Uses for determining distances
    RR Lyrae variables (Section 4.4b)
        All of approximately the same absolute magnitude
        Uses for determining distances
Clusters and stellar populations (Section 4.5)
    Galactic (open) clusters
        Representative of spiral arms
        20 to several hundred members
        Turn-off on H-R diagram gives age
    Globular clusters
        Representative of galactic halo
        10^4 to 10^6 members
        Old enough to have H-R diagrams with horizontal branches
Spectroscopic parallax (Section 4.5a)
    Compare measured H-R diagram with one calibrated in terms of absolute magnitude
X-ray bursts from globular clusters (Section 4.5c)
    Massive black holes may be the cause of the bursts

QUESTIONS

1. Sketch the orbit of a double star that is simultaneously a visual, an eclipsing, and a spectroscopic binary.

2. How much brighter than the sun is a main-sequence star whose mass is 10 times that of the sun?

3. (a) Assume that an eclipsing binary contains two identical stars. Sketch the intensity of light received as a function of time. (b) Sketch to the same scale another curve to show the result if both stars were much larger while the orbit size stayed the same.

4. A Cepheid variable has a period of 30 days. What is its absolute magnitude?

5. Explain briefly how observations of a Cepheid

variable in a distant galaxy can be used to find the distance to the galaxy.

6. (a) Use the mass-luminosity relation to determine about how many times brighter than the sun are the most massive main-sequence stars of which we know. (b) How many times fainter are the least massive main-sequence stars?

7. When we see the light from a distant galaxy, are we seeing mostly low or high mass stars?

8. When we measure the gravitational effects of a distant galaxy on its neighbors, are we measuring the effects of mostly low or high mass stars?

9. If we take two stones, one twice the diameter of the other, and put them in an oven until they are heated to the same temperature and begin glowing, what will be the relationship between the light given off by the stones?

10. (a) What is the absolute magnitude of a Cepheid with a 10-day period? (Refer to Fig. 4–11.) (b) If its apparent magnitude is +12, how far away is it?

11. Cluster X has a higher fraction of main-sequence B stars than cluster Y. Which cluster is probably older?

12. What is the advantage of studying the H-R diagram of a cluster, compared to that of stars in the general field?

13. How does the discovery of x-ray bursts from globular clusters affect your conception of astronomy, which may have been that astronomy involved a leisurely contemplation of the stars?

OBSERVING PROJECTS:

1. (a) Try to see Alcor with your naked eye. (b) With a small telescope, observe Alcor, Mizar A, and Mizar B.

2. Observe Albireo with a small telescope.

3. Follow the brightness variation of Algol over 3 nights. (Each month, *Sky and Telescope* lists the times of minimum of Algol.)

The Sun

We have thus far discussed a range of individual stars of different spectral classes and have discussed groupings of stars in close physical proximity to each other. Studying these distant stars has allowed us to learn a lot about the properties of stars and how they evolve. But not all stars are far away; one is close at hand. By studying the sun, we not only learn about the properties of a particular star but also can study the details of processes that undoubtedly take place in more distant stars as well. We will first discuss the *quiet sun,* the solar phenomena that appear every day. Afterwards, we will discuss the *active sun,* solar phenomena that appear non-uniformly on the sun and vary over time.

5.1 BASIC STRUCTURE OF THE SUN

We think of the sun as the bright ball of gas that appears to travel across our sky every day. We are seeing only one layer of the sun; the properties of the solar interior below that layer and of the solar atmosphere above that layer are very different. The outermost parts of the solar atmosphere even extend through interplanetary space beyond the orbit of the earth.

The layer that we see is called the *photosphere,* which simply means the sphere from which the light comes (from the Greek *photos,* light). As is typical of many stars, about 94 per cent of the atoms and nuclei in the outer parts are hydrogen, about 5.9 per cent are helium, and a mixture of all the other elements make up the remaining one tenth of one per cent. The overall composition of the interior is not very different.

The sun is an average star, in that it is a G2 dwarf and falls right in the middle of the Hertzsprung-Russell diagram. Radiation from the photosphere peaks in the middle of the visible spectrum; after all, our eyes evolved over time to be sensitive to that region of the spectrum because the greatest amount of the solar radiation occurred there. If we lived on a planet orbiting an object that emitted mostly x-rays, we, like Superman, might have x-ray vision.

The disk of the sun takes up about one-half a degree across the sky; we say that it *subtends* one-half degree. This is large enough for us to see detail on the solar surface, and we shall describe that detail in subsequent sections.

The sun, observed in white light on July 4, 1974.

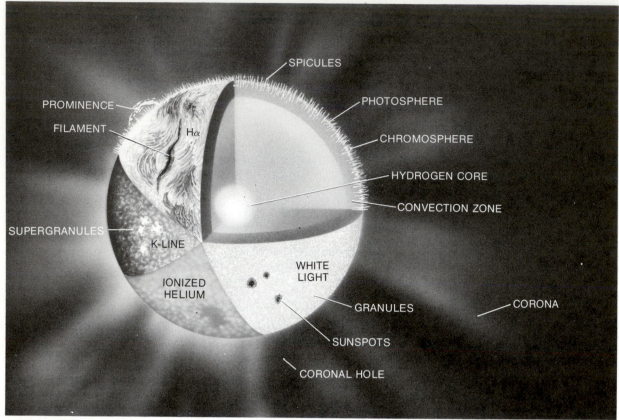

Figure 5-1 The parts of the solar atmosphere and interior. The solar surface is depicted as it appears through unfiltered light (called *white light*), and through filters that pass only light of certain elements in certain temperature stages. These specific wavelengths, counterclockwise from the top, are the $H\alpha$ line of hydrogen, which appears in the red, the K line of ionized calcium, which appears in the part of the ultraviolet that passes through the earth's atmosphere and can be seen with the naked eye, and the line of ionized helium at 304 Å in the extreme ultraviolet, which can be observed only from rockets and satellites.

Beneath the photosphere is the solar *interior*. All the solar energy is generated there at the solar *core*, which is about 10 per cent of the solar diameter at this stage of the sun's life.

The photosphere is the lowest level of the *solar atmosphere* (Fig. 5-1). The parts of the atmosphere above the photosphere are very tenuous, and contribute only a small fraction to the total mass of the sun. These upper layers are very much fainter than the photosphere, and are best seen with the naked eye during a solar eclipse, when the moon blocks the photospheric radiation from reaching our eyes directly. Now we can also study these upper layers with special instruments on the ground and in orbit around the earth.

Just above the photosphere is a jagged, spiky layer about 10,000 km thick, only about 1.5 per cent of the solar radius. This layer glows colorfully pinkish when seen at an eclipse, and is thus called the *chromosphere* (from the Greek *chromos*, color). Above the chromosphere, a pearly-white halo called the *corona* (from the Latin, crown) extends tens of millions of kilometers into space. The corona is continually expanding into interplanetary space and in this form is called the *solar wind*. We shall discuss all these phenomena in succeeding sections.

5.2 THE PHOTOSPHERE

The sun is the only star close enough to allow us to study its surface in detail. The phenomena we discover on the sun presumably also exist on other stars.

5.2a High-Resolution Observations of the Photosphere

One major limitation on observing the sun is the turbulence in the earth's atmosphere. We have seen how turbulence is a limiting factor in stellar observing, as it causes the twinkling of the stars. The problem is even more serious for studies of the sun, since the sun is up in the daytime when the atmosphere is heated by the solar radiation and so is more turbulent than it is at night.

Only at the very best observing sites, specially chosen for their steady solar observing characteristics, can one see detail on the sun subtending an angle as small as 1 second of arc. This corresponds to about 700 km on the solar surface, the distance from Boston to Washington, D.C.

When one studies the solar surface with 1 arc second resolution, one sees a salt-and-pepper texture called *granulation* (Fig. 5–2). The effect is similar to that shown by boiling liquids on earth, which are undergoing what is called *convection*. Convection is a way in which energy can be transported—conduction and radiation are other methods. Granulation on the sun seems to be an effect of convection. Each granule is only about 1000 km across, and represents a volume of gas that is rising from and falling to a shell of convection, called the *convection zone*, located below the photosphere. The granules are convectively carrying energy from the hot solar interior to the base of the photosphere.

It was discovered in the early 1960's that areas of the upper photosphere are oscillating up and down with a 5-minute period. This result, which was totally unexpected, is now thought to be caused by waves of energy coming from the convection zone.

Granulation can be seen in white light—all the radiation from the visible part of the spectrum taken together without being filtered—as distinguished from, say, red light or light of the Hα line of hydrogen.

10 sec

Figure 5–2 Granulation on the solar surface. One arc sec is 700 km on the sun.

5.2b The Photospheric Spectrum

The spectrum of the solar photosphere, as are the spectra of all G2 dwarfs, is a continuous spectrum with absorption lines (Fig. 5–3). Tens of thousands of these absorption lines, which are also called Fraunhofer lines, have been photographed and catalogued. They come from most of the chemical elements, although some of the elements have many lines in their spectra and some have very few. The majority of the lines in the solar spectrum come from iron (Fig. 5–4) and such other elements as magnesium, aluminum, calcium, titanium, chromium, nickel, and sodium. The hydrogen Balmer lines are strong but few in number. Helium is not excited at the low temperatures of the photosphere, and so its spectral lines do not appear.

Fraunhofer, in 1814, labeled the strongest of the absorption lines in the solar spectrum with letters from A through H. His C line, in the red, is now known to be the first line in the Balmer series of hydrogen, and is called Hα (H alpha). We still use Fraunhofer's notation for some of the strong lines. The D lines, a pair of lines close together in the yellow part of the spectrum, are caused by neutral sodium (Na I). The H line and the K line, both in the part of the violet spectrum that is barely visible to the eye, are caused by singly ionized calcium (Ca II). None of the elements other than helium makes up as much as one-tenth of one per cent of the number of hydrogen atoms (Table 5–1). Nevertheless, the D, H, and K lines are among the strongest in the visible spectrum. Of the elements other than hydrogen and helium, sodium and calcium are among the most abundant. Also, most

We are discussing here only the part of the solar spectrum in the visible. The formation of the Fraunhofer lines was discussed in Section 2.7. Note that to absorb is spelled with a "b" but absorption is spelled with a "p."

The temperature of the gas in the photosphere is about 5800 K.

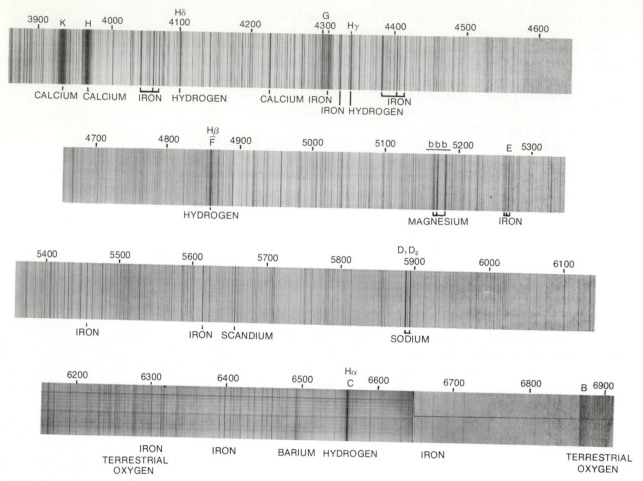

Figure 5–3 The visible part of the solar spectrum.

of the absorbing power of these elements at the temperatures of the photosphere and chromosphere is concentrated in just a few lines. Furthermore, these lines occur in the visible part of the spectrum.

The continuous spectrum, called the *continuum,* and most of the absorption lines, are formed together throughout the photospheric layers. The energy emitted in them is not formed in these layers, but has rather been transported upward from the solar interior.

TABLE 5–1 SOLAR ABUNDANCES OF THE MOST COMMON ELEMENTS

	Symbol	Atomic Number
For each 1,000,000 atoms of hydrogen, there are	H	1
63,000 atoms of helium	He	2
690 atoms of oxygen	O	8
420 atoms of carbon	C	6
87 atoms of nitrogen	N	7
45 atoms of silicon	Si	14
40 atoms of magnesium	Mg	12
37 atoms of neon	Ne	10
32 atoms of iron	Fe	26

4100   4200

Figure 5–4 This region of the solar spectrum (center strip) is enlarged to illustrate that many of the absorption lines result from iron, whose emission lines are visible in the comparison spectrum (above and below).

Figure 5–4 This region of the solar spectrum (center strip) is enlarged to illustrate that many of the absorption lines result from iron, whose emission lines are visible in the comparison spectrum (above and below).

5.3 THE CHROMOSPHERE

Under high resolution, the chromosphere is seen not to be a spherical shell around the sun but rather to be composed of small spikes that rise and fall. They have been compared in appearance to blades of grass or burning prairies. These small spikes are called *spicules* (Fig. 5–5).

Spicules are visualized as cylinders of about 1 arc second in diameter and perhaps ten times that in height, which corresponds to about 700 km across and 7000 km tall. They seem to have lifetimes of about 5 to 15 minutes, and there may be approximately half a million of them on the surface of the sun at any given moment.

Studies of velocities on the sun showed the existence of large organized cells of matter on the surface of the sun called *supergranulation*. Supergranulation cells look somewhat like polygons of approximately 30,000 km diameter. Supergranulation is an entirely different phenomenon from granulation. Each supergranulation cell may contain hundreds of individual granules. Supergranules can be seen in Figure 5–6.

Chromospheric matter appears to be at a temperature of approximately 15,000 K, somewhat higher than the temperature of the photosphere. When we examine the spectra of other, more distant stars in hydrogen or calcium light, we find unmistakable signs of chromospheres in stars of spectral types like the sun. Thus by studying the solar chromosphere we are also learning what the chromospheres of other stars are like—stars of spectral types F, G, K, and M can have chromospheres.

Figure 5–6 Supergranulation is best visible in an image like this one, in which the velocity field of the sun is shown. Dark areas are receding and bright areas are approaching us.

5.3a *The Chromospheric Spectrum at the Limb*

During the few seconds at an eclipse of the sun that the chromosphere is visible, its spectrum can be taken. This type of observation has been performed ever since the first spectroscopes were taken to eclipses in 1868. As the photosphere is completely covered, the chromosphere becomes visible.

Since the chromosphere then appears as hot gas silhouetted against

Figure 5–5 Spicules at the solar limb, observed through a filter passing a narrow band of wavelengths centered 1 Å redward of H α.

Figure 5–7 The flash spectrum 0.6 sec after totality at the African solar eclipse of 1973. We see the chromospheric spectrum.

We have been discussing the visible chromospheric spectrum. In the far ultraviolet, as studied from satellites, the chromospheric and coronal lines are emission lines even when viewed at the center of the disk (Color Plates 11, 12, 15, 16, and 17).

dark sky, the chromospheric spectrum consists of emission lines superimposed on a continuum. (The continuum is usually too faint to study well.)

The chromospheric emission lines appear to flash into view at the beginning and at the end of totality, so the spectrum of the chromosphere is known as the *flash spectrum* (Fig. 5–7 and Color Plate 10). Since the chromospheric gas is hotter than the photospheric gas, most of the chromospheric emission lines are from ions.

5.4 THE CORONA

During total solar eclipses, when first the photosphere and then the chromosphere are completely hidden from view, a faint white halo around the sun becomes visible. This *corona* (Fig. 5–8) is the outmost part of the solar atmosphere, and is the link between the sun and interplanetary and interstellar space. It has been discovered that the corona is at a temperature of about 2,000,000 K.

The actual amount of energy in the solar corona is not large, because there aren't very many coronal particles even though each particle has a high velocity. The corona has less than one billionth the density of the earth's atmosphere, and would be considered to be a very good vacuum in a laboratory on earth. For this reason, the corona serves as a unique and valuable celestial laboratory in which we may study gaseous plasmas in a near vacuum.

Photographs of the corona show that it is very irregular in form. Beautiful long *streamers* extend away from the sun in the equatorial regions. At the poles, delicate thin *plumes* are suspended above the surface. The shape of the corona varies continuously and is thus different at each successive eclipse. The structure of the corona is maintained by the magnetic field of the sun.

5.4a The Coronal Spectrum

The visible region of the coronal spectrum, when observed at eclipses, shows a continuum, and also both absorption lines and emission lines. It is easy to tell which part of the coronal spectrum comes from the emission lines: simply consider everything above the continuum level. The emission lines do not correspond to any normal spectral lines that are known in laboratories on earth or on other stars, and for many years their identification was one of the major problems in solar astronomy. In the late 1930's, it was discovered that they arose in atoms that were multiply ionized. This was the major indication that the corona was very hot. In the photosphere we find atoms that are neutral, singly ionized, or doubly ionized (Ca I, Ca II, and Ca III, for example). In the corona we find ions that are ionized approximately a dozen times (Fe XIV, for example, iron that has lost 13 of its normal quota of 26 electrons). The corona must be very hot indeed, millions of degrees, to have enough energy to strip that many electrons off atoms.

Figure 5–8 The 1973 total solar eclipse, photographed from Africa by an expedition of the High Altitude Observatory. The coronal intensity falls off so rapidly with distance above the sun that photographic film cannot directly record such a large region of the corona; to take this photograph, a special filter was prepared that absorbed a large fraction of light near the center and less and less radially toward the edges. Thus the light that passed through the filter was more uniform in intensity than the incident radiation, and coronal structures are well outlined.

By multiply (mult-i-plē) we mean more than once, that is, twice, three times, and so on.

Figure 5-9 Normally, even on the clearest days, the sky is too bright to allow us to observe the corona. During an eclipse, photospheric light does not reach the terrestrial atmosphere to turn it blue, and we can observe the corona. The K-corona, from the scattering of photospheric light by electrons in the corona, falls off in intensity more rapidly than does the F-corona, from the scattering of photospheric light by interplanetary dust. Although the total energy received in all the emission lines is less than the energy received in the K-corona or F-corona, at the given wavelength of a particular emission line, the emission line is stronger than the K- and F-coronas. The intensity of the sky during an eclipse (horizontal shaded area) is about one-millionth as great as it is on an ordinary day. The amount of earthshine on the moon, sunlight reflected off the earth to the moon, is also graphed. The E-corona is the set of emission lines.

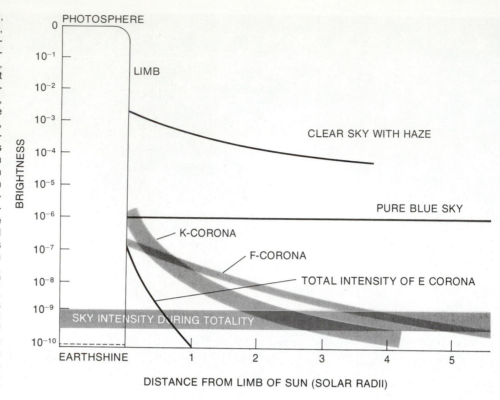

The continuum is caused partly by electrons located near the sun, as they scatter (bounce) photospheric light to us, and partly by a similar scattering by dust particles near the orbit of the planet Mercury. The absorption lines are added in the scattering by the dust. (See Fig. 5-9.)

5.4b Non-Eclipse Observations of the Corona

The corona is normally too faint to be seen except at an eclipse of the sun because it is fainter than the everyday blue sky. But at certain locations on mountain peaks on the surface of the earth, the sky is especially clear and dust free, and the innermost part of the corona can be seen (Fig. 5-10). Special telescopes called *coronagraphs* have been built to study the corona from such sites. Coronagraphs are built with special attention to detail, since their object is not to gather a lot of light but to prevent the strong solar radiation from being scattered about within the telescope. They thus use lenses instead of mirrors, because even the best mirrors are not smooth enough to reflect light without scattering a fraction that may seem tiny but would still be sufficient to overwhelm the coronal light.

Since so many of the problems in observing the faint solar corona are caused by the earth's atmosphere, it is obvious that we want to observe the corona from above the atmosphere. If we stood on the moon, which has no air, we would see the corona rising each lunar morning a bit ahead of the sun.

But it is much less expensive to observe the corona from a satellite in orbit around the earth than it is to do so from the moon. The unmanned seventh Orbiting Solar Observatory, OSO-7 (1971–1974), and the manned Skylab missions (1973–74) used coronagraphs to photograph the corona hour by hour in visible light (Color Plates 13 and 14). These satellites could

A

B

Figure 5–10 From a few mountain sites, the inner-most corona can be photo-graphed without need for an eclipse. (A) The corona-graph at the Haleakala Ob-servatory of the University of Hawaii on Maui. (B) These observations were taken on March 8, 1970, the day after a total eclipse, with this coronagraph. Five exposures were superim-posed to improve the image.

study the corona to much greater distances from the solar surface than can be studied with coronagraphs on earth. Among the major conclusions of the research is that the corona is much more dynamic than we had thought. For example, many blobs of matter were seen to be ejected from the corona into interplanetary space, often in connection with solar flare activity (which will be discussed in Section 5.6).

5.5 SOLAR ECLIPSES

We have discussed the solar chromosphere and corona, parts of the solar atmosphere that are visible to the eye only at the time of a total solar eclipse. Eclipses have played a major role in solar physics, so let us dis-cuss them in detail.

The sun (more precisely, the photosphere) is very large and very far from us. The moon is 400 times smaller in diameter than the sun, but it is also 400 times closer to earth. Because of this, the sun and the moon subtend almost exactly the same angle in the sky—about ½°—which is a happy coincidence (Fig. 5–11).

The moon's position in the sky, at certain points in its orbit around the earth, comes close to the position of the sun. This happens approximately once a month at the time of the new moon, as discussed in Section 10.1. We will see there that since the lunar orbit is inclined with respect to the earth's orbit, the moon usually passes above or below the line joining the earth and the sun. But occasionally the moon passes close enough to the earth-sun line that the moon's shadow falls upon the surface of the earth.

At a total lunar eclipse (see Section 10.2), the earth's shadow is suf-

Figure 5–11 An apple, the Empire State Building, the moon and the sun are very different from each other in size, but here they subtend the same angle because they are different distances from us.

ficiently large that the entire moon fits inside it. Thus anyone on the surface of the earth at whose location the moon has risen sees the lunar eclipse. By contrast, at a total solar eclipse, the lunar shadow barely reaches the earth's surface. As the moon moves through space on its orbit, and as the earth rotates, this lunar shadow sweeps across the earth's surface in a band up to 300 km wide. Only observers stationed within this narrow band can see the total eclipse.

From anywhere outside this band, one sees only a partial eclipse. Sometimes the moon, sun, and earth are not precisely aligned and the darkest part of the shadow—called the *umbra*—never hits the earth. Only a partial eclipse is visible on earth under these circumstances. As long as the slightest bit of photosphere is visible, even as little as 1 per cent, one cannot see the important eclipse phenomena—the chromosphere and corona. Thus partial eclipses are of little value for most serious scientific purposes. After all, the photosphere is 1,000,000 times brighter than the corona; if one per cent is showing then we still have 10,000 times more light from the photosphere than from the corona, which is enough to ruin our opportunity to see the corona (Fig. 5–12). During the initial phase, we are in the *penumbra*.

If you are fortunate enough to be standing in the zone of a total eclipse, you will find excitement all around you as the approaching eclipse is anticipated. Even uneducated people have formed an impression of the cause of an eclipse: it immediately follows the arrival of a horde of astronomers. If the astronomers come, can the eclipse be far behind?

About an hour or an hour and a half before totality begins, the partial phase of the eclipse starts. Nothing is visible to the naked eye immediately, but if you were to look at the sun through a special filter you would see that the moon was encroaching on the sun (Fig. 5–13 and Color Plate 7). At this stage of the eclipse, it is necessary to look through a special filter (or to project the image of the sun with a telescope or a pinhole camera onto a surface such as a piece of cardboard) in order to protect your eyes, for the photosphere is visible and its direct image on your retina could cause burning and blindness (Fig. 5–14).

Figure 5–12 Established observatories are not usually in the path of a total solar eclipse. The partial phase, however, can be seen from a much wider area, and this closeup view in Hα from the Big Bear Solar Observatory shows the irregularities at the edge of the lunar disk projected on the solar surface. The corona could not be seen.

The eclipse progresses gradually, and by 15 minutes before totality the sky grows strangely dark, as though a storm were gathering. During the minute or two before totality begins, bands of shadow race across the ground. These *shadow bands* are caused in the earth's atmosphere, and as a terrestrial rather than a celestial phenomenon are not of much interest to astronomers.

You still need the special filter to watch the final seconds of the partial phase. Only a thin sliver of photosphere is visible. Then the sliver breaks up into a chain of beads along the rim of the moon. These *Baily's beads* are the photosphere shining through valleys that happen to be located at the edge of the moon and oriented so as to make the lunar rim irregular.

The last Baily's bead seems especially bright to the eye. It glistens and dazzles for a few seconds, sparkling like a diamond. This is called the *diamond ring effect* (Color Plate 8).

With the passage of the diamond ring effect, the total phase of the eclipse has begun. For a few seconds, a reddish glow is visible in a narrow band around the leading edge of the moon. This is the chromosphere and perhaps one or more prominences. Scientists studying the chromosphere have to work fast and have their photographic devices operating at a rapid pace in order to make their observations in these few seconds.

Then the corona is visible in all its glory (Figs. 5–8 and 5–20 and Color Plate 9). One can see the equatorial streamers and the polar plumes. At this stage, the photosphere is totally hidden so it is perfectly safe to stare at the corona with the naked eye and without filters. It has approximately the same brightness as the full moon, and is equally safe to look at. Unfortunately, many people are not adequately informed about this most important phase of the total eclipse, and miss the spectacle.

At the end of the eclipse, the diamond ring appears on the other side of the sun, then Baily's beads, and then the final partial phases. All too soon, the eclipse is over.

On the average, a total solar eclipse occurs somewhere in the world every year and a half. There can actually be up to seven solar and lunar eclipses a year, but most of these are either lunar or only partial solar eclipses.

A

B

Figure 5–13 (*A*) The partially eclipsed sun observed from Africa in 1973. (*B*) These children, who visited our eclipse site in Loiengalani, Kenya, just prior to the 1973 total solar eclipse, were practicing the use of special filters to protect their eyes when observing the partial phases.

5.5a Observing Eclipses (Optional)

Scientists prepare for solar eclipses years in advance. The band of totality usually does not cross populated areas of the earth (after all, the surface of the earth is largely covered with water), and astronomers usually have to travel great distances to carry out their observations.

On October 12, 1977, a total eclipse was visible in South America and at sea in the Pacific. The next total solar eclipse to cross the continental U.S. and Canada, and the last one to do so during the 20th century, will occur on February 26, 1979 (Fig. 5–15). The band of totality will include parts of the Pacific Northwest states of Washington and Oregon, and then continue north through Idaho, Montana, North Dakota, and central Canada. The rest of North America will see only a partial eclipse. The next total eclipse visible in the continental United States won't be until August 21, 2017. Annular eclipses (Fig. 5–16) will be visible in the U.S. in 1984 and 1994. Table 5–2 lists future total eclipses.

In these days of orbiting satellites that can study ultraviolet coronal

Figure 5–14 The diamond ring effect at the 1977 total eclipse, observed by the author's expedition from a ship in the Pacific Ocean.

Figure 5–15 In 1979, totality will be visible from sites in the Pacific Northwest of the U.S. and the north central U.S. and Canada.

Figure 5–16 A member of the author's expedition, observing the annular eclipse of December 24, 1974 in the Andes Mountains near Bogotá, Colombia. In an annular eclipse, the moon appears smaller than the sun and a ring (annulus) of photosphere remains visible.

lines or carry coronagraphs above the earth's atmosphere, is it worth traveling to observe eclipses? The matter is controversial, but there is much to be said for the benefits of eclipse observing. Even when you consider that a certain fraction of the eclipses will be ruined by clouds, and that a certain number of experiments will fail at the last minute for a variety of reasons (ranging from the breakdown of an electronic component or a power failure to something trivial like the failure to remove a lens cap), eclipse observations are a relatively inexpensive way—compared to space research—of observing the chromosphere and corona to find out such quantities as the temperature, density, and magnetic field structure.

5.5b An Eclipse Expedition (Optional)

An eclipse expedition is unlike any other kind of observing experience that an astronomer can have. A year or more of preparation is followed by a tense and pressure-packed but fascinating few weeks at the expedition site, all leading up to the inexorable deadline of the eclipse. Then, if the experiment is successful, it may take many months or several years to study the data and draw the conclusions.

On June 30, 1973, a total solar eclipse occurred that lasted over 7 minutes, close to the maximum duration theoretically possible. It was the longest eclipse of the century. Clear weather was predicted and so many teams of astronomers from countries all over the world prepared experiments to carry out during the eclipse.

The path of the eclipse was a narrow band that crossed the continent of Africa from west to east (Fig. 5–17). The point on the map where totality was to have its maximum duration fell in the middle of the Sahara desert; the nearest towns included Timbuktu, which is in Mali, and Agadez, a similar desert outpost in Niger.

Unfortunately, a long drought in the Sahara made the air continually dusty. It did not seem that conditions would be good enough there to observe the eclipse.

The U.S. National Science Foundation decided to coordinate the American efforts and explored other sites. In view of the Saharan dust problem, they sent most of the astronomers to a site in northern Kenya.

TABLE 5–2 TOTAL SOLAR ECLIPSES

Date	Maximum Duration	Location
1977 October 12	2:37	Pacific Ocean, South America (Colombia and Venezuela)
1979 February 26	2:49	U.S.A., Canada, Greenland
1980 February 16	4:08	Central Africa, India, China
1981 July 31	2:03	U.S.S.R.
1983 June 11	5:11	Indian Ocean, Indonesia, New Guinea
1984 November 22	1:59	New Guinea, South Pacific Ocean
1985 November 12	1:59	South Pacific Ocean
1988 March 18	3:46	Indonesia, Philippines
1990 July 22	2:33	Finland, U.S.S.R., U.S.A. (Alaska)
1991 July 11	6:54	U.S.A. (Hawaii), Central and South America

Totality was to be five minutes there, shorter than the 7-minute maximum but still a very respectable length of time.

A month before the eclipse, the astronomers, technicians, and students who were on the expedition left for Africa in a chartered jet. About 70 individuals were along.

After only one day's rest in Nairobi, we set off for the eclipse site, which was in an oasis named Loiengalani, situated in the Northern Frontier District of Kenya, an area that was usually off-limits to tourists because of tribal wars. The eclipse site was a 16-hour drive north of Nairobi to the shore of Lake Turkana (then called Lake Rudolf), a huge fresh-water lake (Fig. 5–18).

At Loiengalani, we spent the next three weeks setting up equipment, checking, testing, and rechecking. Astronomers have to be very versatile at an eclipse site. They may have to construct a small building to shield their instruments from the sand and sun, they may have to pour concrete to make stable platforms for their instruments, and they may have to know details of machine-work or electronics to make last-minute repairs to balky instruments. It helps to be a combination carpenter, mason, machinist, electronics engineer, optical physicist, and more.

Some of the astronomers lived in a newly-enlarged lodge while others lived in rooms in the local mission. Some of us lectured to the children at the mission school to explain what was going to happen and how to watch the eclipse safely (Fig. 5–19).

Tension grew as the day of the eclipse approached. One by one, almost all of the experiments were made to operate. A few days before the eclipse, reporters and tourists from Nairobi swelled the population of the oasis to a few hundred.

The morning of the eclipse, the sun rose in a perfectly clear sky. Then, within minutes, the sky completely clouded over. The eclipse was to take place after noon, and the astronomers went through last-minute checks of their apparatus. The clouds gradually disappeared.

The partial phase began, and the sky grew darker. We had special filters to enable us to follow the progress of the moon across the sun. A few

Figure 5–17 The June 30, 1973 eclipse crossed Africa from west to east at a speed that ranged from about 2000 to 3000 km/hr.

Figure 5–18 The eclipse site alongside Lake Rudolf (now Lake Turkana) near the oasis of Loiengalani, Kenya.

Figure 5–19 Lecturing to students at the mission school in Loiengalani.

Figure 5–18

Figure 5–19

Figure 5-20 The camera was not moved between exposures in this sequence that shows the partial phases and totality of the 1973 eclipse. Note that clouds occasionally covered the sun during the partial phases.

minutes before totality, at about the time that the shadow bands occurred, many of the experiments started recording data on instrumental tape recorders or on automatic devices with pens and moving strips of paper, or started cameras whirring. Only a few small clouds were left in the sky. Fig. 5–20 shows the progress of the eclipse phases.

Then it came. Baily's beads were visible and the diamond ring appeared. "Totality," someone shouted, for many of us were inside buildings or under tents tending our equipment and could not even see the eclipse. For five short minutes everyone worked to try to capture the data that would later on be analyzed to help us understand the solar atmosphere.

The end of the eclipse came, within a few seconds of the time predicted months and years before. The tension was over. The day had been a tremendous success.

Now, years later, the results of the eclipse experiments are being published. Papers are given in meetings of the astronomical societies around the world or appear in scientific journals. Our knowledge of almost all parts of the sun has been considerably advanced.

5.6 SUNSPOTS AND OTHER SOLAR ACTIVITY

In this section we discuss the active sun.

We have discussed the components of the sun as though it were a static object: we described the photosphere, the chromosphere, and the corona, and will describe the interiors of the sun and stars of similar mass in Chapter 6. But a host of other time-varying phenomena are superimposed on the basic structure of the sun. Many of them, notably the sunspots, vary with an 11-year cycle, which is called the *solar activity cycle*.

5.6a Sunspots

Sunspots (Fig. 5–21) are the most obvious manifestation of solar activity. They are areas of the sun that appear relatively dark when seen in white light. They appear dark because they are giving off less radiation than the photosphere that surrounds them. This implies that they are cooler areas of the solar surface, since cooler gas radiates less than hotter gas. Actually, if we could somehow remove a sunspot from the solar surface and put it off in space, it would appear bright against the dark sky and might give off as much light as the full moon.

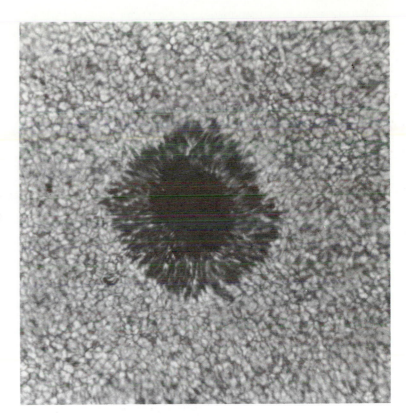

Figure 5-21 A sunspot, showing the dark *umbra* surrounded by the lighter *penumbra*. Granulation is visible in the surrounding photosphere.

A sunspot includes a very dark central region, called the *umbra* from the Latin for "shadow" (pl: *umbrae*). The umbra is surrounded by a *penumbra* (pl: *penumbrae*), which is not as dark (just as during an eclipse the umbra of the shadow is the darkest part and the penumbra is less dark).

To understand sunspots we must understand magnetic fields. When iron filings are put near a simple bar magnet on earth, the filings show a pattern that is illustrated in Figure 5-22. The magnet is said to have a north pole and a south pole, and the magnetic field linking them is characterized by what we call *magnetic lines of force,* or *magnetic field lines* (after all, the iron filings are spread out in what look like lines). The earth (as well as some other planets) has a magnetic field that has many characteristics in common with that of a bar magnet. The structure seen in the solar corona, including polar plumes and equatorial streamers, results from matter being constrained by the solar magnetic field.

We can measure magnetic fields on the sun by using a spectroscopic method. In the presence of a magnetic field, certain spectral lines are split into a number of components and the amount of the splitting depends on the strength of the magnetic field. This is called the *Zeeman effect.*

Measurements of the solar magnetic field were first made by George Ellery Hale. He showed, in 1908, that the sunspots are regions of very high magnetic field strength on the sun, thousands of times more powerful than the earth's magnetic field or than the average solar magnetic field.

Sunspots were discovered in 1610, independently by Galileo in Italy, Fabricius and Christopher Scheiner in Germany, and Thomas Harriot in England. In about 1850, it was realized that the number of sunspots varies with an 11-year-cycle, as is shown in Figure 5-23. This is called the *sunspot cycle,* although we now realize that many related signs of solar activity vary with the same period.

Figure 5-22 Lines of force from a bar magnet are outlined by iron filings.

The next maximum of the sunspot cycle—the time when there are the greatest number of sunspots—should take place in 1980 or 1981. NASA plans to launch a satellite called the Solar Maximum Mission to study solar activity at that time.

5.6b Flares

Violent activity sometimes occurs in the regions around sunspots. Tremendous eruptions called *solar flares* (Fig. 5–24) can eject particles and emit radiation from all parts of the spectrum into space. These solar storms

A

B

Figure 5–24 *(A)* A flare near the solar limb, photographed in hydrogen light on August 20, 1971. *(B)* One of the largest solar flares in decades occurred on August 7, 1972, and led to power blackouts, short wave radio blackouts, and aurorae. The whole sun is seen here in hydrogen light during the peak of the flare.

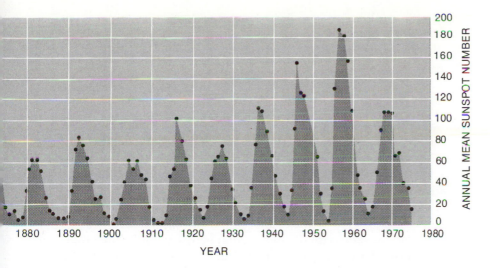

Figure 5–23 The 11-year sunspot cycle is but one manifestation of the solar activity cycle.

begin in a few seconds and can last up to four hours. A typical flare lifetime is 20 minutes. Temperatures in the flare can reach 5 million kelvins, even hotter than the quiet corona. Flare particles that are ejected reach the earth in a few hours or days and can cause disruptions in radio transmission, cause the aurorae (Fig. 5–25 and Color Plate 34)—the *aurora borealis* is the "northern lights" and the *aurora australis* is the "southern lights"—and even cause surges on power lines. Because of these solar-terrestrial relationships, high priority is placed on understanding solar activity and being able to predict it. The U.S. government even has a solar weather bureau to forecast solar storms, just as it has a terrestrial weather bureau.

Figure 5–25 An aurora borealis, beautiful patterns of color in the sky, photographed in Alaska. The aurorae are phenomena of the earth's atmosphere, and are caused by incoming particles from the sun.

Effects in the upper atmosphere of the earth during solar maximum may cause so much radio "skip"—which makes even very distant stations audible, as the signal skips or bounces off the earth's ionosphere (an upper atmospheric level)—that CB radios become harder to use.

No specific model is accepted as explaining the eruption of solar flares. But it seems that a tremendous amount of energy is stored in the solar magnetic fields in sunspot regions. Something unknown triggers the release of the energy.

5.6c Plages, Filaments, and Prominences

Studies of the solar atmosphere in Hα radiation also reveal other types of solar activity. Bright areas called *plages* (from the French word for beach) surround the entire sunspot region (Fig. 5–26). Dark *filaments* are seen threading their way across the sun in the vicinity of sunspots. The longest filaments can extend for 100,000 km. When these filaments happen to be on the limb of the sun, they can be seen to project into space, often in beautiful shapes. Then they are called *prominences* (Fig. 5–27). Prominences can be seen with the eye at solar eclipses, and glow pinkish at that time because of their emission in Hα and a few other spectral lines. They can be observed from the ground even without an eclipse, if an Hα filter is used. Prominences appear to be composed of matter in a similar condition of temperature and density to matter in the quiet chromosphere. Space observations can lead to maps of the temperature structure inside a prominence (Fig. 5–28 and Color Plates 15 and 16).

Sometimes prominences can hover above the sun, supported by magnetic fields, for weeks or months. They are then called *quiescent prominences*, and can extend tens of thousands of kilometers above the limb. A

Figure 5–26 The sun, photographed in hydrogen radiation at the last solar maximum, shows bright areas called plages and many dark filaments around the many sunspot regions.

Figure 5–27 Many prominences appear around the sun, and a giant eruptive prominence ejects matter in this hydrogen-light photograph from the Haleakala Observatory in Hawaii.

A   B   C

quiescent prominence is shown on page 9. Other prominences can seem to undergo rapid changes (Color Plates 11, 12, and 17).

5.6d Solar-Terrestrial Relations

The solar activity cycle is now being carefully examined in order to increase our understanding of solar-terrestrial relations. Although for many years scientists were skeptical of the idea that solar activity could have a direct effect on the earth's weather, scientists presently seem to be accepting more and more the possibility of such a relationship.

An extreme test of the interaction of solar activity with terrestrial weather may be provided by the interesting probability that there most likely were no sunspots at all on the sun from 1645 to 1715! The sunspot cycle may not have been operating for that 70-year period (Fig. 5–29). This was known to Maunder and others in the early years of this century but was largely forgotten until its importance was recently noted and stressed by John A. Eddy of the High Altitude Observatory. Although no counts of sunspots exist for most of that period, there is evidence that people were looking for sunspots; it seems reasonable that there were no counts of sunspots because they were not there and not just because nobody was observing. A variety of indirect evidence has also been brought to bear on the question.

Figure 5–28 Observations in the ultraviolet from the Harvard College Observatory experiment aboard Skylab reveal the temperature structure in and around the prominences, which are themselves at about 15,000 K. (A) Triply ionized oxygen gas (O IV) is at a temperature of about 130,000 K and represents the transition zone from the prominence to the corona. (B) Five-times ionized oxygen gas (O VI) is somewhat hotter (300,-000 K) and represents both gas in the transition zone and gas in the corona. (C) Nine-times ionized magnesium gas (Mg X) represents gas in the corona at a temperature of 1,500,000 K.

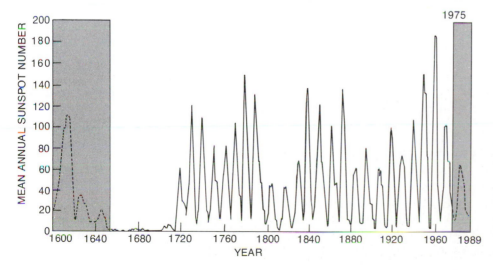

Figure 5–29 The Maunder minimum (1645–1715), when sunspot activity was negligible for decades, may indicate that the sun does not have as regular a cycle of activity as we had thought. Activity is extrapolated into the shaded regions.

Figure 5–30 The solar wind causes the wavy streaming of the tails of comets, as in this view of Comet Mrkos in 1957.

Mg IX 368 Å

Figure 5–31 This Skylab photograph taken on August 14, 1973, shows the coronal hole, a phenomenon that usually appears over one or both poles of the sun and often at other latitudes as well. The ultraviolet line of magnesium studied shows the coronal gas. A coronal hole is also noticeable in visible light at the top on the 1973 eclipse photograph (Fig. 5–8).

No aurorae were reported for many years, for example, and the solar corona appeared very weak when observed at eclipses. It may be significant that the anomalous sunspotless period coincided with a "Little Ice Age" in Europe and with a drought in the southwestern United States. Another important conclusion from the existence of this sunspotless period is that the solar activity cycle may be much less regular than we had thought.

5.7 THE SOLAR WIND AND CORONAL HOLES

At about the time of the launch of the first earth satellites, in 1957, it was realized that the corona must be expanding into space. This phenomenon is called the *solar wind*. The expansion causes comet tails (Fig. 5–30) always to point away from the sun, but the definite existence of a solar wind had not been deduced from this fact.

The solar wind extends into space far beyond the orbit of the earth. Calculations show that it must extend beyond the orbit of Saturn, and possibly even far beyond the orbit of Pluto. The density of particles, always low, decreases with distance from the sun.

The solar wind contains ions and electrons. The relative composition of elements and ions in the solar wind is fixed within a few solar radii of the determined base of the corona, and does not change outside of this volume. We measure the same composition even at the orbit of the earth, which is 200 solar radii from the sun.

From observations made on the Skylab manned missions, it was realized that the corona contains areas that are particularly cool and quiet. The density of gas in those areas is lower than the density in adjacent areas. It is now thought by many astronomers that the solar wind emanates from those areas, which are called *coronal holes* (Fig. 5–31).

The earth's outer atmosphere is bathed in the solar wind. Thus research on the nature of the solar wind is necessary to understand our environment in the solar system.

5.8 OBSERVING WITH THE LATEST ORBITING SOLAR OBSERVATORY *(Optional)*

Even solar astronomers who are observers rather than theoreticians need no longer go to remote mountain peaks to make their observations. They can work at their home institutions with data sent back from satellites in space.

OSO-8, the most recent of the very successful NASA series of Orbiting Solar Observatories, was launched in June 1975 (Fig. 5–32). It carries two major experiments to study the sun. Both take ultraviolet spectra of small areas (1 arc sec × 1 arc sec up to 1 arc sec × 40 arc sec) on the surface of the sun.

Besides these two major experiments, which use telescopes that are kept continuously pointing at the sun, there are several additional experiments located in a part of the spacecraft that rotates every few seconds. These experiments map out swaths across the sky, and measure such things as gamma ray bursts and x-rays from objects outside the solar system.

For the first year after launch, one astronomer for each experiment was designated as chief scientist each day. This provided a clear chain of command. Other tasks rotated on a five-day cycle. The first day, the scientist blocked out the experiments roughly. The second day the scientist fit in the details and made certain that all the different experiments were coordinated. The third day, the detailed plans were changed to computer language that could be transmitted to the satellite. These instructions were sent by the scientist's computer to the computer at NASA's Goddard Space Flight Center at Greenbelt, Maryland, from which they were relayed to the satellite. The fourth day, the scientist supervised the general operation to see that everything was going smoothly, and may have had a chance to get a "quick look" at some of the data that were radioed back to earth by the spacecraft. And the fifth day, the scientist sat all day at a computer console checking a part of the data from each orbit to make certain that all was working well. After the first year, the observing programs were changed weekly instead of daily, but all the steps still had to be followed. Such office work is a far cry from observing at a ground-based telescope, but it carries with it the substantial reward of knowing that the data being received were hitherto unavailable and that these data could be the key to a scientific problem that the astronomer has been considering for many years.

When astronomers go home from such an expedition, they carry with them only a few graphs or sets of numbers that the computer received from the satellite. It may be months or more before the scientists receive all the data they have helped record, and only then can they go to work and study these data.

Figure 5–32 The eighth Orbiting Solar Observatory (OSO-8), launched in June 1975, being tested in the laboratory by being spun.

5.9 THE SOLAR CONSTANT *(Optional)*

Every second, a certain amount of solar energy passes through each square centimeter at the average distance of the earth from the sun. This quantity is called the *solar constant*. Accurate knowledge of the solar constant is necessary to understand the terrestrial atmosphere, for to interpret the atmosphere completely we must know all the ways in which it can gain and lose energy. Further, knowledge of the solar constant enables us to calculate the amount of energy that the sun itself is giving off, and thus gives us an accurate measurement on which to base our quantitative understanding of the radiation of all the stars.

The latest measurements give a value of about 2 calories/cm²/min, which is equivalent to about 135 milliwatts/cm². The value is determined to an accuracy of 1.5 per cent; we have not yet been able to determine the solar constant to higher accuracy. Higher accuracy measurements are necessary, and an experiment for this purpose is scheduled for

The solar constant is the amount of energy per second that would hit each square centimeter of the earth at its average distance if the earth had no atmosphere. 99% of solar energy is in the range from 2760 Å to 49,600 Å (nearly 5 microns), and 99.9% is between 217Å and 10.94 microns.

Calories are a unit of energy, and watts are a unit of power in the metric system.

launch on the Solar Maximum Mission in 1980 or 1981. One interesting aim is to determine if there are long-term changes in the solar constant, perhaps arising from changes in the luminosity of the sun. Certainly our climate would be profoundly affected by small long-term changes. Of course, if the solar constant changed with time, it wouldn't really be a constant after all.

5.10 SOLAR ENERGY FOR EARTH *(Optional)*

Most of the energy that we use on earth can ultimately be traced back to the sun. Oil and natural gas are fossil fuels that release energy received from the sun millions of years ago. Even when we use wood in a fire we must recall that the sun supplied the energy for the growth of trees. Similarly the sun supplies the energy that makes the wind blow and heats the oceans. We are currently investigating the extraction of energy from both wind and ocean water.

The sun is itself a fusion reactor, properly located far from populated areas.

The sun is also important to us as a laboratory in which we can study hot gases in a magnetic field. The knowledge we are gathering from studies of the solar gas is helping us learn to control fusion processes here on earth. If we learn to build magnetic "bottles" to contain hydrogen undergoing fusion at temperatures of millions of degrees, then we can use the hydrogen in the oceans as our energy supply and be self-sufficient in energy for millions of years.

But progress in developing fusion for peaceful purposes has not been as rapid as had been hoped. The energy crisis probably won't allow us to wait. There is increasing interest in gathering energy directly from the sun.

There is now considerable research centering on finding ways of reducing the cost of devices that convert solar radiation into heat and electricity. Consideration is also being given to how we could best utilize such an energy conversion method. Do individuals put solar receptors on every house? Do power companies build giant farms of solar receptors that cover many square miles of countryside? Does one choose to put these farms only in regions like the southwest United States where the amount of sunshine is much higher than it is in New England? How do you store energy for a cloudy day or for the nighttime? Since the energy in sunlight falling on an area only about 100 km in diameter would meet the energy needs of the

Figure 5-33 Gerard O'Neill of Princton has suggested that a self-sufficient space station like this one, permanently located in one of the points in space where the terrestrial and lunar gravity balance each other, could collect solar energy and beam it to earth in the form of microwaves. Although the initial stages would be launched from earth, much of the material used thereafter could be mined on the moon. The concept appears promising, and is under study. (Photo from Science Year, the World Book Science Annual,© 1975 Field Enterprises Educational Corporation.)

entire United States if it could be efficiently converted to electricity, these questions are obviously worth pursuing.

Solar energy can also be used directly to provide heat and hot water. One can build houses to take advantage of the solar radiation by suitably designing and orienting roofs, windows, and overhangs, and by using suitable construction materials that tend to retain heat. Even if only part of a house's energy consumption can be supplied by such methods, the need for outside energy sources is considerably reduced.

The use of solar energy for widespread practical purposes is still in its infancy. For some purposes—providing heating for water and houses—perhaps only technological development and practical experience is necessary rather than scientific research. For other purposes—such as the generation of electricity—we have more hope that research will bring breakthroughs, or at least substantially reduce the cost of the existing devices that directly produce electricity when struck by sunlight. The only thing that can be said for certain is that "solar energy," in the sense of heat or electricity generated from solar radiation, will play an increasingly important part in our national energy budget.

5.11 THE SUN AND THE THEORY OF RELATIVITY

The sun, as the nearest star to the earth, has been very important for testing some of the predictions of Albert Einstein's theory of gravitation, which is known as the general theory of relativity. The theory, which Einstein advanced in final form in 1916, made three predictions that depended on the presence of a large mass like the sun for experimental verification.

Figure 5-34 Albert Einstein visiting Caltech in the 1930's.

These predictions were (1) the gravitational deflection of light, (2) the advance of the perihelion of Mercury, and (3) the gravitational redshift. We shall discuss the gravitational redshift in Section 7.5, and discuss the other two tests here.

Box 5.1   The General Theory of Relativity

Einstein's *general theory of relativity* links three dimensions of space and one dimension of time to describe a four-dimensional space-time. In Einstein's own words, "If you will not take the answer too seriously, and consider it only as a kind of joke, then I can explain it as follows. It was formerly believed that if all material things disappeared out of the universe, time and space would be left. According to the relativity theory, however, time and space disappear together with the things." The general theory explains gravitation. Einstein's earlier *special theory of relativity* (1905) is a theory of relative motion. It holds that the speed of light is an absolute limit, shows how length, mass, and the rate at which time advances depend on the velocity of motion, and leads to the relation between mass and energy, $E=mc^2$.

Actually, according to the general theory of relativity, the presence of the mass warps the space. Thus the light travels in a straight line but space itself is curved, as discussed in Section 1.2 and shown in Figure 5–35. The effect is the same as it would be if space were flat and the light were bent.

Einstein's theory predicts that the light from a star would act as though it were bent toward the sun by a very small amount (Fig. 5–35). We on earth, looking back, would see the star from which the light was emitted as though it were shifted slightly away from the sun. Only a star whose radiation grazed the edge of the sun would seem to undergo the full deflection; the effect diminishes as one considers stars farther away from the solar limb. To see the effect, one has to look near the sun at a time when the stars are visible, and this could be done only at a total solar eclipse.

The British astronomer Sir Arthur Eddington and other scientists observed the total solar eclipse of 1919 from a number of sites in Africa and South America. The effect for which they were looking was a very delicate one, and it was not enough merely to observe the stars at the moment of eclipse. One had to know what their positions were when the sun was not present in their midst, so the astronomers had already made photographs of the same field of stars six months earlier when the same stars were in the nighttime sky. Even though his observations had been limited by clouds, Eddington detected that light was deflected by an amount that agreed with Einstein's revised predictions. When Einstein eventually heard of the result, he was so pleased that he wrote his mother a postcard to tell her of his success. ("Dear Mother! Joyful news today," Einstein wrote. "H. A. Lorentz has telegraphed me that the English expeditions have actually proved the deflection of light near the sun.") The scientists hailed this confirmation of Einstein's theory, and from the moment of its official announcement, Einstein was recognized by scientists and the general public alike as the world's greatest scientist.

Although the experiment has been repeated at many solar eclipses, most recently in Mexico in 1970 and Africa in 1973, the precision of even the best results is only about 5 per cent (that is, we cannot tell from the data

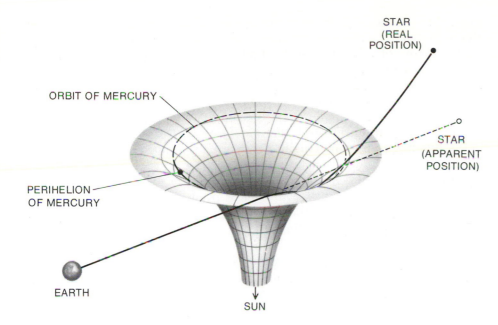

STAR
(REAL
POSITION)

ORBIT OF MERCURY

STAR
(APPARENT
POSITION)

PERIHELION
OF MERCURY

EARTH

SUN

Figure 5-35 Under the general theory of relativity, the presence of a massive body essentially warps the space nearby. This can account for both the bending of light near the sun and the advance of the perihelion point of Mercury by 43 arc sec per century more than would otherwise be expected. The diagram shows how a two-dimensional surface warped into three dimensions can change the direction of a "straight" line that is constrained to its surface; the warping of space is analogous, although with a greater number of dimensions to consider.

whether the actual value is as much as 5 per cent more or less than the value we specify as the "measured" value). The data agree with Einstein's prediction, but are not accurate enough to distinguish between Einstein's theory and newer, more complicated, rival theories of gravitation. Fortunately, the effect of gravitational deflection is constant through the electromagnetic spectrum, and the test can now be performed more accurately by observing the bending of radiation from radio sources, especially quasars (see Chapter 24), as the sun passes in front of them. This is done with radio interferometers (which are described in Chapter 23). The interferometer results agree with Einstein's theory to within 1 per cent, enough to make the competing theories very unlikely.

A related test involves not deflection but a delay in time of signals passing near the sun. This can now be performed with signals from interplanetary spacecraft, most recently with a Viking orbiter near Mars, and these data also agree with Einstein's theory.

Another of the triumphs of Einstein's theory, even in its earliest versions, was that it explained the advance of the perihelion of Mercury. The orbit of Mercury, as are the orbits of all the planets, is elliptical; the point at which the orbit comes closest to the sun is called the *perihelion* (Fig. 5-36). The elliptical orbit is pulled around the sun over the years, mostly by the gravitational attraction on Mercury by the other planets, so that the perihelion point is at a different orientation in space. Each century (!) the perihelion point moves around the sun by approximately 5600 seconds of arc (which is less than 2°). Subtracting the effects of precession and of the other planets leaves 43 seconds of arc per century whose origin had not been understood before Einstein.

Einstein's general theory predicts that the presence of the sun's mass warps the space near the sun. Since Mercury is sometimes closer to and sometimes farther away from the sun, it is sometimes traveling in space that is warped more than the space it is in at other times. This should have the effect of changing the point of perihelion by 43 seconds of arc per century. The agreement of this prediction with the measured value was an important observational confirmation of Einstein's theory. More recent re-

Figure 5-36 The advance of the perihelion of Mercury.

SUN

PERIHELION

NEXT PERIHELION

finements of solar system observations have shown that the perihelions of Venus and of the earth also advance by the even smaller amounts predicted by the theory of relativity. In the 1960's, it was suggested that there was another contribution to the perihelion advance that had been previously overlooked. But the effect reported then seems to have been accounted for in other ways. So Einstein's theory survived the challenge.

Historically, in explaining the perihelion advance, general relativity was explaining an observation that had been known. On the other hand, in the bending of light it actually predicted a previously unobserved phenomenon.

As a general rule, scientists try to find theories that not only explain the data that are at hand, but also make predictions that can be tested. This is an important part of the *scientific method,* which is discussed further in Section 19.5. Because the bending of electromagnetic radiation by a certain amount was a prediction of the general theory of relativity that had not been anticipated, the verification of the prediction was a more convincing proof of the theory's validity than the theory's ability to explain the perihelion advance.

SUMMARY AND OUTLINE OF CHAPTER 5

Aims: To study the sun, which is the nearest star and an example of all the other stars whose surfaces we can not observe in such detail

Definitions: parts of the sun (interior, photosphere, chromosphere, corona, spicules, granulation, solar wind, coronal holes) (Sections 5.1, 5.2, 5.3, 5.4, and 5.7)

Spectra:
  Photospheric spectrum: continuum with Fraunhofer lines; 5800 K; abundances of the elements (Section 5.2b)
  Chromospheric spectrum: emission at the limb during eclipses in the visible part of the spectrum (flash spectrum); typical temperature is 15,000 K (Section 5.3a)
  Coronal spectrum (Section 5.4a): high temperature of 2,000,000 K

Space observations: Orbiting Solar Observatories, Skylab (Sections 5.4b and 5.8)

Eclipse phenomena: band of totality, Baily's beads, diamond ring (Section 5.5)

Solar activity cycle: sunspots, magnetic fields, flares, plages, filaments, prominences (Section 5.6)

Solar wind flows from coronal holes (Section 5.7)

Solar constant and its possible variability (Section 5.9)

Solar energy (Section 5.10)

Using the solar mass to test Einstein's general theory of relativity (Section 5.11)
  Gravitational deflection
    Predicted in advance of detection
    First detected at eclipses
    Now also tested in radio region of spectrum
  Advance of the perihelion of Mercury
    Phenomenon was known in advance of Einstein's theory

QUESTIONS

1. Sketch the sun, labeling the interior, the photosphere, the chromosphere, the corona, sunspots, and prominences. Give the approximate temperature of each.

2. What elements make most of the lines in the Fraunhofer spectrum? What elements make the strongest lines? Why?

3. Explain why the photospheric spectrum is an absorption spectrum and the chromospheric spectrum seen at an eclipse is an emission spectrum.

4. Graph the temperature of the interior and atmosphere of the sun as a function of distance from the center.

5. (a) Why isn't there a solar eclipse once a month? (b) Whenever there is a total solar eclipse, a lunar eclipse occurs either two weeks before or two weeks after. Explain.

6. Why can't we observe the corona every day from any location on earth?

7. Describe the series of phenomena one observes at a total solar eclipse.

8. Why does the chromosphere appear pinkish at an eclipse?

9. How do we know that the corona is hot?

10. Describe the sunspot cycle and a mechanism that explains it.

11. What are the differences between the solar wind and the normal "wind" on earth?

12. (a) What is the solar constant? (b) Why is it difficult to measure precisely?

13. (a) Of what part of the spectrum is it most important to make accurate measurements in order to determine the solar constant? (b) What would the solar constant be if we lived on Mars?

14. What are the advantages and disadvantages of solar energy as a source of energy for the decade of the 1980's?

15. In what tests of general relativity does the sun play an important role? Describe the current status of these investigations.

TOPICS FOR DISCUSSION

1. Discuss the relative importance of solar observations (a) from the ground, (b) at eclipses, (c) from unmanned satellites, and (d) from manned satellites.

2. (a) Why might solar energy become a more important part of the solution to our energy problem around the year 2000 than it could be in the next few years? Discuss the advantages and disadvantages it might have relative to (b) nuclear fission, (c) nuclear fusion.

6

Stellar Evolution

We have seen how the Hertzsprung-Russell diagram for a cluster of stars allows us to deduce the age of the cluster and the ages of the stars themselves (see Section 4.5). Our human lifetimes are very short compared to the billions of years that a typical star takes to form, live its life, and die. Thus our hope of understanding the life history of an individual star depends on studying large numbers of stars, for presumably we will see them at different stages of their lives. In this chapter we will discuss the birth of stars, then the processes that go on in a stellar interior during a star's life on the main sequence, and then begin the story of the evolution of stars when they finish this stage of their lives. The next three chapters will continue the story.

6.1 STARS IN FORMATION

The process of star formation starts with a region of gas and dust of slightly higher density than its surroundings. Perhaps this results only from a random fluctuation in density. If the density is high enough, the gas and dust begin to contract under the force of gravity. As they contract, energy is released, and it turns out (from a basic theorem) that half of that energy heats the matter. As the gas and dust are heated, they begin to give off an appreciable amount of radiation.

The "track" of this protostar,* the path it takes on a Hertzsprung-Russell diagram, is shown in Figure 6–1. At first, the protostar brightens, and the track moves upward and toward the left on the diagram. While this is occurring, the central part of the protostar continues to contract and the temperature rises. The higher temperature results in a higher pressure, and

---

*"Proto-" is a prefix of Greek origin meaning "primitive."

A photograph in red light of the region near Herbig-Haro object number 100 in the dark cloud in the constellation Corona Australis. *Herbig-Haro objects* are bright nebulous regions of gas and dust that may be stars in formation. The Herbig-Haro object is just to the lower left of the center of the photograph, near the center of the black patch that is the high-density cloud from which the H–H object was formed.

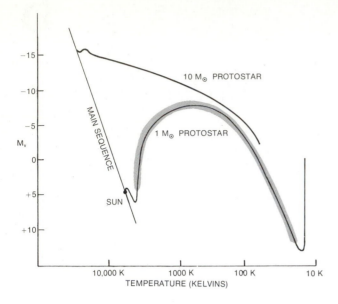

Figure 6–1 Evolutionary tracks for two protostars, one of 1 solar mass and the other of 10 solar masses. ($M_\odot$ is the symbol for the mass of the sun.) The evolution of the shaded portion of the track for the 1 solar mass star is described in the text. These stages of its life last for about 50 million years. More massive stars whip through their protostar stages more rapidly; the corresponding stages for a 10 solar mass star may last only 200,000 years. These very massive stars wind up being very luminous.

thus acts as a force in the outward direction. Eventually a point is reached where this outward force balances the inward force of gravity for the central region.

Box 6.1   Evolutionary Tracks

We can plot the position of a star on the Hertzsprung-Russell diagram for any particular instant of its life. As a protostar, for example, it would have a specific luminosity and temperature a million years before it began nuclear fusion, and another specific luminosity and temperature half a million years later. Each of the pairs of luminosity and temperature would correspond to a point on the H-R diagram. When hydrogen burning began, the star would also correspond to a specific point on the H-R diagram. If we connect the points representing the entire lifetime of the star, we have an *evolutionary track*. A star, of course, is only at one point of its track at any given time.

By this time the dust has vaporized and the gas is opaque, so that energy emitted from the central region does not escape directly. The outer layers continue to contract. Since the surface area is decreasing, the luminosity decreases and the track of the protostar begins to move downward on the H-R diagram. As the protostar continues to heat up, it also continues to move toward the left on the H-R diagram (that is, it gets hotter).

The time scale of this gravitational contraction depends on the mass of gas in the protostar. Very massive stars contract to approximately the size of our solar system in only ten thousand years or so. These massive objects become O and B stars, and are located at the top of the main sequence. They are sometimes found in groups, which are called *O and B associations*.

Less massive stars contract much more leisurely. A star of the same mass as the sun may take tens of millions of years to contract, and a less massive star may take hundreds of millions of years to pass through this stage.

Theoretical analysis shows that the dust surrounding the stellar embryo we call a protostar should absorb much of the radiation that the protostar emits. Do we detect any objects in the sky that meet the characteristics we expect from this scenario?

Theoretically, the radiation from the protostars should heat the dust to temperatures that produce primarily infrared radiation. Indeed, infrared astronomers have found several objects that are especially bright in the infrared but that have no known optical counterparts. These objects seem to be located in regions where the presence of a lot of dust and gas and other young stars indicates that star formation might be going on. Some of these infrared objects may contain stars in formation.

In the visible part of the spectrum, several classes of stars that vary erratically are found. One of these classes, called *T Tauri stars,* includes stars that have a wide range of spectral types. Their visible radiation can vary by as much as several magnitudes, and they are known to be strong infrared emitters. Presumably these are stars that have not quite reached the main sequence and have not completely settled down to a steady and reliable existence.

6.2 STELLAR ENERGY GENERATION

All the heat energy in stars that are still contracting toward the main sequence results from the gravitational contraction itself. If this were the only source of energy, though, stars would not shine for very long on an astronomical time scale—only about 30 million years. Yet we know that even rocks on earth are older than that, since rocks billions of years old have been found. We must find some other source of energy to hold the stars up against their own gravitational pull.

The gas in a protostar will continue to heat up until the central portions become hot enough for *nuclear fusion* to take place. Using this process, which we discuss later, the star can generate enough energy inside itself to support it during its entire lifetime on the main sequence.

The basic fusion processes fuse four hydrogen nuclei into one helium nucleus, just as hydrogen atoms are combined into helium in a hydrogen bomb here on earth. In the process, tremendous amounts of energy are released.

In the interiors of stars, we are dealing with nuclei instead of the atoms we have discussed in stellar atmospheres, because the high temperature strips the electrons off the nuclei. The electrons are mixed in with the nuclei through the center of the star; there can be no large imbalance of positive and negative charges, or strong repulsive forces would arise inside the star.

A hydrogen nucleus is but a single proton. A helium nucleus is more complex. It consists of two protons and two neutrons (Fig. 6–3). Protons and neutrons have essentially the same mass and are similar in many of their properties, though the proton has a charge and the neutron has none.

The mass of the helium that is the final product of the fusion process is slightly less than the sum of the masses of the four hydrogen nuclei that went into it. A small amount of mass "disappears" in the process: .007 (a

Figure 6–2 T Tauri itself is embedded in a Herbig-Haro object known as Hind's variable nebula. It has changed in brightness considerably in the last century, first fading from view and then brightening considerably. This probably results from changing illumination from T Tauri on a fixed dust cloud.

Nuclear fusion *is the joining of two nuclei to form a single, larger, nucleus.*

H¹ NUCLEUS
PROTON

He⁴ NUCLEUS

Figure 6–3 The nucleus of hydrogen's most common form is a single proton, while the nucleus of helium's most common form consists of two protons and two neutrons.

number that is easy to remember for James Bond aficionados) of the mass of the four hydrogen nuclei.

The mass does not really simply disappear, but is rather converted into energy according to Albert Einstein's famous formula: $E = mc^2$. Now c, the speed of light, is a large number, and c^2 is even larger. Thus even though m is only a small fraction of the original mass, the amount of energy released is prodigious. The loss of only .007 of the mass of the sun, for example, is enough to allow the sun to radiate as much as it does at its present rate for a period of at least 10^{10} years. This fact, not realized until 1920, solved the long-standing problem of where the sun and the other stars got their energy.

All the main-sequence stars are approximately 90 per cent hydrogen (that is, 90 per cent of the atoms are hydrogen), so there is lots of raw material to stoke the nuclear "fires." We speak colloquially of *nuclear burning*, although, of course, the processes are quite different from the chemical processes that are involved in the "burning" of logs or of autumn leaves. In order to be able to discuss these processes, we must first discuss the general structure of nuclei and atoms.

6.3 ATOMS

An atom consists of a small *nucleus* surrounded by *electrons*. Most of the mass of the atom is in the nucleus, which takes up a very small volume in the center of the atom. The effective size of the atom is set by the electrons, and the chemical interactions of atoms to form molecules are also determined by the electrons.

The nuclear particles with which we need be most familiar are the *proton* and *neutron*. Both these particles have nearly the same mass, only 1.6×10^{-24} gram, 1836 times greater than the mass of an electron.

While the neutron has no electric *charge*, the proton has one unit of positive electric charge. The electrons, which surround the nucleus, have one unit each of negative electric charge, so when an atom has the same number of protons and electrons it is electrically neutral. This is called a neutral atom, and is denoted with the Roman numeral I. For example, neutral helium is denoted He I.

When an atom has lost an electron, it has a net positive charge of 1 unit. As we stated in Section 2.7, the atom is now a type of ion (Fig. 6–4); this state of the atom is denoted with a Roman numeral II. Thus an individual ion of helium that has lost one electron is in the He II state. If the atom is twice ionized, that is, has lost two electrons, it is in state III.

The definition of ions does not concern the nucleus; for the nucleus, we need keep track only of the numbers of protons and neutrons and not of

Figure 6–4 Hydrogen and helium ions. The sizes of the nuclei are greatly exaggerated with respect to the sizes of the orbits of the electrons.

H I   H II

He I

He II

He III

$_1H^1$ $_1H^2 = D$ = DEUTERIUM $_1H^3 = T$ = TRITIUM $_2He^3$ $_2He^4$

Figure 6–5 Isotopes of hydrogen and helium. $_1H^2$ (deuterium) and $_1H^3$ (tritium) are much rarer than the normal isotope, $_1H^1$. $_2He^3$ is much rarer than $_2He^4$.

orbiting electrons. The number of protons in the nucleus determines the quota of electrons that the neutral state of the atom must have, since it determines the charge of the nucleus. Each element is defined by the specific number of protons in its nucleus. The element with one proton is hydrogen, that with two protons is helium, that with three protons is lithium, and so on. A table of the elements appears as Appendix 4.

The number of neutrons in a nucleus is not fixed for a given element, although the number is always somewhere between 1 and 2 times the number of protons. (Hydrogen, which need have no neutrons, and helium are the only exceptions to this rule.) The possible different forms of the same element have only slightly different numbers of neutrons. These different forms are called *isotopes*.

For example, the nucleus of ordinary hydrogen contains one proton and no neutrons. An isotope of hydrogen (Fig. 6–5) called deuterium (and sometimes "heavy hydrogen") has one proton and one neutron. Another isotope of hydrogen called tritium has one proton and two neutrons.

Most isotopes do not have specific names, and one keeps track of the numbers of protons and neutrons with a system of superscripts and subscripts. Though there are several alternative ways of writing the form, we shall use the convention that a subscript before the symbol denoting the element is the number of protons, and a superscript following the symbol is the total number of protons and neutrons together. The number of protons is called the *atomic number,* and the number of protons and neutrons together is called the *mass number.* For example, $_1H^2$ is deuterium, since deuterium has 1 proton, which gives the subscript, and an atomic mass of 2, which gives the superscript. Deuterium has atomic number equal 1 and mass number equal 2. Similarly $_{92}U^{238}$ is an isotope of uranium with 92 protons (atomic number $= 92$) and a mass number of 238, which is divided into 92 protons and $238–92 = 146$ neutrons.

Each element exists only in certain isotopic forms. For example, most of naturally occurring helium is in the form $_2He^4$, with a lesser amount as $_2He^3$. Sometimes an isotope is not stable, in that after a time it will spontaneously change into another isotope or element; we say that such an isotope is *radioactive*.

During a certain type of radioactive decay, a particle called a neutrino is given off. A neutrino is a neutral particle (its name, conferred by the physicist Enrico Fermi, comes from the Italian for "little neutral one"). It has the very interesting property, for a particle, of always traveling at the speed of light. Now, ordinary matter cannot travel at the speed of light, according to Einstein's theory of relativity, because its mass gets larger and larger and approaches infinity as its speed approaches the speed of light. This problem does not arise in the case of the neutrino because the neutrino, like the photon, would have no mass if it were at rest.

Neutrinos have a very useful property for the purposes of astronomy: they do not interact very much with matter. Thus when a neutrino is formed deep inside a star, it can usually escape to the outside without interacting

By convention, *we mean a usage that is agreed upon.*

with any of the matter in the star. This means that the neutrino heads right out of the star, without bumping into any of the matter in the star along the way. Electromagnetic radiation, on the other hand, does not escape from inside a star so easily. A photon of radiation can travel only about 1 cm in a stellar interior before it is absorbed, and it is millions of years before a photon zigs and zags its way to the surface.

The elusiveness of the neutrino makes it a valuable probe of the conditions inside the sun at the present time, but also makes it very difficult for us to detect on earth. Later in this chapter we shall discuss the major experiment that is being performed to study solar neutrinos.

6.4 STELLAR ENERGY CYCLES

Several chains of reactions have been proposed to account for the fusion of four hydrogen atoms into a single helium atom. Hans Bethe, now at Cornell University, suggested some of these procedures during the 1930's. The different possible chains that have been proposed are important at different temperatures, so chains that are dominant in the centers of very hot stars may be different chains from the ones that are dominant in the centers of cooler stars.

When the center of a star is at a temperature less than 15×10^6 K, the *proton-proton* chain is dominant (Fig. 6–6). In the first stage, two ordinary nuclei of hydrogen fuse to become a deuterium nucleus, a positron, and a neutrino. The neutrino immediately escapes from the star, but the positron soon collides with an electron. They annihilate each other, forming gamma rays.

Figure 6–6 The proton-proton chain; e+ stands for a positron (a particle like an electron except for having positive charge), ν (nu) is a neutrino, and γ (gamma) is radiation at a very short wavelength.

$$_1H^1 + _1H^1 \longrightarrow _1H^2 + e^+ + \nu$$

$$_1H^1 + _1H^1 \longrightarrow _1H^2 + e^+ + \nu$$

$$_1H^2 + _1H^1 \longrightarrow _2He^3 + \gamma$$

$$_1H^2 + _1H^1 \longrightarrow _2He^3 + \gamma$$

$$_2He^3 + _2He^3 \longrightarrow _2He^4 + 2 _1H^1$$

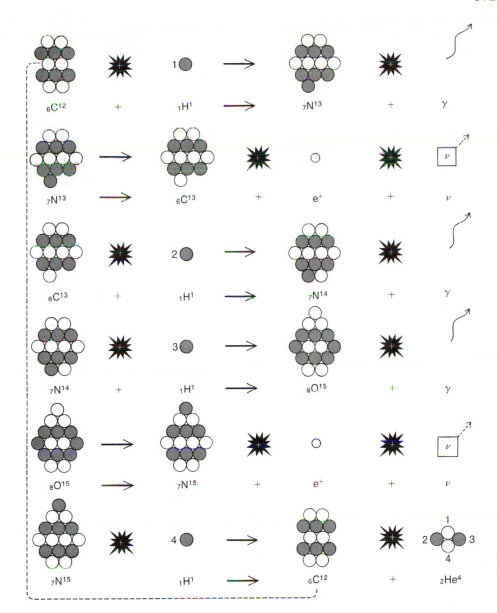

$_6C^{12}$ + $_1H^1$ $\longrightarrow$ $_7N^{13}$ + γ

$_7N^{13}$ $\longrightarrow$ $_6C^{13}$ + e^+ + ν

$_6C^{13}$ + $_1H^1$ $\longrightarrow$ $_7N^{14}$ + γ

$_7N^{14}$ + $_1H^1$ $\longrightarrow$ $_8O^{15}$ + γ

$_8O^{15}$ $\longrightarrow$ $_7N^{15}$ + e^+ + ν

$_7N^{15}$ + $_1H^1$ $\longrightarrow$ $_6C^{12}$ + $_2He^4$

Figure 6–7 The carbon cycle. Note that the carbon is left over at the end, ready to enter into another cycle.

Next, the deuterium nucleus fuses with yet another nucleus of ordinary hydrogen to become an isotope of helium with two protons and one neutron. High-energy radiation is released at this stage in the form of more gamma rays.

Finally, two of these helium isotopes fuse to make one nucleus of ordinary helium plus two nuclei of ordinary hydrogen. We have put in six hydrogens and wind up with one helium plus two hydrogens, a net transformation of four hydrogens into one helium. The small fraction of mass that disappears in the process is converted into energy according to the formula $E = mc^2$.

For hotter stellar interiors than that of the sun, the *carbon cycle* would dominate (Fig. 6–7). The carbon cycle begins with the fusion of a hydrogen nucleus with a carbon nucleus. After many steps, and the insertion of

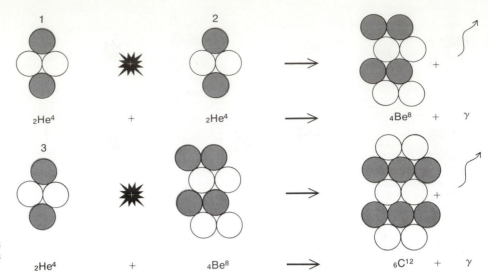

Figure 6–8 The triple-alpha process, which takes place only at temperatures above about 10^8 K.

four hydrogen nuclei, we are left with one helium nucleus plus a carbon nucleus. Thus as much carbon remains at the end as there was at the beginning, and the carbon can start the cycle again. Again, four hydrogens have been converted into one helium, .007 of the mass has been transformed, and an equivalent amount of energy has been released according to $E = mc^2$. Since the cycle can also start with nitrogen and oxygen, the process is sometimes called the *carbon-nitrogen cycle* or the *CNO cycle*.

At even higher temperatures, as occur in very hot stars where the interior temperatures are above 10^8 K, we can fuse helium nuclei to make carbon nuclei. The nucleus of a helium atom is called an "alpha particle" for historical reasons, as the Greek characterizations derive from a former confusion of radiation and particles. (We now know that α particles, formerly called α rays, are helium nuclei, β particles, formerly called β rays, are electrons, and γ rays, as we have seen, are radiation of short wavelength.) Since three helium nuclei ($_2\text{He}^4$) go into making a single carbon nucleus ($_6\text{C}^{12}$) [with beryllium ($_4\text{Be}^8$) as an intermediate step], the procedure is known as the *triple-alpha process* (Fig. 6–8). A series of other processes can build some of the heavier elements inside stars. This is the study of *nucleosynthesis*.

The theory can account for the abundances we observe of the elements heavier than helium. Currently, we think that synthesis of isotopes of hydrogen and helium took place in the few minutes after the origin of the universe (see Section 25.5), and that the heavier elements were formed, along with additional helium, in stellar processes.

6.5 THE STELLAR PRIME OF LIFE

Now that we have discussed basic nuclear processes, let us return to an astronomical situation. We last discussed a protostar in a collapsing phase, with its internal temperature rapidly rising.

One of the most common definitions of temperature describes the temperature in terms of the velocities of individual atoms or other particles. Since this type of temperature depends on the kinetics (motions) of particles, it is called the *kinetic temperature*. A higher kinetic temperature (we shall call it simply "temperature" from now on) corresponds to higher particle velocities.

For a collapsing protostar, the energy from the gravitational collapse goes into giving the individual particles greater velocities; that is, the temperature rises. For nuclear fusion to begin, atomic nuclei must get close enough to each other so that the nuclear force, technically called the *strong force*, can play its part. But all nuclei have positive charges, because they are composed of protons and neutrons. The positive charges on any two nuclei cause an electrical repulsion between them, and this force tends to prevent fusion from taking place. However, at the high temperatures typical of a stellar interior, some nuclei have enough energy to overcome this electrical repulsion and to come into sufficiently close proximity for the strong force to take over. The electrical repulsion that must be overcome is the reason why hydrogen nuclei, which have net positive charges of 1, will fuse at lower temperatures than will helium nuclei, which have net positive charges of 2.

Box 6.2   Forces in the Universe

There are four known types of forces in the universe:

The *strong force*, also known as the *nuclear force*, is the strongest. It is the binding force of nuclei. Although it is very strong close up, it grows weaker rapidly with distance. Actually, the strong force is no longer thought to be a basic force. We now think that both mesons and nuclear particles are composed of 3 kinds of particles called *quarks* (from a line of James Joyce's *Finnegan's Wake*). The quarks are linked together by the *color force*; each kind of quark can have one of three properties called *colors*. There also seem to be other types of quarks.

The *electromagnetic force*, 1/137 the strength of the strong force, leads to electromagnetic radiation in the form of photons. It is the force involved in chemical reactions.

The *weak force*, important only in the decay of certain elementary particles, is currently under careful study by particle physicists. It is very weak, only 10^{-13} the strength of the strong force, and also has a very short range.

The *gravitational force* is the weakest of all over short distances, only 10^{-39} the strength of the strong force. But the effect from the masses of all the particles is cumulative—it adds up—so that on the scale of the universe gravity dominates the other forces.

In the center of a star, the fusion process is a self-regulating one. If the nuclear energy production rate increases, then an excess pressure is generated that would tend to make the star grow larger if it were not for the fact that this expansion in turn would cool down the gas. Thus the temperature can remain roughly constant for a very long time. This is the main-sequence phase of a stellar lifetime. Fusion processes are well regulated in stars. When we learn how to control fusion in power-generating stations on earth, which currently seems a long way off, our energy crisis will be over.

The more mass a star has, the hotter its core becomes before it generates enough pressure to counteract the gravity. Thus more massive stars use their nuclear fuel at a much higher rate than less massive stars, and even though the more massive stars have more fuel to burn, they go through it relatively quickly. This explains the mass-luminosity relation (Section 4.2).

Depending on the temperature and density inside the star, and thus on how massive the star is, the helium that is formed inside the star may itself fuse into carbon via the triple-alpha process. In the interiors of the most massive stars, the carbon itself burns, and then the products of the carbon combustion burn, and so on, until eventually an iron core is built up.

Because of basic properties of nuclei, iron is the heaviest element that can form in a stellar interior. All the fusion processes that form nuclei lighter than and including iron are processes that release energy. On the other hand, one must add energy to the system in order to form nuclei heavier than iron. Since there is no source for such extra energy, iron accumulates at the center of the star.

6.6 THE NEUTRINO EXPERIMENT

Since the stellar interior lies under opaque layers of gases, we cannot observe directly any electromagnetic radiation it might emit. All the gamma rays, x-rays, ultraviolet, visible, infrared, and radio radiation that we receive come from the surface layers or atmospheres of stars.

Only the neutrino, a nuclear particle mentioned in Section 6.3, can escape directly from a stellar interior. It interacts so weakly with matter that it is hardly affected by the presence of the rest of the solar mass, and zips right out into space at the speed of light.

Raymond Davis, Jr., a chemist at the Brookhaven National Laboratory, has spent the last ten years trying to detect the neutrinos from the solar interior that on theoretical grounds should be created by nuclear reactions there. Since the experiment is our only direct opportunity to test the theoretical work on stellar interiors, the results are exceedingly important. Because they raise serious questions about the state of our knowledge, we shall discuss them in some detail.

How do we detect the neutrinos? Neutrinos, after all, pass through the earth and sun, barely affected by their mass. At this instant, neutrinos are passing through your body. Davis makes use of the fact that very occasionally a neutrino will interact with the nucleus of an atom of chlorine, and transform it into an isotope of the noble gas argon. (The "noble" elements do not readily combine with other elements to form molecules; the atoms of argon are thus not hidden in molecules.)

This transformation takes place very rarely, so Davis had to accumulate

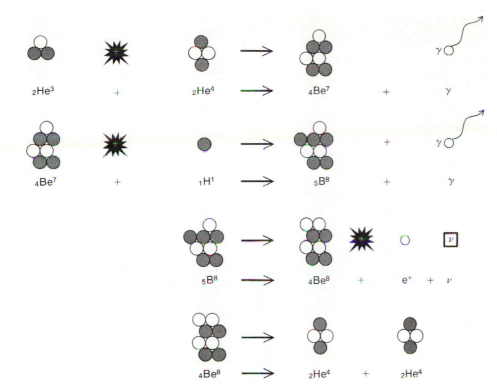

$$_2He^3 + \quad _2He^4 \longrightarrow \quad _4Be^7 + \quad \gamma$$

$$_4Be^7 + \quad _1H^1 \longrightarrow \quad _5B^8 + \quad \gamma$$

$$_5B^8 \longrightarrow \quad _4Be^8 + \quad e^+ + \quad \nu$$

$$_4Be^8 \longrightarrow \quad _2He^4 + \quad _2He^4$$

Figure 6–9 The chain of fusion via the boron-beryllium route, through a version of the proton-proton chain, leads to the ejection of neutrinos of an energy that should be detectable by Davis's apparatus.

a large number of chlorine atoms. He found it best to do this by filling a large tank with liquid cleaning fluid, C_2Cl_4, where the subscripts represent the number of atoms of carbon and chlorine in this perchloroethylene molecule. One fourth of the chlorine is the Cl^{37} isotope, which is able to interact with a neutrino. Davis now has a large tank containing 400,000 liters (100,-000 gallons) of this cleaning fluid.

Even with this huge tankful of chlorine atoms, calculations show that Davis should expect only about one neutrino-chlorine interaction every day. This experiment is surely one of the most difficult ever attempted.

When an interaction occurs, a radioactive argon atom is formed. Any such argon atoms can be removed from the tank by standard chemical techniques involving bubbling helium gas through the fluid. Then the radioactivity of the few resulting argon atoms can be measured, also by standard means.

Davis set up his apparatus far underground, in order to shield it from other particles from space that could also interact with the chlorine. Thus 1.5 km underground, deep in the Homestake Gold Mine in Lead, South Dakota, you will find the most incredible of the world's "telescopes"—this 400,000 liter tank of cleaning fluid (Fig. 6–10).

Davis runs the experiment for 1 to 3 month intervals by leaving the tank undisturbed for that length of time. Then he flushes the tank with helium to remove the argon, and measures the amount of argon. In the meantime, he prepares to "observe" for the next one to three month cycle.

It is fair to say that Davis's results have astounded the scientific community, which had confidently expected to hear reports of a small but

Figure 6–10 The neutrino telescope, deep underground in the Homestake Gold Mine in Lead, South Dakota, consists mainly of a tank containing 400,000 liters of perchloroethylene. The tank is now surrounded by water.

Figure 6–11 Davis tried out the water that helps the kilometer of earth overhead shield his neutrino telescope from particles other than neutrinos.

measurable number of neutrino interactions. But Davis reported: no neutrinos.

Theoreticians went back to work, and tried to carry out complicated calculations including more realistic assumptions than they had previously used. They included, for example, the fact that the sun is not just a stationary ball of gas but that it is rotating. Still, the new calculations did not give predictions of appreciably lower value.

As time went on, Davis's experiment was refined and the total amount of observing time increased. His latest value is not necessarily zero, but less than one-third the best theoretical prediction. His measurement is at the level that represents the minimum that standard theories are able to predict.

Several questions immediately come to mind. The first deals with Davis's apparatus, and whether there are some experimental effects that could explain the results in some normal manner. Davis has carried out a series of careful checks of his apparatus, and most astronomers and chemists are convinced that the experimental setup is not the cause of the problem.

Next, perhaps we do not understand the neutrino as well as we thought. Perhaps it disintegrates during the 8 minutes it takes to travel from the sun to the earth. That would indeed account for the failure to detect neutrinos, but in itself it would be an important result for physics. There is no experimental evidence for this situation, and not many scientists are still seriously considering it.

One of the most interesting possibilities is that we really don't understand the basic physics of nuclear reactions as well as we had thought.

Another interesting possibility is that our understanding of stellar interiors is not satisfactory. If, for example, stars are cooler inside than we have predicted, fewer neutrinos would be produced. Thus the neutrino experiment has clear and direct ramifications for the study of astrophysics.

Another possibility now being considered with increasing respect is the idea that maybe we are doing everything right but that the sun is simply not generating neutrinos at the moment! The neutrino experiment tells us what the sun is doing now, or really what it did eight minutes earlier. The other way we can detect the effect of the nuclear furnace in the solar interior is to observe the solar surface, but it takes millions of years for energy generated inside to work its way up to the surface. Perhaps ice ages in the past resulted from other periods long ago when the sun was turned "off"; perhaps the current episode will lead to ice ages hundreds of thousands of years from now. We have thought of main-sequence stars as sedentary objects, regularly transforming hydrogen into helium at a steady rate. But they may be actually cyclic, and we would be merely observing the sun at a period when no nuclear fusion is going on in its interior.

These explanations may sound desperate, but the situation is not yet hopeless. Research is continuing to try to produce theoretical explanations of Davis's neutrino results. In the meantime, Davis continues to run his apparatus.

Though some of the more recent runs gave values somewhat higher than average, Davis views these as mere fluctuations of a statistical nature. Only time will tell.

The neutrino experiment is one of the most interesting to be carried out in astronomy in recent years, and seems to be giving the most profound and unexpected results. The least that we can conclude is that until the matter is settled, we must treat all the theoretical predictions about stellar interiors with a bit of caution.

6.7 DYING STARS

We have been considering what we know and how much we may not really know about stars as they spend their middle age on the main sequence as dwarfs. After this phase of life, we have seen that the stars evolve upward on the H-R diagram. Equivalently, they become brighter. Also, their surfaces become cooler and they move to the right on the H-R diagram. On the main sequence, most stars behave similarly, regardless of their mass. Of course, their lifetimes are very different, and so are their temperatures. But once they evolve off the main sequence, mass differences play an even more important role.

Note that stars do not actually move along the main sequence; they stay for a long time at essentially the same place on the H-R diagram.

We shall devote the next three chapters to the various end stages of stellar evolution. First we shall discuss the less massive stars, such as the sun. In Chapter 7, we shall see how such stars swell in size to become giants, possibly become planetary nebulae, and then end their lives as white dwarfs.

More massive stars come to more explosive ends. In Chapter 8 we shall see how some stars are blown to smithereens, and how strange objects called "pulsars" may be the remnants.

In Chapter 9, we shall discuss the strangest kind of stellar death of all. The most massive stars may become "black holes," and effectively disappear from view.

SUMMARY AND OUTLINE OF CHAPTER 6

Aims: To study the formation of stars, their main-sequence lifetimes, and
the mechanisms of stellar energy generation

Protostars (Section 6.1)
  Their evolutionary tracks on the H-R diagram
  They may form in dark clouds
  Infrared observations: may be observing dust shells around protostars
T Tauri stars (Section 6.1)
  Irregular variations in magnitude
  May not have reached the main sequence
Stellar energy generation (Sections 6.2, 6.3, and 6.4)
  Nuclear fusion
  Definition of nuclear particles, isotopes
  The proton-proton chain: dominates in cooler stars, including the sun
  The carbon cycle: for hotter stars
  The triple-alpha process: takes place after helium has been formed and
    at high temperatures
  Stellar nucleosynthesis
The main sequence: balance of pressure and gravity (Section 6.5)
A test of the theory: the neutrino experiment (Section 6.6)
  The mechanism
  The results: too few neutrinos
  The possible explanations
  The consequences of our faith in the theory
Mass as the determining factor in evolution (Section 6.7)

QUESTIONS

1. Since individual stars can live for billions of years, how can observations taken at the current time tell us about stellar evolution?

2. What is the source of energy in a protostar? At what point does a protostar become a star?

3. What is the *evolutionary track* of a star?

4. Arrange the following in order of development: OB associations; T Tauri stars; dark clouds; pulsars.

5. What is the net charge (in terms of proton charges) of Ca I, Mg II, Fe II, Fe XII. (Consult Appendix 4 for a list of the elements. The number of protons an element has is called its atomic number, and is tabulated there.)

6. (a) If you remove one neutron from helium, the remainder is what element?
   (b) Now remove one proton. What is left?
   (c) Why is He IV not observed?

7. (a) Explain why nuclear fusion takes place only in the centers of stars rather than on their surfaces as well.
   (b) What is the major fusion process that takes place in the sun?

8. If you didn't know about nuclear energy, what possible energy source would you suggest for stars? What is wrong with these alternative explanations?

9. (a) If all the hydrogen in the sun were converted to helium, what fraction of the solar mass would be lost?
   (b) How many times the mass of the earth would that be?

10. (a) How does the temperature in a stellar core determine which nuclear reactions will take place?
    (b) Why do more massive stars have shorter main-sequence lifetimes?

11. Why is the gravitational attraction between you and the earth stronger than the electrical attraction?

12. In what form is energy carried away in the proton-proton chain?

13. In the proton-proton chain, the products of some reactions serve as input for the next and therefore don't show up in the final result. Identify these intermediate products. What are the **net** input and output of the proton-proton chain?

14. What do you think would happen if nuclear reactions in the sun stopped? How long would it be before we noticed?

15. Why do neutrinos give us different information about the sun than does light?

16. Why are the results of the solar neutrino experiment so important?

The Death of Stars Like the Sun

The sun is just an average star, in that it falls in the middle of the main sequence. But the majority of stars have less mass than the sun. So when we discuss the end of the main-sequence lifetimes of such low-mass dwarfs, we are discussing the future of most of the stars in the universe. In this chapter we will discuss the late stages of evolution of all stars that, when they are on the main sequence, contain up to about 4 solar masses. We shall call these *lightweight stars*. Let us consider a star like the sun, in particular, remembering that all stars up to 4 solar masses go through similar stages but at different rates.

7.1 RED GIANTS

During the main-sequence phase of lightweight stars, as with all stars, hydrogen in the core gradually fuses into helium. By about 10^{10} years after a one-solar-mass star first reaches the main sequence, no hydrogen is left in its core, which is thus composed almost entirely of helium. Though there is no longer any hydrogen in the core, hydrogen is still undergoing fusion in a shell around the core.

Since no fusion is taking place in the core, there is no ongoing nuclear process to replace the heat that flows out of this hot central region. The core no longer has enough pressure to hold up both itself and the overlying layers against the force of gravity. As a result, the core then begins to contract under the force of gravity (we say it contracts *gravitationally*). This gravitational contraction not only replaces the heat lost by the core, but also in fact heats the core up further. (Half the energy from gravitational contraction always goes into kinetic motion in the interior. An example of such increased kinetic motion is the rise in temperature that we call heat. This is an important theorem that we met before in Section 6.1 in discussing the contraction of protostars.) Thus, paradoxically, soon after the hydrogen burning in the core stops, the core becomes hotter than it was before because of the gravitational contraction.

The Bubble Nebula, NGC 7635, often described as a planetary nebula (although some recent work indicates that it might receive its energy from outside rather than from inside, unlike ordinary planetary nebulae).

Figure 7–1 An H-R diagram showing the evolution of a 1 solar mass star toward its red giant stage.

As the core becomes hotter, the hydrogen-burning shell around the core becomes hotter too, and the reactions proceed at a higher rate. Part of the increasing energy production goes into expanding the outer parts of the star. In fact, the surface temperature of the star decreases. Thus, as soon as a main-sequence star forms a substantial core of helium, the outer layers grow slowly larger and redder. The star winds up giving off more energy (because the nuclear reaction rates are increasing). The total luminosity, which is the emission per unit area times the total area, is increased. The luminosity increases even though each bit of the surface is less luminous than before (as a result of its decreasing temperature) because the total surface area that is emitting increases rapidly.

The process proceeds at an ever accelerating pace, simply because the hotter the core gets, the faster heat flows from it and the more rapidly it contracts and heats up further. The hydrogen shell gives off more and more energy. The layers outside this hydrogen-burning shell continue to expand.

The sun is presently still at the earlier stage where changes are stately and slow. But in a few billion years, when the hydrogen in its core is exhausted, the time will come for the sun to brighten and redden faster and faster until it eventually swells and engulfs Mercury and Venus. At this point, the sun will no longer be a dwarf but will rather be what is called a *red giant* (Fig. 7–1). Its great luminosity will sear and char everything on earth. When the sun is sufficiently bloated to be nearly the size of the earth's orbit (Fig. 7–2), we won't be around to admire the show.

The gravity of a spherical mass acts, to a body outside the spatial limits taken up by that mass, as though all the mass were concentrated at its center. A red giant is so bloated that its outer layers are relatively far from its center, so the force of gravity at the surface is not very great. Thus some matter can escape from the star.

While the outer layers are expanding, the inner layers continue to heat up, and eventually reach 10^8 K. At this point, the triple-alpha process begins for all but the least massive lightweight stars, as groups of three helium nuclei fuse into single carbon nuclei. (This triple-alpha process was described in Section 6.4.) For stars about the mass of the sun, the onset of the triple-alpha process happens rapidly. The development of this *helium flash* produces a very large amount of energy in the core, but only for a few years. The consequences of this input of energy at the center are to reverse the evolution that took place prior to the helium flash: now the core expands and the outer portion of the star contracts. As the star adjusts to this situation it becomes smaller and less luminous, moving back down and to the left on the H-R diagram. When the effect of the helium flash is stabilized, the star is burning helium in the core and hydrogen in the shell.

The helium flash is the rapid onset of the conversion of helium to carbon through the triple-alpha process.

7.2 PLANETARY NEBULAE

Sometime after the helium burning sets in, a substantial carbon core forms. This carbon core contracts and heats up as did the helium core before it, driving up the helium reaction rates in the shell surrounding the core. The star then moves up and to the right on the H-R diagram, becoming a luminous red giant once again. This time, however, the swelling and cooling continue on and on until, after a time, the outer layers grow sufficiently cool that the nuclei and electrons combine and form neutral atoms. As electrons that are free have more energy than electrons that are bound up as parts of atoms, the system as a whole experiences an increase in energy when the previously free electrons recombine with the nuclei.

The excess energy is given off as photons, which are absorbed by the gas and blow the outer layers outward. But as the outer layers expand, they in turn become cooler, so that still more nuclei and electrons recombine and still more energy is released. In this way a vicious circle quickly sets in, which is broken in many cases when the outer layers of the star are blasted off. By the time the blown-off layers have moved a few astronomical units away from the star, they have spread into a shell thin and cool enough to be transparent. (1 astronomical unit is the average earth-sun distance.)

We know of a thousand such objects in our galaxy that can be explained by the above theoretical picture, and there may be 10,000 more. They were named *planetary nebulae* because hundreds of years ago, when they were discovered, they appeared similar to the planets Uranus and Neptune when viewed in a small telescope. Both planets and "planetary nebulae" appeared as small, greenish disks. We have since discovered what planetary nebulae really are; they have nothing at all to do with planets, but they retain their old name for historical reasons. Their greenish color is caused by the presence of certain strong emission lines of multiply ionized oxygen (that is, oxygen that has lost more than one electron) and other elements. These lines happen to fall in the green region of the visible spectrum.

Planetary nebulae are exceedingly beautiful objects (Figs. 7–3, 7–4, and 7–5, and Color Plates 19, 20, and 21). These nebulae are actually semi-transparent spheres or cylinders of gas. When we look at their edges, we are looking obliquely through the shells of gas (Fig. 7–6), and there is enough gas along our line of sight to be visible. But when we look through the center of the spherical shell, there is less gas along our line of sight, and it appears transparent.

Astronomers are particularly interested in planetary nebulae because they want to study the various means by which stars can eject mass into interstellar space. This tells them both how a star can reduce its own mass and also about the origin of some of the interstellar matter. Each planetary nebula represents the ejection of 10 or 20 per cent of a solar mass, which is only a small fraction of the mass of the star.

We can measure the ages of planetary nebulae by tracing back the shells at their current rate of expansion and calculating when they would have been ejected from the star. Most of the planetary nebulae we can observe appear to be less than 50,000 years old. This is consistent with our picture of planetary nebulae as a transient phase in the lifetimes of stars of about one solar mass; after about 100,000 years the nebulae will have expanded so much that the gas will be invisible.

7.3 WHITE DWARFS

When a star of approximately one solar mass (and other low-mass stars) has burned all the hydrogen in its core into helium and then all the helium in its core into carbon, it does not heat up sufficiently to allow the carbon to fuse into still heavier elements. Accordingly, there comes a time when

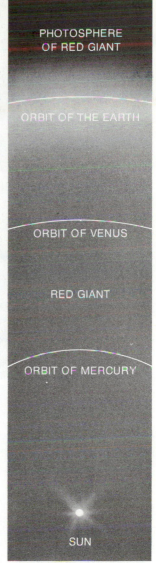

PHOTOSPHERE OF RED GIANT

ORBIT OF THE EARTH

ORBIT OF VENUS

RED GIANT

ORBIT OF MERCURY

SUN

Figure 7–2 A red giant swells so much it can be the diameter of the earth's orbit.

Figure 7–3 NGC 2392, a planetary nebula in the constellation Gemini, the Twins. This object is sometimes known as the Eskimo Nebula.

Figure 7–4   Figure 7–5

Figure 7–4 The Dumbbell Nebula, M27, a planetary nebula in the constellation Vulpecula, the Fox.

Figure 7–5 NGC 7293, a planetary nebula in the constellation Aquarius, the Water Bearer.

Figure 7–6 A hollow shell, when viewed from outside, can give the appearance of a ring, since when looking at the center we see perpendicularly through nearly-transparent gas. Only near the edges of the shell, where we are looking through the gas at a very oblique angle, does the gas become readily visible.

nuclear reactions are no longer generating energy to maintain or increase the internal pressure that balances the force of gravity.

All stars that at this stage in their lives contain less than 1.4 solar masses have the same fate. Many or all stars in this low-mass group come from the red giant phase, and may pass through the phase of being central stars of planetary nebulae. Those that do not go through a planetary nebula stage lose their mass in some other way, perhaps during the giant stage. In any case, they lose large fractions of their original masses. Stars that contained one solar mass when they were on the main sequence now contain 0.5 solar mass. Stars that contained 4 solar masses when they were on the main sequence may now contain only 1.4 solar masses.

When their nuclear fires die out for good, the stars with less than 1.4 solar masses remaining shrink in size, and reach a stable condition that we shall describe below. As they shrink they grow very faint (in the opposite manner to that in which red giants grew brighter as they grew bigger). Because the Planck curve of radiation from many of the stars peaks to the left of the center of the visible part of the spectrum, most appear whitish in color. (Since the intensity falls off slowly on the long wavelength side of the Planck curve, even those stars whose radiation peaks in the uv give off amounts of radiation that vary only slightly across the visible, and so also appear whitish.) Some stars of this type are cooler, though, and so appear yellow or even red. Whatever their actual color, all these stars are called *white dwarfs*. The white dwarfs occupy a region of the Hertzsprung-Russell diagram that is below and to the left of the main sequence (Fig. 7–7).

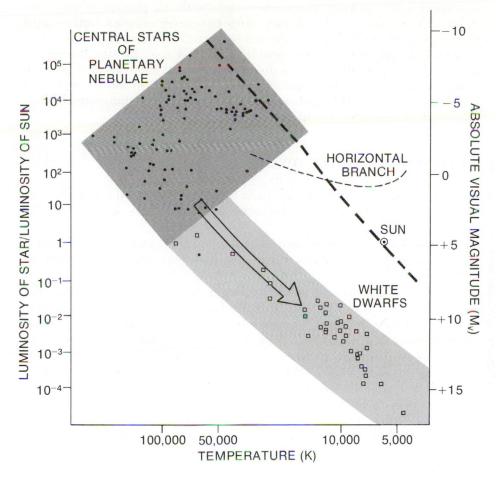

Figure 7-7 When the planetary nebula blows off a red giant, the core of the star remains visible and can be detected at the center of the nebula. Since this *central star* has a high temperature, it appears to the left of the main sequence. These stars eventually shrink and cool to become white dwarfs, as shown on this H-R diagram.

White dwarfs represent a stable phase in which stars of less than 1.4 solar masses live out their old age. Obviously, something must be holding up the material in the white dwarfs against the force of gravity; nuclear reactions generating thermal pressure no longer take place in their interiors. The property that holds up the white dwarfs is a condition called *electron degeneracy;* we thus speak of "degenerate white dwarfs."

Electron degeneracy is a condition that arises in accordance with certain laws of quantum mechanics, and is not something that is intuitively obvious. As the star contracts and the electrons get closer together, there is a continued increase in their resistance to being pushed even closer. This manifests itself as a pressure. There comes a time when the pressure generated in this way exceeds the normal thermal pressure. This only happens at very great densities. When this pressure from the degenerate electrons is sufficiently great, it balances the force of gravity and the star stops contracting.

The effect of the degenerate electron pressure is to stop the white dwarf from contracting at a point when the gas is in a very compressed state. In a white dwarf, a mass approximately that of the sun is compressed into a volume only the size of the earth (Fig. 7-8). A single teaspoonful of a white dwarf weighs 5 tons; it would collapse a table if you somehow had some in your hand and tried to put it down. A white dwarf contains matter so dense that it is in a truly incredible state.

Do not confuse the term "white dwarf" with the term "dwarf." The former refers to the dead hulks of stars in the lower left of the H-R diagram, while the latter refers to normal stars on the main sequence.

What will happen to the white dwarfs with the passage of time? Because of their electron degeneracy, they can never contract further. Still, they have some heat stored in the nuclear particles present, and that heat will be radiated away over the next billions of years. Then the star will be a burned-out hulk called a *black dwarf,* though it is probable that no white dwarfs have yet lived long enough to reach that final stage. It will be billions of years before the sun becomes a white dwarf and then many billions more before it reaches the black dwarf stage.

7.4 OBSERVING WHITE DWARFS

White dwarfs, as we can see from their positions on the H-R diagram, are very faint and thus should be correspondingly difficult to detect. This is indeed the case.

White dwarfs are discovered by looking for blue stars with high proper motions or by their gravitational effect on their companion stars if they are in binary systems.

In the former case, by studying proper motions we find the stars that are close to the sun. The ones that are fainter than main-sequence stars would be at their distances must be white dwarfs. To understand the latter case, we must realize that for any system of masses orbiting each other, we can define an imaginary point called the *center of mass* around which each individual mass appears to orbit. This center of mass, in its proper motion, moves in a straight line, though the path in the sky of any of the individual bodies appears wavy (Fig. 4–6).

At least three of the 40 stars within 5 parsecs of the sun—Sirius, 40 Eridani, and Procyon—have white dwarf companions. Another nearby object, known as van Maanen's star, is a white dwarf, although not in a multiple system. So even though we are not able to detect white dwarfs at great distances from the sun, there seems to be a great number of them. Hundreds of white dwarfs have been discovered.

Even though the theory that explains white dwarfs has seemed consistent with actual observations, their size had never before been measured directly until 1975 when a set of observations was made in the ultraviolet from the Copernicus satellite. The ultraviolet part of the spectrum of Sirius B could be distinguished from that of its much brighter neighbor, and was found to peak at 1100 Å. This corresponds to a temperature of 30,000 K. Since the absolute magnitude of Sirius B is known, its surface area and thus its radius can be found. The radius of Sirius B is indeed only 4200 km, smaller than the earth's. Even if we allow for an uncertainty of 1000 or 2000 km, that does not change the import of the result that this is direct confirmation of the picture of white dwarfs that we drew in this chapter.

7.5 WHITE DWARFS AND THE THEORY OF RELATIVITY

One special reason to study white dwarfs is to make use of them as a laboratory to test extreme physical conditions. It is impossible to create such strong gravity in a laboratory on earth. We have already discussed two tests of general relativity in Section 5.11: the deflection of starlight by the sun, and the advance of the perihelion of the planet Mercury. A third test is the *gravitational redshift* of light.

Some black dwarfs come from stars that were not massive enough to begin hydrogen burning; others are cooled white dwarfs.

SURFACE OF SUN

EARTH

SIRIUS B
(WHITE DWARF)

Figure 7–8 The sizes of the white dwarfs are not very different from that of the earth. A white dwarf contains about 300,000 times more mass than does the earth, however.

Figure 7–9 Sirius B can be seen as the faint dot alongside the much brighter Sirius A image in this photograph from the Lick Observatory. The spikes of light seemingly radiating from Sirius A are an optical effect known as diffraction. Judicious arrangement of the position of the supports can cause much of the light of Sirius A to go in non-obstructing directions.

The sun is bright enough that we can spread out its spectrum and measure the wavelengths of spectral lines from the photosphere with tremendous accuracy. The redshift that Einstein predicted has been successfully detected, though the effect is minuscule. Furthermore, turbulence in the solar photosphere, such as the rise and fall of granules, distorts and confuses the results. One would like to find a body with stronger gravity than the sun, and white dwarfs are just such objects.

Unfortunately, one must know the mass and size of the white dwarf to perform the test accurately, and the masses and sizes of white dwarfs are not known very well. The observational part of the test is not difficult, because strong redshifts are found. The results agree with the predictions of the theory of relativity to within the possible error that results from our uncertain knowledge of the masses and sizes of white dwarfs. The test has best been carried out with 40 Eridani B.

A gravitational redshift is a redshift caused by gravity, as predicted by the general theory of relativity. Einstein's theory predicts that light leaving a mass would be slightly redshifted. The greater the amount of gravity at the place where the light is emitted, the greater the redshift.

7.6 NOVAE

Although the stars are generally thought to be unchanging on a human time scale, occasionally a "new star," a *nova* (pl: *novae*), becomes visible. Such occurrences have been noted for thousands of years; ancient Oriental chronicles report many such events.

A nova is a newly visible star rather than a really new star. (It is certainly not new in the manner of the new stars detected in the infrared, as discussed in Section 6.1.) A nova, rather, represents a brightening of a star by 5 to 15 magnitudes or more, which is equivalent to a brightening by hundreds or millions of times. Often, while brightening by this large factor, a nova passes from the realm of objects that cannot be seen with the naked eye to that of objects that can be seen with the naked eye, or from objects that are too faint to be seen even with the largest telescopes to objects that can be observed.

A nova (Fig. 7–10) may brighten within a few days or weeks. It ordinarily fades drastically within months, and then continues to fade gradually over the years. Several "recurrent" novae that appear to brighten at intervals of years or decades are known.

The spectra of novae show absorption lines that are Doppler shifted to the blue. This implies that gas between the star and earth is moving toward the earth. From this we deduce that gas is being thrown off the star. Soon

Figure 7–10 Nova Herculis 1934, showing its rapid fading from 3rd magnitude on March 10, 1935 to below 12th magnitude on May 6, 1935.

Figure 7–11 In 1951, a shell of gas could be seen to surround Nova Herculis 1934.

In this century, exceptionally bright novae appeared in 1901, 1918, 1925, 1934, 1942, and 1975 (Fig. 7–12).

Figure 7–12 These photographs are part of a unique series of observations covering the eruption of Nova Cygni 1975 during the period of its brightening. A Los Angeles amateur astronomer, Ben Mayer, was repeatedly photographing this area of the sky at the crucial times to search for meteors. Never before had a nova's brightening been so well observed. On August 28 (left), no star was visible at the arrow. By August 30 (right), the nova had reached 2nd magnitude, the brightness of Deneb, which is seen at the right.

the photospheric material that has been ejected thins out enough to become transparent, and we begin to see emission lines from the gas that has expanded in all directions from the star. Only 10^{-4} or so of a solar mass is thrown off in a nova outburst, so there is no reason that recurrent novae can't repeat their outbursts many times. Months or years after the outburst of light, the shell of gas sometimes becomes detectable through optical telescopes (Fig. 7–11).

Many astronomers believe that most if not all novae occur when a binary system has one member that has evolved into a white dwarf and another member that is en route to becoming a red giant. Though the details are not certain, the nova might occur through the following processes. We have mentioned that the outer layers of a red giant are not held very strongly by the star's gravity. If a white dwarf is nearby, some of the matter originally from the red giant can surround one or both of the stars. Some may fall on the surface of the white dwarf and perhaps trigger the proton-proton chain on its surface for a brief time. The energy thus produced blows off the material. Obviously, this mechanism could recur every few years as more new matter falls on the white dwarf.

Objects in other parts of the spectrum besides the visible can also brighten suddenly. An already strong x-ray source in the constellation Monoceros (the Unicorn), formerly known only by its catalogue number as A0620–00, brightened in August 1975 until it was more than five times as bright as any other x-ray source. Radio emission was detected from the source, and an 8th magnitude optical star was identified with the x-ray object using a telescope newly moved from an urban site to Kitt Peak by the University of Michigan, Dartmouth, and M.I.T. for the purpose of providing coordinated visible and x-ray observations.

A search of the library of old photographic plates at the Harvard Observatory showed that the star had been an optical nova before, in 1917. Thus the object, now called Nova Monocerotis 1975, is obviously a recurrent nova. This is the first recurrent nova to erupt since x-ray astronomy developed sufficiently to allow observations to be made in that part of the spectrum.

SUMMARY AND OUTLINE OF CHAPTER 7

Aims: To understand what happens to stars of up to 4 solar masses when they have finished their time on the main sequence of the H-R diagram

Red giants (Section 7.1)
  They follow the end of hydrogen burning in the core
    Core contracts gravitationally

August 28, 1975   11:30 U.T.

August 30, 1975   6:45 U.T.

Core and hydrogen-burning shell become hotter
The increased energy expands the outer layers
The total luminosity is the luminosity of each bit of surface times
the total surface area
Star is red, so each bit of surface has relatively low luminosity
Surface area is very greatly increased, so total luminosity is greatly
increased
Helium flash
Rapid onset of triple-alpha process
Planetary nebulae (Section 7.2)
Their appearance
Ages: less than 50,000 years old
The gas is blown off when it absorbs photons
Mass loss: 0.1 or 0.2 solar mass
White dwarfs (Sections 7.3, 7.4, and 7.5)
End result of lightweight stars
Less than 1.4 solar masses remaining
Electron degeneracy
Detecting white dwarfs via proper motion or by their presence in a
binary system
Use in testing the gravitational redshift
Einstein's general theory of relativity endorsed
Novae (Section 7.6)
Mass loss: only 10^{-4}
Binary model: interaction of a red giant and a white dwarf
X-ray novae

QUESTIONS

1. What event signals the end of the main-sequence life of a star?

2. When hydrogen burning in the core stops, the core contracts and heats up again. Why doesn't hydrogen burning start again?

3. When the core starts contracting, what eventually halts the collapse?

4. If you are outside a spherical mass, the force of gravity just varies inversely as the square of the distance from the center. What is the ratio of the force of gravity at the surface of the sun to what it will be when the sun has a radius of one astronomical unit?

5. Why is helium "flash" an appropriate name?

6. If you compare a photograph of a nearby planetary nebula taken 80 years ago with one taken now, how would you expect them to differ?

7. Why is a star brighter after it sheds a planetary nebula?

8. What keeps a white dwarf from collapsing further?

9. What are the differences between the sun and a one solar mass white dwarf?

10. When the sun becomes a white dwarf, approximately how much mass will it have? Where will the rest of the mass have gone?

11. Which has a higher surface temperature, the sun or a white dwarf?

12. Sketch an H-R diagram indicating the main sequence, and the region of white dwarfs. Now indicate the position of an ordinary dwarf that appears yellowish.

13. Why does the study of the gravitational effects of a system often give a better estimate of the mass of the system than the direct observation of all the components?

14. What evidence do we have that material is being ejected from novae?

15. When the proton-proton chain starts at the center of a star, it continues for billions of years. When it starts at the surface (as in a nova) it only lasts a few weeks. How can you explain the difference?

Supernovae, Neutron Stars, and Pulsars

We have seen how the run-of-the-mill lightweight stars end, not with a bang but a whimper. More massive stars, stars that contain more than 4 solar masses when they are on the main sequence, put on a more dazzling display. They cook the heavy elements deep inside themselves and then blow themselves almost to bits, forming still more heavy elements in the process. Then the matter that is left behind settles down to even stranger states of existence than that of a white dwarf. In this chapter we shall discuss the death of stars of 4 to 8 solar masses, which we may call *middle-weight stars*. Let us consider one of the possible versions of the story, though we shall not attempt here to discuss all the different models that have been advanced to explain supernovae. In Chapter 9, we shall discuss the death of *heavyweight stars*, stars of more than 8 solar masses.

The division into light-weight (Chapter 7), middle-weight, and heavyweight stars was suggested by Martin Schwarzschild of Princeton. Note that we are really discussing mass rather than weight.

8.1 RED SUPERGIANTS

The stars that are much more massive than the sun whip through their main-sequence lifetimes at a rapid pace. These prodigal stars use up their store of hydrogen very quickly. A star of 15 solar masses may take only 10 million years from the time it first reaches the main sequence until the time when it has exhausted all the hydrogen in its core. This is a lifetime a thousand times shorter than that of the sun. When the star exhausts the hydrogen in its core, the outer layers expand and the star becomes a red giant.

For these massive stars, the core can then gradually heat up to 100 million degrees, and the triple-alpha process begins to transform helium into carbon. However, for massive stars the helium burns steadily after the helium ignition, unlike the case for less massive stars.

By the time helium burning is concluded, the outer layers have expanded even further, and the star has become much brighter than even a red giant. We call it a *red supergiant* (Fig. 8–1); Betelgeuse, the star that

Figure 8–1 An H-R diagram showing in very general terms the evolution of a massive star off the main sequence to a red supergiant.

Filaments at the north end of the supernova remnant known as the Cygnus Loop.

Figure 8–2 Betelgeuse is the star labeled α in the shoulder of Orion, the hunter, shown here in Johann Bayer's *Uranometria*, first published in 1603.

marks the shoulder of Orion, is the best-known example (Fig. 8–2). Supergiants are inherently very luminous stars, with absolute magnitudes of up to −10. The sun, with an absolute magnitude of +5, is only one-millionth as luminous as the most brilliant of the red supergiants. A supergiant's mass is spread out over such a tremendous volume that its average density is less than one millionth that of the sun.

8.2 SUPERNOVAE

After reaching supergiant status, a very massive star will eventually explode in a glorious burst called a *supernova* (pl: *supernovae*). The conditions in the center of the star change so quickly at this stage that it becomes very difficult to model them satisfactorily in sets of equations or on computers. Many different models have been advanced to describe what happens in a supernova; none is universally accepted.

Let us describe an example of a specific set of events that could lead to a supernova explosion. We present the details of the process, which most likely occurs for heavyweight rather than middleweight stars, in order to illustrate how the advance of stellar evolution can run away with itself and go out of control. Other specific processes can also explain the supernova phenomenon, and may be more important in middleweight stars.

This model explains Type II supernovae. Type I supernovae, on the other hand, may result from the addition of mass to a white dwarf, which leads to the star's collapse. Type I and Type II supernovae can be distinguished from each other by their spectra.

In the case we are considering, a substantial iron core has been formed and begins to shrink and heat up. The iron represents the ashes of the previous stage of nuclear burning. Eventually the temperature becomes high enough for the iron to undergo nuclear reactions. The stage is now set for disaster because iron nuclei have a fundamentally different property from other nuclei when undergoing nuclear reactions. Unlike the other nuclei, iron absorbs rather than produces energy in order to undergo either fusion or fission. Thus iron takes up energy when it is being transformed to other elements, either heavier or lighter. This energy is no longer available to heat the core.

The core responds to this loss of energy by shrinking and heating up still more. The iron can be broken up into lighter nuclei by the high-energy photons of radiation that are generated, and in so doing absorbs still more energy. The process goes out of control. Within seconds—a fantastically short time for a star that has lived for millions and millions of years—the core collapses and heats up catastrophically. It was formerly thought that matter falling in upon the collapsing core from the layers above would undergo violent nuclear reactions in the intense heat of the core, but this process is now thought to be less likely and less important than the following process:

As matter falls in upon the collapsing core from the layers above, electrons and protons fuse into neutrons. In the process, huge numbers of neutrinos, the elusive particles we discussed in Sections 6.3 and 6.6, are given off. At this stage of stellar collapse, the outer layers of the star are so dense that they stop the neutrinos and in so doing are pushed away from the core with great force. Shock waves—which compress gas, similarly to the sonic booms that we experience on earth—race through these outer layers as they are blown off, possibly causing heavy elements to form. With a tremendous explosion, the star is destroyed. Only the core is left behind. This explosion of the star is called its supernova phase. To summarize, the essential part of the story is that once the core is made up of iron, the star is likely to explode.

The heavy elements formed either in the center of the star or in the supernova explosion itself are spread out into space, where they enrich the interstellar gas. When a star forms out of such enriched gas, these heavy elements are present to make up some of any planets that may coalesce. Thus supernovae provide the heavy elements that are necessary for life to arise. Many of the atoms in each of us have been through such supernova explosions.

In the days or weeks following the explosion, the amount of radiation emitted by the supernova can equal that emitted by the rest of its entire galaxy. Most of the supernovae that we observe (Figs. 8–3 and 8–4) are in distant galaxies, and we can see a single star outshine its galaxy for a period of weeks. It may brighten by over 20 magnitudes, a factor of 10^8 in luminosity.

Optical astronomers have photographed two dozen of these stellar shreds, which are known as *supernova remnants* (Fig. 8–5, the photograph that opens this chapter, and Color Plates 23 and 24), in our galaxy alone. The gas gives off not only optical radiation but also strong radio radiation, and the supernova remnants are best studied by means of radio astronomy (which is described in Chapter 21). About 100 supernova remnants are known from observations made in the radio part of the spectrum. Some of these have now also been observed in the x-ray part of the spectrum (Color Plate 22).

Supernovae were named at a time when it was thought that they were merely unusually bright novae. But now we know that novae and supernovae are very different phenomena. A supernova explosion represents the death of a star and the scattering of most of its material, while a nova uses up only a small fraction of a stellar mass and can be recurrent.

In our own galaxy, we have historical records of only five supernovae in the last thousand years, so we can conclude that many of the 24 optical remnants and over 100 radio remnants known must come from stars that exploded before such events were recorded.

Figure 8–3 Views of the central part of the galaxy NGC 5252, taken in 1959 and in 1972. The supernova that appeared in 1972 was nearly as bright as the rest of the galaxy.

Figure 8–4 Views of the galaxy NGC 4725 in the constellation Coma Berenices. There was an 8 month interval between the photographs—May 10, 1940 to January 2, 1941. The line points to the supernova.

Tycho's supernova, observed by the great astronomer in 1572, is the bright object at the left edge of Cassiopeia's chair on page 12. (Tycho and Kepler, who observed a supernova in 1604, will be discussed in Section 10.3.)

In 1054 A.D., Chinese chronicles recorded the appearance of a "guest star" in the sky that was sufficiently bright that it could be seen in the daytime. No one is certain why no European records of the supernova were made, for the Bayeux tapestry illustrates a comet and thus shows that even in the Middle Ages people were aware of celestial events. Certain cave and rock paintings made by Indians in the American southwest may show the supernova (Fig. 8–6), though this interpretation is controversial.

When we look at the reported position in the sky, in the constellation Taurus, we see an object that clearly looks as though it is a star torn to shreds (Fig. 8–14 and Color Plate 23). It is called the Crab Nebula. The Crab appears to be about 1/5 of the distance from the sun to the center of our galaxy. There is no doubt that the Crab is the remnant of the supernova whose radiation reached earth in 1054 A.D., twelve years before William the Conqueror invaded England.

Figure 8–5 S147, the remnant of a supernova explosion in our galaxy. The long delicate filaments shown cover an area over 3° × 3°, about 40 times the area of the moon.

From a study of the rate at which supernovae seem to appear in distant galaxies, it is estimated that supernovae should appear in our galaxy about once every 30 to 50 years. There have been only a few known supernovae in our galaxy in the past 900 years, though any that occurred on the far side of our galaxy may have been hidden from our view by all the interstellar matter. There could be a supernova in our part of the galaxy any day. We've already been waiting for hundreds of years, since the last one was observed in 1604.

8.3 COSMIC RAYS

Most of the information we have discussed thus far in this book has been gleaned from the study of electromagnetic radiation. But certain high-energy particles have been discovered to be traveling through space in addition to the photons. These particles are nuclei of atoms moving at tremendous velocities. They are called *cosmic rays.*

To best capture cosmic rays from outer space, one must travel above most of the earth's atmosphere in a rocket, satellite, or balloon. (Only a few of the most energetic cosmic rays reach the ground.) Often balloons bear stacks of photographic emulsion to altitudes of 50 kilometers; the cosmic rays leave marks as they travel through the film, and a three-dimensional picture of the track can be built up because of the three-dimensional nature of the stack of emulsion. The cosmic rays formed deep in space, called *primary cosmic rays,* often interact with atoms in the earth's atmosphere, at which time *secondary cosmic rays* are given off. It is possible to detect many such secondary rays without leaving the earth's surface.

The origin of the primary cosmic rays has long been a matter of controversy. It seems likely to many astronomers that the cosmic rays are particles accelerated to tremendous velocities in supernova explosions, and that they have been traveling through space since they were ejected.

8.4 NEUTRON STARS

We have discussed the fate of the outer layers of a star that explodes as a supernova. Now let us discuss the fate of the core.

As iron fills the core of a massive star, the temperatures are so high that the iron nuclei begin to break apart into smaller units like alpha particles (helium nuclei). The pressure is no longer high enough to counteract gravity, and the outer layers collapse.

In the supernova explosion that ensues, the matter left behind in the core has been compressed to the point where the iron has broken apart. Protons (which have positive charge) and electrons (which have negative charge) are forced together, and neutrons (which have no charge) result.

The core may contain as little as a few tenths of a solar mass or possibly as much as two or three solar masses following the explosion. This remainder is at an even higher density than that at which electron degeneracy holds up a white dwarf. At this density a condition called *neutron degeneracy,* in which the neutrons cannot be packed any more tightly, appears. This condition is completely analogous to electron degeneracy, with the difference that neutron degeneracy comes from the application of quantum mechanical rules to the neutrons. The pressure caused by neutron degen-

Figure 8–6 An American Indian cave painting discovered in northern Arizona that may depict the supernova explosion 900 years ago that led to the Crab Nebula. It had been wondered why only Oriental astronomers reported the supernova, so searches have been made in other parts of the world for additional observations. Note that the drawing is obviously distorted if it is indeed the supernova, as the star could not appear over the dark disk of the unlit part of the moon.

Figure 8–7 A neutron star may be only 20 kilometers in diameter (10 km in radius), the size of a city, even though it may contain a solar mass or more.

eracy balances the gravitational force that tends to collapse the core, and as a result the core reaches equilibrium as a *neutron star*.

If more than two or three solar masses remain, then the force of gravity will overwhelm even neutron degeneracy. We shall discuss that case in Chapter 9.

Whereas a white dwarf packs the mass of the sun into a volume the size of the earth, the density of a neutron star is even more extreme. A neutron star may be only 20 kilometers or so across (Fig. 8–7), in which space it may contain the mass of about two suns. In its high density, it is like a single, giant nucleus. A teaspoonful of a neutron star could weigh a billion tons. This state of matter may seem simply inconceivable, but on theoretical grounds it is possible for it to exist even on such a large scale.

Before it collapses, the core (like the sun) has only a weak magnetic field. But as the core collapses, the magnetic field is concentrated, and grows stronger as a result. By the time the core shrinks to neutron star size, it has an extremely powerful magnetic field, much stronger than any we can produce on earth.

A neutron star may be the strangest type of star of which we can conceive. When neutron stars were discussed in theoretical analyses in the 1930's, there seemed to be no hope of actually observing one. Nobody had a good idea of how to look for a neutron star. Let us leave this story here for a moment, and jump to consider events of 1967, which will later prove to be related to the search for neutron stars.

8.5   THE DISCOVERY OF PULSARS

By 1967, radio astronomy had become a flourishing science. Dozens of large radio telescopes were in existence all around the world, and were being used to observe radio emission from objects in space. The details of radio astronomy will be described in Chapter 21, but for the current discussion all we need know is that various objects in space emit electromagnetic signals in the radio part of the spectrum, and that for the last forty years astronomers have been able to build giant antennas called *radio telescopes* to collect and focus these radio waves. Radio telescopes, just like radio receivers in our living rooms, are subject to static—rapid variations in the strength of the signal. It is difficult to measure the average intensity of a signal if the signal strength is jumping up and down many times a second, so radio astronomers usually adjust their instruments so that they do not record any variation in signal shorter than a second or so.

Figure 8–8 The radio telescope—actually a field of aerials—with which pulsars were discovered.

This should have had the effect of merely "smoothing" the incoming signal, while not distorting it. Of course it would mask any rapid variations in signal, but none were expected to arise in the source itself, as most astronomical sources were thought to be relatively steady in their emission of electromagnetic radiation. The only rapid variations expected besides terrestrial static were those that correspond to the twinkling of stars.

An astronomer at Cambridge University in England, Antony Hewish, wanted to study this "twinkling" of radio sources, which is called *scintillation*. The light from stars twinkles because of effects of our earth's atmosphere. The radio waves from radio sources scintillate not because of any terrestrial effects but rather because radio signals are affected by clouds of electrons in the solar wind. Electrons in the earth's ionosphere and in interstellar space contribute lesser amounts to the observed scintillation of

|————| **10 sec**

Figure 8–9A A series of pulses of radio waves observed at the National Radio Astronomy Observatory at Green Bank, West Virginia.

radio sources. To study scintillations, Hewish built a radio telescope that was uniquely able to study rapid variations of faint signals.

One day in 1967, Jocelyn Bell Burnell, then a student working on the project, noticed that a set of especially strong variations in the signal appeared in the middle of the night, when scintillations caused by the solar wind are usually weak. After a month of observation, it became clear that the position of the source of the signals remained fixed with respect to the stars (that is, according to sidereal time, which was discussed in Section 3.11), a sure sign that the object was not terrestrial or solar.

Detailed examination of the signal showed that, surprisingly, it was a rapid set of pulses, with one pulse every 1.3373011 seconds (Fig. 8–9A). The pulses were very regularly spaced.

One immediate thought was that the signal represented an interstellar beacon sent out by extraterrestrial life on another star. For a time the source was called an LGM, for Little Green Men. Soon, Burnell located three other sources, pulsing with regular periods of .253065, 1.187911, and 1.2737635 seconds, respectively.

They were briefly called LGM 1, LGM 2, LGM 3, and LGM 4, but by this time it was obvious that the signals had not been sent out by extraterrestrial life. For one thing, it seemed unlikely that there would be four such beacons at widely spaced locations in our galaxy. Besides, any beings would probably be on a planet orbiting a star, and no effect of a Doppler shift from any orbital motion was detected.

When the discovery was announced to an astonished astronomical community, it was immediately apparent that the discovery of these pulsating radio sources, called *pulsars,* was one of the most important astronomical discoveries of the decade. But what were they?

Other observatories turned their radio telescopes to search the heavens for new pulsars. This often required the purchase of new equipment to be able to follow signals that varied in time so quickly. Dozens of new pulsars were found. The periods range from hundredths of a second to four seconds.

8.6 PROPERTIES OF PULSAR SIGNALS

A typical pulse shape is shown in Figure 8–9B. The time from one pulse to the next pulse is called the *period.*

When the positions of all the known pulsars are plotted on a chart of the heavens, it can easily be seen that they are concentrated along the plane of our galaxy (Fig. 8–10) rather than being uniformly distributed over the sky. Thus they are clearly objects in our galaxy, for if they were extragalactic objects (objects located outside our galaxy), we would expect them to be distributed uniformly.

Figure 8–9B A schematic series of pulses.

Figure 8–10 The distribution of the first 154 pulsars on a projection that maps the entire sky, with the plane of the Milky Way along the zero degree line on the map. The concentration of pulsars near 60° galactic longitude on this map merely represents the fact that this section of the sky has been especially carefully searched for pulsars because it is the area of the Milky Way best visible from the Arecibo Observatory. From the concentration of pulsars along the plane of our galaxy, we can conclude that pulsars are members of our galaxy; had they been extragalactic, we would have expected to see as many near the poles of this map. We can extrapolate that there are at least 100,000 pulsars in our galaxy.

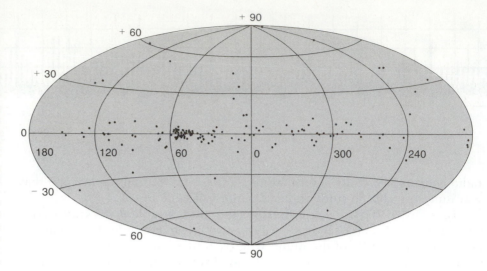

Figure 8–11 Even if a large star were to turn off all at one time, the size of the star is such that it would appear to us to darken over a measurable period of time because radiation travels at the finite speed of light. Pulsars pulse so rapidly that we know that they cannot be large objects flashing.

8.7 WHAT ARE PULSARS?

Theoreticians went to work to try to explain the source of the pulsars' signal. The problem had essentially two parts. The first part was to explain what supplies the energy of the signal. The second part was to explain what causes the signal to be regularly timed, that is, what the clock mechanism is.

From studies of the second part of the problem, the choices were quickly narrowed down. The signals could not be coming from a pulsation of a whole normal main-sequence star because such a star would be too big. If the sun, for example, were to turn off all at one instant, we on earth would not see it go dark all at once. The point nearest to us would disappear first, and in the next seconds the darkening would spread farther back around the side of the sun (Fig. 8–11). Pulsars pulsed too rapidly to be the size of a normal dwarf star.

That left white dwarfs and neutron stars as the likely candidates. Remember, at that time neutron stars were merely objects that had been predicted theoretically but had never been detected observationally. Could the pulsars be the more ordinary objects, namely, special kinds of white dwarfs?

There were two basic mechanisms astronomers could conceive of as possibly controlling the pulsar timing. It was thought that the pulses might be coming from a star that was actually oscillating in size and brightness, or that they were coming from a star that was rotating. Another less likely alternative was that the emitting mechanism involved two stars orbiting around each other in a binary system, and it was soon shown that systems of orbiting stars would not have the characteristics observed for pulsars. This left the rotation or oscillation of white dwarfs or neutron stars.

On theoretical grounds, a white dwarf would oscillate in less than a minute. However, it could not oscillate as rapidly as once every second, as would be required if pulsars were oscillating white dwarfs. Neutron stars, however, would indeed oscillate more rapidly, but they would oscillate once every 1/100 second or so, too rapidly to account for the pulsars. Thus oscillating neutron stars are ruled out as well.

What about rotation? Consider a lighthouse whose beacon casts its powerful beam many miles out to sea. As the beacon goes around, the beam sweeps past any ship very quickly, and returns again to illuminate that ship after it has made a complete rotation. Perhaps pulsars do the same thing:

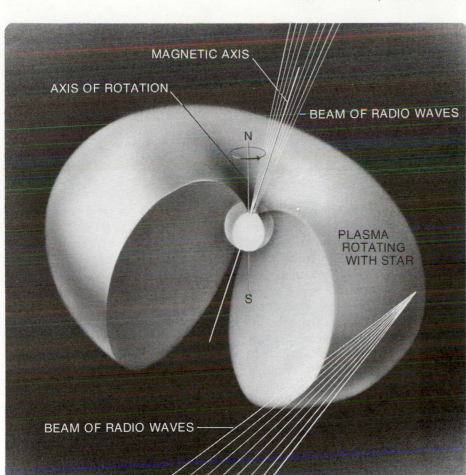

MAGNETIC AXIS

AXIS OF ROTATION

BEAM OF RADIO WAVES

N

S

PLASMA ROTATING WITH STAR

BEAM OF RADIO WAVES

Figure 8-12 In the light-house model for pulsars, which is now commonly accepted, a beam of radiation flashes by us once each pulsar period, just as a lighthouse beam appears to flash by a ship at sea. It is believed that the generation of a pulsar beam is related to the neutron star's magnetic axis being aligned in a different direction than the neutron star's axis of rotation. The mechanism by which the beam is generated is not currently understood. Two possible variations of the lighthouse theory are shown here. The beam may be emitted along the magnetic axis, as shown in the top half of the figure. Alternatively, the beam could be generated on the surface of a doughnut of magnetic field (shown cut away) that surrounds and rotates with the neutron star, as shown in the bottom half of the figure.

they emit a beam of radio waves that sweeps out a path in space. We see a pulse each time the beam passes the earth (Fig. 8–12).

But what type of star could rotate at the speed required to account for the pulsations? Could a white dwarf be rotating fast enough? Recall that a white dwarf is approximately the size of the earth. If an object that size were to rotate once every 1/4 second, the centrifugal force would overcome even the immense gravitational force of a white dwarf, and the outer layers would begin to be torn off. This indicated that pulsars were probably not white dwarfs; if a pulsar with a period shorter than 1/4 second were to be found, then the white dwarf model would be completely ruled out.

Neutron stars, on the other hand, are much smaller, so the centrifugal force would be weaker and the gravitational force would be stronger. As a result, neutron stars can indeed rotate four times a second. There is nothing to rule out their identification with pulsars. Since no other reasonable possibility has been found, astronomers accept the idea that pulsars are in fact neutron stars that are rotating. This is called the *lighthouse model.* Thus by discovering pulsars, we have also discovered neutron stars.

Note that for this argument to be complete, and convincing, we must be as certain as we can be that there are no other possibilities to consider. Whenever we argue from elimination of other possibilities, we must be certain that we have a complete list (Table 8–1).

At the moment, about 300 pulsars have been discovered. But the light-house model implies that there are many more, for we can see only those

The first pulsars discovered all had periods longer than 1/4 sec. Later, a shorter period pulsar was discovered (as will be described in the next section); others are now also known.

TABLE 8–1 POSSIBLE EXPLANATIONS OF PULSARS

Hypothesis	Probability
Regular dwarf or giant	Ruled out
System of orbiting white dwarfs	Ruled out
System of orbiting neutron stars	Ruled out
Oscillating white dwarf	Ruled out
Oscillating neutron star	Ruled out
Rotating white dwarf	Ruled out
Rotating neutron star	Most likely

pulsars whose beams happen to strike the earth. If a pulsar were spinning at a different angle, then we would not know that it was there.

We have dealt above with only the second part of the explanation of the emission from pulsars: the clock mechanism. We understand much less well the mechanism by which the radiation is actually emitted in a beam. Presumably it has something to do with the extremely powerful magnetic field of the neutron star. Energy may be generated far above the surface of the neutron star in the magnetic field, perhaps near the location where the magnetic field lines, which are swept along with the star's rotation, attain velocities that approach that of light. The radio waves may be generated by charged particles that have escaped from the neutron star near its magnetic poles. These magnetic poles may not coincide with the neutron star's poles of rotation. After all, the earth's north magnetic pole is not at the north pole but near Hudson's Bay, Canada (Fig. 8–13).

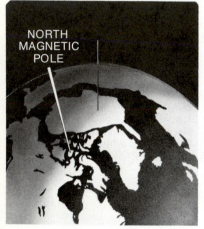

Figure 8–13 Note that the earth's magnetic axis, like that of a pulsar, is aligned in a different direction from that of the earth's axis of rotation; our magnetic north pole lies in the vicinity of Hudson's Bay in Canada. The earth rotates in 24 hours; a neutron star may rotate in 1/30 sec, which is a period 2,500,000 times shorter. Thus the effect of the magnetic field being carried around at a skew angle is intensified many-fold.

8.8 THE PULSAR IN THE CRAB NEBULA

Several months after the first pulsars had been discovered, strong bursts of radio energy were discovered coming from the direction of the Crab Nebula (Fig. 8–14). The source was named NP 0532. The bursts were sporadic rather than periodic, but there was hope that the astronomers were seeing only the strongest pulses and that a periodicity could be found.

Within two weeks of feverish activity at several radio observatories, it was discovered that the pulsar in the Crab Nebula had a period of 0.033 second, the shortest period by far of all the known pulsars. The fastest pulsar previously known pulsed four times a second, while NP 0532 pulsed at the very rapid rate of thirty times a second. No white dwarf could possibly rotate that fast.

Furthermore, the Crab Nebula is a supernova remnant, and theory predicts that neutron stars should exist at the centers of supernova remnants. Thus the discovery of a pulsar there, exactly where a neutron star would be expected, was the discovery that clinched the identification of pulsars with neutron stars. Presumably, we do not detect pulsars in the positions of Tycho's and Kepler's supernovae because their spin axes are oriented such that their beams of radiation do not strike the earth.

Optical astronomers, of course, had searched very carefully at the reported positions of pulsars in order to try to discover optical objects there. A substantial amount of observing time with the largest telescopes in the world had been devoted to these projects. But no optical objects had been found, and at this time the position of the Crab Nebula had not been examined since it was not yet known to contain a pulsar. After several failures, the search was abandoned.

Figure 8–14 The Crab Nebula, the remnant of a supernova explosion that became visible on earth in 1054 A.D. The pulsar, the first pulsar detected to be blinking on and off in the visible part of the spectrum, is marked with an arrow. In the long exposure necessary to take this photograph, the star turns on and off so many times that it appears to be a normal star. Only after its radio pulsation had been detected was it observed to be blinking in optical light. It had long been suspected, however, of being the leftover core of the supernova because of its unusual spectrum, which has only a continuum and no spectral lines.

Soon after NP 0532 was discovered, three astronomers at the University of Arizona decided to examine its position to look for an optical pulsar. They turned their telescope towards a faint star in the midst of the Crab Nebula, and it very soon became apparent that the star was pulsing! Essentially they did something that many astronomers had come to think impossible—they found an optical pulsar.

For about ten years after that discovery, no other optical pulsar was found. During that time there were 150 objects known to be pulsing in the radio spectrum, and only one—which has the shortest period yet discovered and which is thus presumably the youngest known pulsar—in the optical spectrum. In 1977, flashes from the Vela pulsar, the third fastest, were discovered. Several x-ray pulsars are known, and will be discussed in a subsequent section. The Crab and Vela pulsars are also gamma-ray pulsars.

8.9 THE CLOCK SLOWS DOWN

When pulsars were first discovered, their most prominent feature was the extreme regularity of the pulses. It was hoped that they could be used as precise timekeepers, perhaps even more exact than atomic clocks on earth.

After they had been observed for some time, however, it was noticed that the pulsars were slowing down very slightly. Their periods were all gradually increasing.

The filaments in the Crab Nebula still glow brightly, even though over 900 years have passed since the explosion. Furthermore, the Crab gives off tremendous amounts of energy across the spectrum from x-rays to radio waves. It had long been wondered where the Crab got this energy. Theoretical calculations show that the amount of rotational energy lost by the Crab's neutron star as its rotation slows down is just the right amount to provide the energy radiated by the entire nebula. Thus the discovery of the Crab pulsar solved a long-standing problem in astrophysics: where the energy originates that keeps the Crab Nebula shining.

Figure 8–15 The x-ray pulsar SMC X-1 in the Small Magellanic Cloud has a period of 0.716 sec. These observations were taken with instrumentation of the Naval Research Laboratory during the joint Soviet-American Apollo-Soyuz project in 1975.

Note that when we say "pulsar," we mean an object pulsing radio waves. When we mean "x-ray pulsar," we say the whole phrase.

8.10 X-RAY PULSARS

Telescopes in orbit that are sensitive to x-rays have detected a number of strong x-ray sources. While these telescopes are not able to make the high-speed observations that are necessary to see if pulsations are taking place, x-ray telescopes sent up for brief periods in rockets have been able to make such observations.

Three x-ray pulsars have been found, that is, objects that pulse only in the x-ray region of the spectrum. Two of them, Hercules X-1 and Centaurus X-3 (the first x-ray source to be discovered in the constellation Hercules and the third in Centaurus), are located in our galaxy. The third is not in our galaxy, but rather in the Small Magellanic Cloud, a companion galaxy to our own system (and is thus called SMC X-1) (Fig. 8–15).

The 1.24 second period of the x-ray pulses from Hercules X-1, unlike the period of an ordinary pulsar, is growing shorter, that is, the rotation of the neutron star is speeding up. Theoreticians think that the x-ray pulsars are part of binary systems, and that they are radiating because mass from the companion star is being funneled towards the poles of the pulsar by the pulsar's strong magnetic field. Thus energy is being added to the rotating neutron star, and unlike regular (radio) pulsars, it speeds up instead of slowing down.

Hercules X-1 is the companion to a long-observed variable star (a single-line spectroscopic binary, a type of double star described in Section 4.1) named HZ Herculis (variable star HZ in the constellation Hercules); they orbit each other every 1.7 days. The mass of Hercules X-1 has been carefully evaluated. The observations indicate that the companion has 1.3 solar masses, within the limit that neutron degeneracy can support (as discussed in Section 7.3). The latest electronic techniques were used to detect individual photons emitted from the optically visible star when the beam of x-rays from the dark companion, according to the lighthouse theory, strikes it every 1.24 seconds. These observations make it plausible that Hercules X-1 is indeed a neutron star like the other pulsars.

The discovery of pulsars was a complete surprise that provided great excitement for astronomers over the last decade. What will the next decade bring?

SUMMARY AND OUTLINE OF CHAPTER 8

Aims: To see how stars of medium and heavy mass become supernovae, with the cores of the stars of medium mass condensing to become neutron stars. We observe the neutron stars as pulsars.

This entire chapter concerns the evolution of stars of more than 4 solar masses.

When they exhaust their hydrogen in the core, they become red giants.

Later, heavier elements are built up in the core, and the stars become supergiants (Section 8.1).

After iron cores form, the stars explode as supernovae. Only a few optical supernovae have been seen in our galaxy in the last 900 years. Supernova remnants can be studied with techniques of radio astronomy and x-ray astronomy (Section 8.2).

Cosmic rays, high-energy particles in space, probably come from supernovae (Section 8.3).

After a star's fusion stops, neutron degeneracy can support a remaining mass of up to 2 or 3 solar masses; more massive stars will be discussed in the next chapter.

A neutron star may be only 10 km in radius, and so is fantastically dense (Section 8.4).

Pulsars were discovered when a radio telescope was built that did not mask rapid time variations in the signal (Section 8.5).

Pulsars have very regular signals, although they have been discovered to be slowing down very slightly. Pulse periods range from .033 to 4 sec (Section 8.6).

Pulsars are considered to be objects that emit in the radio spectrum, although the Crab and the Vela pulsars also pulse light, x-rays, and γ-rays. Three sources pulse only x-rays (Section 8.10).

Scientists have identified pulsars with rotating neutron stars by a process of elimination (Section 8.7):

They can't be full-size stars, because the pulse width is too small.

They can't be a binary pair of white dwarfs or neutron stars.

They can't be oscillating white dwarfs or neutron stars, because the periods of oscillation would be too long or too short, respectively.

They can't be rotating white dwarfs, because white dwarfs can barely rotate fast enough to account for the first known pulsars and surely can't rotate fast enough to account for the pulsar in the Crab Nebula.

Rotating neutron stars are all that are left, and could satisfy the observations. Thus we accept the lighthouse model.

A pulsar is observed in the center of the Crab Nebula, a supernova remnant, where we would expect to observe a neutron star (Section 8.8).

Pulsars are gradually slowing down as they grow older (Section 8.9).

An x-ray pulsar has been observed to be speeding up gradually, opposite to the ordinary pulsars (Section 8.10).

QUESTIONS

1. Why are red supergiants so bright?

2. What are the basic differences between a nova and a supernova?

3. If we see a massive main-sequence star (a heavyweight star), what can we assume about its age, relative to most stars? Why?

4. What do we know about the core of a star when it leaves the main sequence?

5. In a supernova explosion of a 20 solar mass star, about how much material is blown away?

6. A supernova can brighten by 20 magnitudes. Verify that this is a factor of 10^8.

7. A typical galaxy has a luminosity 10^{11} times that of the sun. If a supernova equals this luminosity, what was its luminosity before it exploded.

8. Would you expect the appearance of the Crab Nebula to change in the next 500 years?

9. If the light from a supernova 2000 parsecs from us was received in 1054, when did the star actually explode?

10. What are cosmic rays?

11. In your own words, fill in Table 8–1 with the reasons why each of the explanations for pulsars involving single stars was ruled out except for the case of rotating neutron stars.

12. How can you explain the fact that x-ray pulsars seem to be speeding up while ordinary pulsars are slowing down?

Black Holes

The strange forces of electron and neutron degeneracy support dying lightweight and middleweight stars against the force of gravity. The strangest case of all occurs at the death of heavyweight stars, which contained 8 to 60 or more solar masses when they were on the main sequence. These stars generally undergo supernova explosions like the middleweight stars, but some of the heavyweight stars may retain cores of over 2 or 3 solar masses. Nothing in the universe is strong enough to hold up the remaining masses against the force of gravity. The masses collapse, and continue to collapse forever. We call the result of such a collapse a *black hole*.

9.1 THE FORMATION OF A STELLAR BLACK HOLE

When nuclear fusion ends in the core of a star, gravity causes the star to contract. We have seen in Chapter 7 that a star that has a final mass of less than 1.4 solar masses will end its life as a white dwarf. In Chapter 8 we saw that a heavier star will become a supernova, and that if the remaining mass of the core is less than 2 or 3 solar masses it will wind up as a neutron star. We are now able to observe both white dwarfs and neutron stars and so can study their properties directly.

It seems reasonable that in some cases more than 2 or 3 solar masses remain after the supernova explosion. The star collapses through the neutron star stage, and we know of no force—not even neutron degeneracy—that can stop the collapse. In some cases, the matter may have become so dense as the star collapsed that it passed through the neutron star stage before any explosion took place; in these cases, no supernova resulted.

The value for the maximum mass that a neutron star can have is a result of theoretical calculation. We must always be aware of the limits of accuracy of any calculation, and particularly a calculation that deals with matter in a state that is very different from states that we have been able to study experimentally. Still, the best modern calculations show that the limit of a

Do not confuse a black hole with a black body. A black body is merely a radiating body whose radiation follows Planck's law. A black hole, essentially, does not radiate (though an exception to this rule—really an alternate way of looking at the whole picture—will be discussed in Section 9.5).

The overexposed dark object in the center of this negative print is a blue supergiant star HDE 226868, which is thought to be the companion of the first black hole to be discovered, Cygnus X-1.

neutron star mass is about 2 solar masses, and is probably less than 3 solar masses, though some theoreticians set higher limits.

We may then ask what happens to a 5 or 10 or 50 solar mass star as it collapses, if it retains more than 3 solar masses. It must keep collapsing, getting denser and denser. We have seen that Einstein's general theory of relativity predicts that a strong gravitational field will redshift radiation, and that this prediction has been verified both for the sun and for white dwarfs (as discussed in Section 7.5). Also, radiation will be bent by a gravitational field, or at least appear to us on earth as though it were bent. This prediction has been verified for the sun (see Section 5.11).

As the mass contracts, radiation is continuously redshifted more and more, and radiation leaving the star other than perpendicularly to the surface is bent more and more. Eventually, when the mass has been compressed to a certain size, radiation from the star can no longer escape into space. The star has withdrawn from our observable universe, in that we can no longer receive radiation from it. We say that the star has become a *black hole*.

Why do we call it a black hole? We think of a black surface as a surface that reflects none of the light that hits it. Similarly, any radiation that hits the surface of a black hole continues into the black hole and is not reflected. In this sense, the object is perfectly black.

Figure 9–1 As the star contracts, a light beam emitted other than radially outward will be bent.

Figure 9–2 Light can be bent so that it falls back onto the star.

9.2 THE PHOTON SPHERE

Let us consider what happens to radiation emitted by the surface of a star as it contracts. Although what we will discuss affects radiation of all wavelengths, let us simply visualize standing on the surface of the collapsing star with a flashlight.

If we stand on the surface of a supergiant star, we note only very small effects of gravity on the light from our flashlight. We can shine the beam straight up, or at any angle, and it seems to go straight out into space.

As the star collapses, two effects begin to occur. Although we on the surface of the star cannot notice them ourselves, a friend of ours on a planet revolving around the star can detect the effects and radio back information to us about them. For one thing, our friend will see that our flashlight beam is redshifted; he could tell this either by observing a spectral line, if any, or by noting the shift of the peak of the Planck curve of the radiation toward the red. Second, our flashlight beam would be bent by the gravitational field of the star (Fig. 9–1). If we shined the beam straight up, it would continue to go straight up. But if we shined it away from the vertical, the beam would be bent slightly away from the vertical (Fig. 9–2).

Only if the flashlight is pointed within a certain angle of the vertical does the light continue outward. This angle forms a cone, with its apex at the flashlight, and is called the *exit cone*. As the star grows smaller yet, we find that the flashlight has to be pointed more directly upward in order for its light to escape. The exit cone grows smaller as the star continues to shrink.

When we shine our flashlight upward in the exit cone, the light escapes. When we shine our flashlight in a direction outside the exit cone, the light is bent sufficiently that it falls back to the surface of the star. When we shine our flashlight exactly along the side of the exit cone, the light goes into orbit around the star, neither escaping nor returning to the surface (Fig. 9–3).

The sphere around the star in which the light can orbit is called the *photon sphere*. Its size can be calculated theoretically. It is 13.5 km in radius for a star of 3 solar masses, and its size varies linearly with the mass; that is, it is 27 km in radius for a star of 6 solar masses, and so on.

As the star continues to contract, the exit cone gets narrower and narrower. Light emitted within the exit cone edge still escapes. The photon sphere remains at the same height even though the matter inside it has contracted further.

9.3 THE EVENT HORIZON

If we were to depend on our intuition, which is based on classical physics as advanced by Newton, we might think that the exit cone would simply continue to get narrower. But when we apply the general theory of relativity, we find that at a certain radius the cone vanishes. Light no longer can escape into space, even when it is traveling straight up, away from the center of the gravitational mass.

The solution to Einstein's equations that predicts this was worked out by Karl Schwarzschild in 1916, shortly after Einstein advanced his general theory. The radius of the star at the time at which light can no longer escape is called the *Schwarzschild radius* or the *gravitational radius*, and the spherical surface at that radius is called the *event horizon* (Fig. 9–4). The radius of the photon sphere is exactly 3/2 times this Schwarzschild radius.

We can visualize the event horizon in another way, by considering a classical picture (that is, one based on the Newtonian theory of gravitation). The picture is essentially that conceived in 1796 by Laplace (who was also responsible for the nebular hypothesis of planetary formation as described in Section 10.5). A body must have a certain velocity, called the *escape velocity*, to escape from the gravitational pull of another body. For example, we have to launch rockets at 11 km/sec (25,000 miles/hr) in order for them to escape from the earth's gravity. For a more massive body of the same size, the escape velocity would be higher. Now imagine that this body contracts, and we are drawn closer to the center of the mass with all the mass concentrated in a sphere below us. As this happens, the escape velocity would rise. As the body crosses its Schwarzschild radius, the escape velocity becomes equal to the speed of light. Thus even light cannot escape. If we then begin to apply the special theory of relativity, we might then reason that since nothing can go faster than the speed of light, nothing can escape. Now let us return to the picture according to the general theory of relativity.

The size of the Schwarzschild radius depends linearly on the amount of mass that is collapsing. A star of three solar masses, for example, would have a Schwarzschild radius of 9 km. A star of six solar masses would have a Schwarzschild radius of twice 9 km, or 18 km. One can calculate the Schwarzschild radii for less massive stars as well, although the less massive stars would be held up in the white dwarf or neutron star stages and not collapse to their Schwarzschild radii. The sun's Schwarzschild radius is only 3 km.

Once the star passes inside its event horizon, it continues to contract. Nothing can ever stop its contraction. In fact, the mathematical theory

Figure 9–3 When the star has contracted enough, only light emitted within the exit cone escapes. Light emitted on the exit cone goes into the photon sphere.

The Schwarzschild solutions *of equations from Einstein's general theory of relativity are the theoretical basis for the existence of black holes.*

"Now, here, you see, it takes all the running you can do, to keep in the same place. If you want to get somewhere else, you must run twice as fast as that."
The Red Queen,
Through the Looking Glass
by Lewis Carroll, 1871

Figure 9–4 When the star becomes smaller than its Schwarzschild radius, we can no longer observe it. We say that it has passed its *event horizon*, by analogy to the statement that we cannot see a thing on earth once it has passed our horizon.

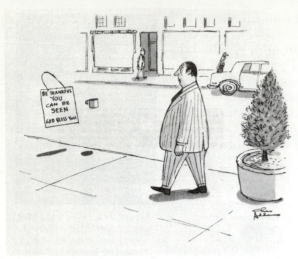

Figure 9–5 This drawing by Charles Addams is reprinted with permission of *The New Yorker Magazine, Inc.,* © 1974.

predicts that it will contract to zero radius, a situation that seems physically impossible to conceive of. The point at which it will have zero radius (it has infinite density there) is called a *singularity*. Strange as it seems, theory predicts that a black hole contains a singularity.

Even though the mass that causes the black hole has contracted further, the event horizon doesn't change. It remains at the same radius forever, as long as the amount of mass inside doesn't change.

9.4 ROTATING BLACK HOLES *(Optional)*

Once matter is inside a black hole, it loses its identity in the sense that from outside a black hole, all we can tell is the mass of the black hole, the rate at which it is spinning, and what electric charge it has. These three quantities are sufficient to completely describe the black hole. Thus, in a sense, black holes are simple objects to describe physically, because we only have to know three numbers to characterize each one. By contrast, for the earth we have to know shapes, sizes, densities, motions, and other parameters for the interior, the surface, and the atmosphere. The theorem that describes the simplicity of black holes is often colloquially stated by astronomers active in the field as "a black hole has no hair"; that is, a black hole has no basic properties at all that can be described aside from mass, spin, and charge. All other properties, such as size, can be derived from these three basic properties.

Most of the theoretical calculations about black holes, and the Schwarzschild solutions in particular, are based on the assumption that black holes do not rotate. But this assumption is only a convenience; we think, in fact, that the rotation of a black hole is one of its important properties. It was not until 1963 that Roy P. Kerr solved Einstein's equations for a situation that was later interpreted in terms of the notion that a black hole is rotating. In this more general case, there are two event horizons with somewhat different properties. Unlike the situation for a non-rotating black hole, in a rotating black hole (Fig. 9–6) it is not necessary that matter inside the event horizons reach the singularity. The matter might be able to reappear at some other point in our universe, or even reappear in our universe at some other time in the past or future!

An object can be shot into the ergosphere of a rotating black hole at such an angle that, after it is made to split into two, one part continues into the event horizon while the other part is ejected from the ergosphere with more energy than the incoming particle had. This energy must have come from somewhere; it could only have come from the rotational energy of the black hole because there are no other possibilities. Thus by judicious use of the ergosphere, we can tap the energy of the black hole. We are always looking for new energy sources and this is the most efficient known, although it is obviously very far from being practical.

9.5 NON-STELLAR BLACK HOLES *(Optional)*

We have discussed how black holes can form by the collapse of a star of greater than 3 solar masses. But theoretically a black hole will result if a mass of any amount is sufficiently compressed. (A neutron star, for example, could become a black hole if enough mass were added to it.) No mass less than 3 solar masses will contract sufficiently under the force of its own gravity in the course of stellar evolution. But the density of matter was so high at the time of the origin of the universe (see Chapter 26) that smaller masses may have been sufficiently compressed to form what are called mini black holes.

There is no good evidence for a mini hole, but some scientists predict that there could be some the size of pinheads (and thus masses equivalent of those of asteroids) floating around in space.

Stephen Hawking, the English astrophysicist who suggested the existence of mini black holes, has further deduced that small black holes can seem to emit energy in the form of elementary particles (neutrinos and so forth). The mini holes would thus evaporate and disappear. This may seem to be a contradiction to the concept that mass can't escape from a black hole. Actually, Hawking suggests that a black hole so affects space near it that a pair of particles—a nuclear particle and its antiparticle—can form simultaneously. Since a particle and its antiparticle completely annihilate each other—wipe each other out—this is basically not adding matter. (Similarly, we can both add and subtract from the same side of an equation without changing anything.) The antiparticle disappears into the black hole, and the remaining particle reaches us. This situation is important only for the smallest mini black holes, for the amount of radiation decreases sharply as we consider black holes of greater and greater masses. Only mini black holes up to the mass of an asteroid—far short of stellar masses—would have had time to disappear since the origin of the universe. Hawking's ideas set a lower limit on the size of black holes now in existence, since we think the mini black holes were formed only at the time of the big bang (which is discussed in Chapter 26).

On the other extreme of mass, we can consider what a black hole would be like if it contained a very large number, i.e., thousands or millions, of solar masses. The density is very high when a black hole of the mass of even a large star is formed. For a black hole of larger mass, the density that it had when it went through its event horizon was smaller. If we consider a black hole of 3 hundred million solar masses, its density would be quite low when this mass disappeared within its event horizon, even approaching that of water. The radius of this black hole would be a hundred million times the radius of a 3 solar mass black hole, namely, 900 million km. This is only $1/10,000$ of a light year, or 6 A. U., approximately the size of the orbit of Jupiter.

Thus if we were traveling through the universe in a spaceship, we couldn't count on detecting a black hole by noticing a volume of high density. We could pass through the event horizon of a high-mass black hole

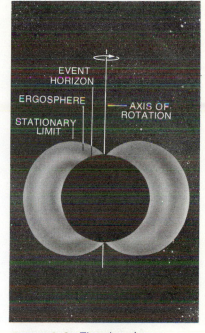

Figure 9–6 The description of a black hole so far in this chapter has assumed that the black hole is not rotating. But it seems reasonable, or even likely, that a black hole would rotate (just as a neutron star is rotating rapidly); a set of solutions to Einstein's equations discovered by Roy P. Kerr can be interpreted to describe a rotating black hole. From the top, the object looks like an ordinary non-rotating black hole. But from the side, one finds a region called the *ergosphere* because, in principle, work (which in Greek is *ergon*) can be extracted from it. The ergosphere is bounded on the outside by the *stationary limit;* within this limit nothing can be at rest. Light directed in the sense of the black hole's rotation will escape from the ergosphere, and an astronaut could too.

without even noticing. We would never be able to get out, but it would be hours before we would notice that we were being drawn into the center at an accelerating rate.

Where could such a supermassive black hole be located? The existence of bursters may be evidence that black holes of about 1000 solar masses are at the center of globular clusters. The center of our galaxy is one possibility for an even more massive black hole. Strange bursts of x-ray emission have been detected, and it is reasonable that there could be a concentration of mass there (see Chapter 22). Though we do not expect to observe radiation from the black hole itself, we hope to observe radiation from the gas surrounding. it.

9.6 DETECTING A BLACK HOLE

What if we were watching the black hole from the outside? Photons that were just barely inside the exit cone when emitted would leak very slowly away from the black hole. Nevertheless, these few photons would be spread out in time, and the image would be very faint. Further, as the star went through its event horizon, it would get very red and then blink out. It would do this in a fraction of a second, so the odds are low that we would actually be watching a collapsing star as it went through the event horizon.

But all hope is not lost for detecting a black hole, even though we can't hope to see the photons from the moments of collapse. The black hole disappears, but it leaves its gravity behind. It is a bit like the Cheshire Cat from *Alice's Adventures in Wonderland,* which fades away, leaving only its grin behind (Fig. 9–7).

The black hole attracts matter, and the matter accelerates toward it. Some of the matter will be pulled directly into the black hole, never to be seen again. But other matter will go into orbit around the black hole, and will orbit at a high velocity.

It seems likely that the gas in orbit will be heated. Theoretical calculations of the friction that will take place between adjacent filaments of gas in orbit around the black hole show that the heating will be so great that the gas will radiate strongly in the x-ray region of the spectrum. Thus, though we cannot observe the black hole itself, we can hope to observe x-rays from the gas surrounding it.

In fact, a large number of x-ray sources are known in the sky. They have been best surveyed from a NASA satellite called Uhuru. Some of these sources have been identified with galaxies or quasars. Others pulse regularly, like Hercules X-1, and are undoubtedly neutron stars (see Section 8.10). Some of the rest, which pulse sporadically, may be related to black holes.

It is not enough to find an x-ray source that gives off sporadic pulses, for one can think of other mechanisms besides revolution of matter around a black hole that can lead to such pulses. One would like to show that a collapsed star of greater than 3 solar masses is present.

In Section 4.2, we discussed how masses are determined from studies of binary stars. Our hope of pinning down the mass of a collapsed object, and thus identifying it with a black hole, rests on such a study.

When we search the position of the x-ray sources we are interested in finding a spectroscopic binary (that is, a star whose spectrum shows a Doppler shift that indicates the presence of an invisible companion). Then, if we can show that the companion is too faint to be a normal, main-

The Uhuru satellite was launched in 1970 in an international effort by an Italian team from a site in Kenya; Uhuru means "freedom" in Swahili, and is the Kenyan national motto.

Figure 9–7 Lewis Carroll's Cheshire Cat, from *Alice's Adventures in Wonderland,* shown here in John Tenniel's drawings, is analogous to a black hole in that it left its grin behind when it disappeared while a black hole leaves its gravity behind when its mass disappears. Alice thought that the Cheshire Cat's persisting grin was "the most curious thing I ever saw in all my life!" We might say the same about the black hole and its persisting gravity.

sequence star, it must be a collapsed star. If, further, the mass of the unobservable companion is greater than 3 solar masses, it must be a black hole.

The most persuasive case is named Cygnus X-1, and was the first x-ray source that was found to vary in intensity on a time scale of milliseconds. In 1971, radio radiation was found to come from the same direction. A 9th magnitude star called HDE 226868 was found at its location (Fig. 9–8 and page 162).

HDE 226868 has the spectrum of a blue supergiant, and thus has a mass of about 15 times that of the sun. Its spectrum is observed to vary in radial velocity with a period of 5.6 days, indicating that the supergiant and the invisible companion are orbiting each other with that period. From the orbit, it is deduced that the invisible companion must certainly have a mass greater than 4 solar masses; the best estimate is 8 solar masses. Because this is so much greater than the limit of 2 or 3 solar masses above which neutron stars cannot exist, it seems that even allowing for possible errors in measurement or in the theoretical calculations, too much mass is present to allow the matter to settle down as a neutron star. Thus many, and probably most, astronomers believe that a black hole has been found in Cygnus X-1.

Calculations have been carried out, especially by Kip Thorne of Caltech, to predict how the supergiant and the black hole would interact if we accept the binary theory. Thorne finds that the binary theory predicts that the gravity of the black hole would attract matter from the outer layers of the supergiant, and that this matter would go into orbit around the black hole. Because of the orbital motion of the matter, it would form a disk around the black hole, not unlike a large-scale version of Saturn's rings (Fig. 9–9). As the matter is heated, it gives off x-rays.

Matter in the disk around the black hole would orbit very quickly. If a hot spot developed somewhere on the disk, it might beam a cone of radiation into space, similar to the lighthouse model of pulsars. If the hot spot lasted for several rotations, we could detect a pulse of x-rays every time the cone swept past the earth. Thorne predicts that the period of the x-ray pulses would be extremely short, only a few milliseconds.

The Uhuru x-ray telescopes are neither large enough nor sensitive enough to variations of the signal with time to detect such periodic structure. But in 1977 NASA launched HEAO-1 (High-Energy Astronomy Observatory 1), with a large, sensitive x-ray telescope (page 60). Perhaps

The study of spectroscopic binaries, which we discussed in Section 4.1, turns out to be important for our hope of discovering a black hole.

Figure 9–8 A section of the image reproduced in the photograph opening this chapter of the blue supergiant star HDE 226868. The black hole Cygnus X-1 that is thought to be orbiting the supergiant star is not visible. Note that the fact that the image of the supergiant appears so large is entirely an overexposure effect in the film and does not represent the actual angular diameter subtended by the star, which is too small to resolve from the earth.

Figure 9–9 An artist's conception of the disk of swirling gas that would develop around a black hole like Cygnus X-1 (*right*) as its gravity pulled matter off the companion supergiant (*left*). The x-radiation would arise in the disk. The painting is by Lois Cohen and is used courtesy of the Griffith Observatory.

the rapid pulses will be detected. Probably the results will be inconclusive. Even if the pulses are not detected, it would not argue against Cygnus X-1 being a black hole. Only the hot spot theory would be tested.

It still looks as though Cygnus X-1 represents the first observational detection of a black hole. The new generation of x-ray telescopes should tell us more.

9.7 GRAVITATIONAL WAVES *(Optional)*

Most of the data we have thus far discussed in this book have come from studies of the electromagnetic spectrum. But Einstein's general theory of relativity predicts the existence of still another type of signal that we might be able to detect: gravitational waves.

Whereas electromagnetic waves affect only charged particles, we would expect gravitational waves to affect all matter. But gravitational waves, if they exist, would probably be very weak. They would be emitted only in situations where large masses were being accelerated rapidly. In the previous chapter, we mentioned one possible example: two neutron stars revolving in close orbits around each other. Another possible example would be a supernova or the collapse of a massive star to become a black hole. Even a simple close binary system containing a white dwarf or a neutron star will give out a reasonable amount of gravitational radiation.

Joseph Weber of the University of Maryland was the first to build sensitive apparatus to try to search for gravitational waves. His detector (Fig. 9–10) is a large metal cylinder, delicately suspended from wires so as to insulate it from restraint and from local motion. When a gravitational wave hits the cylinder, it should begin vibrating. Weber has sensitive apparatus to measure the vibrations. Starting in 1969, Weber reported success in detecting gravitational waves.

Weber's results seemed exciting, but they were treated with skepticism by many scientists because the frequency of events is much higher than we would have expected. Nobody had predicted, for example, that black holes would form as often in our galaxy as Weber's results would indicate, or that so much mass would fall into a giant black hole even if there were one in the center of our galaxy.

Eventually other scientists built other gravitational wave apparatus, some in configurations that they had calculated would be more sensitive than Weber's. None of the other scientists has detected any gravitational waves, and most astronomers feel that Weber's detectors must have been affected by other kinds of events, perhaps of terrestrial origin. Nonetheless, Weber gets credit for having stimulated this exciting field.

Figure 9–10 Joseph Weber and his gravitational wave detector, a large aluminum cylinder, weighing 4 tons, delicately suspended so that gravitational waves would set it vibrating 1660 times per second.

SUMMARY AND OUTLINE OF CHAPTER 9

Aims: To understand what black holes are, how they form, and how they might be detected

Gravitational collapse occurs if more than 2 or 3 M$_\odot$ remains (Section 9.1)
Critical radii of a black hole
   Photon sphere: exit cones form (Section 9.2)
   Event horizon: exit cones close (Section 9.3)
   Schwarzschild radius (gravitational radius) defines the limit of the black hole, the event horizon
   Singularity
Rotating black holes (Section 9.4)
   Can get energy out of the ergosphere.
Non-stellar black holes (Section 9.5)
   Mini black holes could have formed in the big bang
   Very massive black holes do not have high densities
Detecting a black hole (Section 9.6)
   X-radiation expected
   Detection in a spectroscopic binary, invisible high-mass companion
   Cygnus X-1: a black hole?
Gravitational waves (Section 9.7)
   They are not electromagnetic waves
   They might set matter on earth vibrating, though it is difficult to show that the detectors have been set vibrating by gravitational waves rather than by other causes
   Current situation: probably not yet detected

QUESTIONS

1. Why doesn't electron or neutron degeneracy prevent a star from becoming a black hole?

2. Why is a black hole blacker than a black piece of paper?

3. (a) Is light acting more like a particle or more like a wave when it is bent by gravity?
   (b) Can you explain the bending of light as a property of a warping of space, as discussed in Section 5.11?

4. (a) How does the escape velocity of the moon compare with the escape velocity of the earth? Would a larger rocket engine be necessary to escape from the gravity of the earth or from the gravity of the moon?
   (b) How will the velocity of escape from the surface of the sun change when the sun becomes a red giant? A white dwarf?

5. (a) What is the Schwarzschild radius for a 10 solar mass star?
   (b) What is your Schwarzschild radius?

6. What is the relation in size of the photon sphere and the event horizon? If you were an astronaut in space, could you escape from within the photon sphere of a black hole? Could you escape from within its event horizon?

7. (a) Why does the mass of a black hole that results from a collapsed star tend to increase rather than decrease with time?
   (b) What property would show up in mini black holes, if they exist, that would allow them to eventually lose mass?

8. Would we always notice when we reached a black hole by its high density? Explain.

9. Could we detect a black hole that was not part of a binary system?

10. (a) Under what circumstances does the presence of an x-ray source associated with a spectroscopic binary suggest to many astronomers the possible presence of a black hole? Why?
    (b) For what additional properties of the objects in the binary system do we search?

Part III
The Solar System

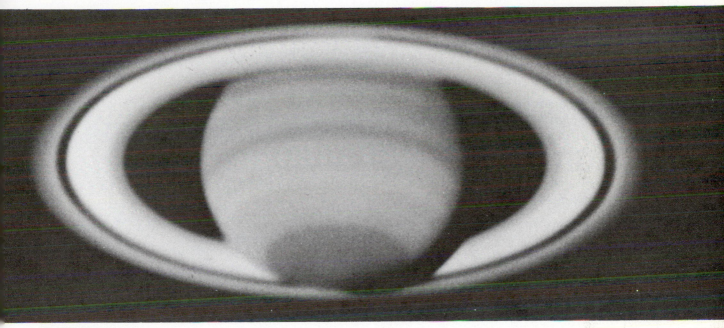

Saturn

(left) Viking 1 pushes aside a rock on Mars, in order to sample soil that has been protected from solar ultraviolet radiation.

The Solar System . . .

The Earth and the rest of the solar system may be important to us, but they are only minor companions to the stars. In *Captain Stormfield's Visit to Heaven,* by Mark Twain, the Captain races with a comet and gets off course. He comes into heaven by a wrong gate, and finds that nobody there has heard of "the world" ("**the** world, there's billions of them!" says a gatekeeper). Finally, the gatekeepers send a man up in a balloon to try to detect "the world" on a huge map. The balloonist has to travel so far that he rises into clouds and after a day or two of searching he comes back to report that he has found it: an unimportant planet, more properly named "the Wart."

We too must learn humility as we ponder the other objects in space. And while it is no doubt the case that the heavens are filled with a vast assortment of suns and planets more spectacular than our own, still, the solar system is our own local environment. We would like to understand it and come to terms with it as best we can. Besides, in understanding our own solar system, we may even find some keys to understanding the rest of the Universe.

To get an idea of its scale, imagine that the solar system is scaled down and placed on a map of the United States. Let us say that the sun is a hot ball of gas taking up all of Rockefeller Center, more than a kilometer across, in the center of New York City.

We would then find Mercury to be a ball 4 meters across at the distance of mid–Long Island, and Venus to be a 10 meter ball one and a half times farther away. The Earth is only slightly bigger and is located at the distance of Trenton, New Jersey. Mars is half that size, 5 meters across, located past Philadelphia.

Only for the planets beyond Mars would the planets be much different in size from the Earth, and the separations become much greater. Jupiter is 100 meters across, the size of a baseball stadium, past Pittsburgh at the Ohio line. Saturn without its rings is a little smaller than Jupiter (including the rings it is a little larger), and is past Cincinnati toward the Indiana line. Uranus and Neptune are each about 30 meters across, about the size of a baseball infield, and are at the distances of Topeka and Santa Fe, respectively. And Pluto, as small as the inner planets, is as far away as Los Angeles, 40 times farther away from the sun than the Earth is. Occasionally a comet sweeps in from Alaska, or some other random direction, passes around the Sun, and returns in the general direction from which it came.

The planets fall naturally into two groups. The first group, the *terrestrial planets,* consists of Mercury, Venus, Earth, and Mars. All are rocky in nature. The terrestrial planets are not very large, and have densities about five times that of water. (When the metric system was set up, the gram was defined so that water would have a convenient density of exactly 1 gram/cubic centimeter. Thus the terrestrial planets have densities of about 5 grams/cm^3.)

The second group, the *giant planets,* consists of Jupiter, Saturn, Uranus, and Neptune. All these planets are much larger than the terrestrial planets, and also much less dense, ranging down to slightly below the density of water. Jupiter and Saturn are largely gaseous in nature, similar to the Sun in composition. Uranus and Neptune have large gaseous atmospheres but have rocky or icy surfaces and interiors. Both Saturn and Uranus have systems of rings. Pluto, planet number nine, is anomalous in several of its properties, as we shall discuss in Section 17.3, and so may have had a very different history from the other planets.

Between the orbits of the terrestrial planets and the orbits of the giant planets are the orbits of thousands of chunks of small "minor planets." These *asteroids* range up to 1000 kilometers across. Sometimes much smaller chunks of interplanetary rock penetrate the Earth's atmosphere and hit the ground. We shall discuss these *meteorites* and where they came from together with asteroids and comets in Chapter 18.

Many people are interested in the planets in order to study their history — how they formed, how they have evolved since, and how they will change in the future. Others are more interested in what the planets are like today. Still others are interested in the planets mainly to consider whether they are harboring intelligent life. The study of the origin of the solar system and the study of the origin of the Universe have the same name: *cosmogony.* And because the Moon, planets, and other objects in the solar system have undergone less water erosion on their surfaces than has the Earth, in observing these celestial bodies we actually see them as they appeared eons ago. In this way the study of the planets and, in particular, their surfaces, provides information about the solar system's early stages and its cosmogony. So even though the planets together contain only about one tenth of one percent of the mass of the Sun, their study may help us discover the origins of the solar system.

PROMINENCE

SUN

PROMINENCE

JUPITER

SATURN

URANUS

MERCURY

EARTH

NEPTUNE

VENUS

MARS

PLUTO

SUNSPOT
GROUP

10

The Planets and Their Origins

10.1 THE PHASES OF THE MOON AND PLANETS

From the simple observation that the apparent shapes of the Moon and planets change, we can make conclusions that are important for our understanding of the mechanics of the solar system. The fact that the Moon goes through a set of *phases* approximately once every month is perhaps the most familiar everyday astronomical observation (Fig. 10–1). In fact, the name month comes from the word moon. The actual period of the phases, the interval between a particular phase of the Moon and its next repetition, is approximately 29½ Earth days (Fig. 10–2). This period can vary by as much as 13 hours. The explanation of the phases is quite simple.

The Moon is a sphere, and at all times the side that faces the Sun is lighted and the side that faces away from the Sun is dark. The phase of the Moon that we see from the Earth, as the Moon revolves around us, depends on the relative orientation of the three bodies: Sun, Moon, and Earth. The situation is simplified by the fact that the plane of the Moon's revolution around the Earth is nearly, although not quite, the plane of the Earth's revolution around the Sun.

Basically, when the Moon is almost exactly between the Earth and the Sun, the dark side of the Moon faces us. We call this a "new moon." A few days earlier or later we see a sliver of the lighted side of the Moon, and call this a "crescent." As the month wears on, the crescent gets bigger, and about 7 days after new moon, half the face of the Moon that is visible to us is lighted. We sometimes call this a "half moon." Since the lighted half we see is one fourth of the whole moon, this is more properly called a "first-quarter moon." (Instead of apologizing for the fact that astronomers call the same phase both "quarter moon" and "half moon," I'll just continue with a straight face and try to pretend that there is nothing strange about it.)

When over half the Moon's disk is visible, we have a "gibbous" moon.

"The Moon" is often capitalized to distinguish it from moons of other planets; in general writing, it is usually written with a small "m." In Part III, we shall capitalize "Earth" to put it on a par with the other planets and the Moon. For consistency, we shall also capitalize "Sun" and "Universe."

The phases of moons or planets are the shapes of the sunlighted areas as seen from our vantage point.

The relative sizes of the planets and the Sun. The sizes of the planets are exaggerated with respect to the Sun by a factor of 2 in this drawing.

Figure 10–1 The phases of the Moon depend on the Moon's position in its orbit around the Earth.

FIRST QUARTER

WAXING GIBBOUS

WAXING CRESCENT

6:00 PM

9:00 PM   3:00 PM

FULL   MIDNIGHT   NOON   NEW

3:00 AM   9:00 AM

SUN'S RAYS

6:00 AM

WANING GIBBOUS

WANING CRESCENT

THIRD QUARTER

One week after the first-quarter moon, the Moon is on the opposite side of the Earth from the Sun, and the entire face visible to us is lighted. This is called a "full moon." One week later, we have a "third-quarter," and then we go back to "new moon" again and repeat the cycle of phases.

Note that since the phase of the Moon is related to the position of the Moon with respect to the Sun, one can tell when the Moon will rise at a certain phase. For example, since the Moon is 180° across the sky from the Sun when it is full, a full moon is always rising just as the Sun sets (Fig. 10–3). Each day thereafter, the moon rises approximately one hour later (24 hours divided by the period of revolution of the Moon around the

Figure 10–2 The phases of the Moon.

4 days

7 days
1st quarter

10 days

14 days
full

EAST   WEST   EAST   WEST

Figure 10–3 Because the phase of the Moon depends on its position in the sky with respect to the Sun, we can see why a full moon is always rising at sunset while a crescent moon is either setting at sunset, as shown here, or rising at sunrise.

Earth). The third-quarter moon, then, rises at midnight, and at temperate latitudes is high in the sky at sunrise. The new moon rises with the Sun in the east at dawn. The first-quarter moon rises at noon and is high in the sky at sunset. It is often visible in the late afternoon.

The Moon is not the only object in the solar system that is seen to go through phases. Mercury and Venus both orbit inside the Earth's orbit, and so sometimes we see the side that faces away from the Sun and sometimes we see the side that faces toward the Sun. Thus at times Mercury and Venus are seen as crescents, though it takes a telescope to observe their shapes.

The outer planets, though, from Mars on outward, are never between the Sun and the Earth. Thus they never appear as crescents, although sometimes they can appear gibbous.

10.2 ECLIPSES

Because the Moon's orbit around the Earth, and the Earth's orbit around the Sun are not precisely in the same plane (Fig. 10–4), the Moon usually passes slightly above or below the Earth's shadow at full moon, and the Earth usually passes slightly above or below the Moon's shadow at new moon. But every once in a while, up to seven times a year, the Moon is at the part of its orbit that crosses the Earth's orbital plane at full moon or new moon. When that happens, we have a lunar or a solar eclipse.

We have already discussed eclipses of the Sun, when the Moon comes directly between the Earth and the Sun. Eclipses of the Moon, when the Moon goes into the Earth's shadow, are more commonly seen.

20 days   22 days   24 days   26 days
                     third quarter

5 DEGREES

EARTH

MOON

SUN'S RAYS

Figure 10–4 The plane of the Moon's orbit is tipped with respect to the plane of the Earth's orbit, so the Moon usually passes above or below the Earth's shadow.

Figure 10–5 (A) When the Moon is between the Earth and the Sun, we observe an eclipse of the Sun. When the Moon is on the far side of the Earth from the Sun, we see a lunar eclipse. The part of the Earth's shadow from which the Sun is only partially shielded from the Moon's view is called the *penumbra*; the part of the Earth's shadow from which the Sun is entirely shielded from the Moon's view is called the *umbra*. Note that the distances in this diagram are not to scale. (B). A partial phase of a lunar eclipse; the Moon remains visible even during totality, though it is relatively faint and longer exposures must be used. Here the bright overexposed part of the moon is in the penumbra, while the left part of the moon is in the umbra.

Many more people see a lunar eclipse when one occurs, because the Moon lies entirely in the Earth's shadow and the sunlight is entirely cut off from it (Fig. 10–5). So anywhere on the Earth that the Moon has risen, the eclipse is visible. In a solar eclipse, on the other hand, the alignment of the Moon between the Sun and the Earth must be precise, and only those people in a narrow band on the surface of the Earth see the eclipse.

A lunar eclipse is a much more leisurely event to watch than a solar eclipse. The partial phase, when the Earth's shadow gradually covers the Moon, lasts for hours. And then the total phase, when the Moon is entirely within the Earth's shadow, can itself last for over an hour. During this time, the sunlight is not entirely shut off from the Moon (Fig. 10–6). A small amount is refracted around the edge of the Earth by our atmosphere. Most of the blue light is taken out during the sunlight's passage through our atmosphere; this is how blue skies are made for the people one fourth of the way around the globe from the point at which the Sun is overhead. The remaining light is reddish, and this is the light that falls on the Moon. Thus, the eclipsed Moon appears reddish.

Scientists can make observations at lunar eclipses to study the properties of the lunar surface. For example, one can monitor the rate of cooling of areas on the lunar surface with infrared radiometers. But now we have landed on the Moon and have made direct measurements at a few specific places, so the scientific value of lunar eclipses has been greatly diminished but not eliminated. There remains little of scientific value to do during lunar eclipses, quite a different situation from that at a solar eclipse.

EARTH

MOON

SUN

UMBRA MOON

EARTH

SOLAR ECLIPSE   A

LUNAR ECLIPSE

PENUMBRA

B

REDDISH LIGHT

RED LIGHT REFRACTED TOWARD MOON

ALL LIGHT

EARTH

MOON

BLUE LIGHT SCATTERED MORE BY ATMOSPHERE

Figure 10-6 The Earth's atmosphere scatters shorter wavelengths much more than it scatters longer wavelengths, so red light is scattered much less than blue light. Some of the red light survives its passage through the Earth's atmosphere and is bent, or refracted, toward the Moon. Thus the Moon often appears reddish during the total phase of a lunar eclipse.

It has been known for thousands of years that some of the bright dots that we see in the sky at night are unchanging (fixed) in position—the fixed stars—while others appear to move among them—the wandering stars.

10.3   THE MOTIONS OF THE PLANETS

When we observe the planets in the sky, we notice that their positions vary from night to night with respect to each other and with respect to the stars. The stars, on the other hand, are so far away that their positions are relatively fixed with respect to each other.

The motion of the planets in the sky is not continuously in the same direction with respect to the stars. Most of the time they appear to drift eastward, that is, move across the sky slightly slower than the stars, but sometimes they drift backwards, that is, move slightly faster than the stars and thus westward (Fig. 10–7). We call the backward motion *retrograde motion.*

Figure 10-7 The retrograde loop of Mars in 1977 and 1978. Opposition occurs in Cancer on January 22, 1978; Gemini and Canis Minor are among the other constellations visible in this simulation photographed at the Vanderbilt Planetarium by George Lovi. Pollux is the bright star just above and to the right side of the retrograde loop. Regulus is the bright star to the left that Mars passes close by.

Figure 10–8 Ptolemy. (Burndy Library, photograph by Owen Gingerich).

10.3a A Sense of History

One of the earliest and greatest philosophers, Aristotle, who lived in Greece about 350 B.C., summarized the astronomical knowledge of his day into a qualitative cosmology that remained unchallenged for 1800 years. On the basis of what seemed to be very good evidence—what he saw—Aristotle thought, and actually believed that he **knew,** that the Earth was at the center of the universe and that the planets, Sun, and stars revolved around it. The universe was made up of a set of 55 celestial spheres that fit around each other and that had rotation as their natural motion. Each of the heavenly bodies was carried around the heavens by a sphere, whose motion was still further compounded by that of other spheres. These motions combined to account for the various observed motions, including revolution around the Earth, retrograde motion, and variation in latitude. Each planet had several spheres in order to account for its various motions. The outermost sphere was that of the fixed stars, beyond which lay the prime mover, *primum mobile,* that caused the general rotation of the stars overhead.

Aristotle's theories ranged through much of science. He held that all bodies below the spheres of the Moon were made of four basic "elements": earth, air, fire, and water. The fifth "essence"—the quintessence—was a perfect, unchanging, transparent element of which the celestial spheres around that of the Moon are formed.

Aristotle's theories dominated scientific thinking for almost two millennia, until the Renaissance. Unfortunately, most of his theories were far from what we now consider to be correct, so we tend to think that the widespread acceptance of Aristotelian physics impeded the development of science.

Almost 500 years after Aristotle, in about 140 A.D., the Greek astronomer Claudius Ptolemy (Fig. 10–8) elaborated on Aristotle's thoughts and presented a detailed theory of the universe that explained the retrograde

Figure 10–9 Aristotle's Earth-centered theory. The Earth is at the center, orbited in larger and larger circles by the Moon, Mercury, Venus, the Sun, Mars, Jupiter, Saturn, and the stars.

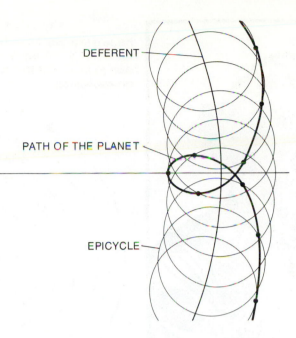

DEFERENT

PATH OF THE PLANET

EPICYCLE

Figure 10–10 In the Ptolemaic theory, retrograde motion was caused by motion of the planets on small circles called epicycles that moved along a larger circle, centered at the Earth, called the deferent.

motion. Ptolemy's model, as was Aristotle's, was Earth-centered (Fig. 10–9). To account for the retrograde motion of the planets, the planets had to be moving not just on large circles around the Earth but rather on smaller circles, called *epicycles*, whose centers moved around the Earth on larger circles, called *deferents* (Fig. 10–10). It seemed natural that the planets should follow circles in their motion since circles were thought to be "perfect" figures. Sometimes the center of the deferent was not centered at the Earth. The planets moved at a constant rate of angular motion but another complication was that the point around which the planet's angular motion moved uniformly was neither at the center of the Earth nor at the center of the deferent. Ptolemy's views were very influential in the study of astronomy, for versions of his ideas and of the tables of planetary motions that he computed were accepted for nearly 15 centuries.

The credit for the breakthrough in our understanding of the solar system is commonly given to Nicholas Copernicus (Fig. 10–11), a 16th-century Polish astronomer whose five-hundredth birthday was celebrated by the astronomical community in 1973. Copernicus advanced a *heliocentric— sun centered—*theory (Fig. 10–12) that the retrograde motion of the planets could be readily explained if the Sun rather than the Earth was at the center of the universe, that the Earth is a planet, and that the planets move around the Sun in circles. (Aristarchus of Samos, a Greek scientist, had also suggested a heliocentric theory 18 centuries earlier, though we do not know how detailed a picture of planetary motions he presented. In any case, his heliocentric suggestion required the apparently ridiculous notion that the Earth moved, in contradiction to our senses and to the theories of Aristotle. Aristarchus's theory was lost in antiquity.)

Copernicus used his own version of the heliocentric theory to calculate tables of planetary positions. His detailed predictions were not in much better agreement with the existing observations than tables based on Ptolemy's model, because Copernicus still used Ptolemy's observations. Also, the errors in the observations were very large. The heliocentric theory

Helios was the Sun god in Greek mythology.

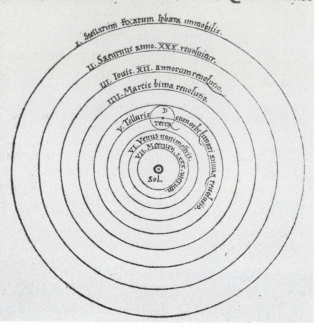

Figure 10–11

Figure 10–12

Figure 10–11 Copernicus, in a 16th century woodcut.

Figure 10–12 The heliocentric theory, as illustrated in Copernicus's *De Revolutionibus,* published in 1543.

Figure 10–13 Copernicus's signature from the Uppsala University Library in Sweden, photographed by Charles Eames and reproduced courtesy of Owen Gingerich.

appealed to Copernicus primarily on the philosophical grounds of simplicity.

Copernicus's heliocentric theory, published in 1543 in a book known as *De Revolutionibus (On the Revolutions),* explained the retrograde motion of the planets in the following manner.

Let us consider, first, an outer planet like Mars as seen from the Earth (Fig. 10–14). As the Earth approaches the part of its orbit that is closest to Mars (which is orbiting much more slowly), the projection of the Earth-Mars line to the stars (which are essentially infinitely far away compared to the planets) moves slightly against the stellar background. As the Earth in its orbit comes closest to Mars, and then passes it, the projection of the line joining the two planets can actually seem to go backwards since the Earth is going at a greater speed than Mars. Then, as the Earth continues around its orbit, Mars appears to go forward again. A similar explanation can be demonstrated for inner as well as outer planets.

The Italian scientist Galileo Galilei began to believe in the Copernican system in the 1590's, and later provided important observational confirmation of the theory. In 1610, simultaneously with the first settlements in the American colonies, Galileo was the first to use a telescope for astronomical observation. His discovery of the moons of Jupiter (Fig. 10–15) showed that not everything in the solar system orbited the Earth, and his discovery of sunspots showed that the Sun had "blemishes" on it. This observation was a key disagreement with the religious and philosophical Aristotelian view that the Sun was perfect. Galileo's discovery that Venus went through an

RETROGRADE
MOTION

STATIONARY
POINTS

DIRECT MOTION OF MARS
AMONG THE STARS

MARS' ORBIT

SUN

EARTH'S ORBIT

Figure 10–14 The Copernican theory explains retrograde motion as an effect of projection. The drawing shows the explanation of retrograde motion for Mars or for the other *superior planets,* that is, planets whose orbits lie outside that of the Earth. Similar drawings can explain retrograde motion for *inferior planets,* that is, planets whose orbits lie inside that of the Earth (namely, Mercury and Venus).

entire set of phases (Fig. 10–16), like the Moon, could not be explained with the Ptolemaic Earth-centered system. If Venus traveled in an epicycle located between the Earth and the Sun, Venus should always appear as a crescent since it is observed never to stray far in the sky from the Sun. This rationale was contrary to Galileo's observations. But while the Roman Catholic Church was not concerned with anyone's making models to explain observations, they were concerned with assertions that those models represent physical truth. Galileo was forced, in his old age, to recant his belief in the Copernican theory.

In the last part of the 16th century, not long after Copernicus's death, Tycho Brahe had made a series of observations of Mars and other planets at his observatory, named Uraniborg, on an island off the mainland of Denmark (Fig. 10–17). These positional observations were considerably more precise than any observations that had been made up to that time, even though the telescope had not yet been invented. In 1597, Tycho lost his

Figure 10–15 A translation (*left*) of Galileo's original notes (*right*) summarizing his first observations of Jupiter's moons in January 1610. The shaded areas were probably added later. It had not yet occurred to Galileo that the objects were moons in revolution around Jupiter.

On the 7th of January Jupiter is seen thus

On the 8th thus   east *   it was therefore direct and not retrograde

On the 12th day it is seen in this arrangement

The 13th are seen very close to Jupiter 4 stars   or better so

On the 14th it is cloudy

The 15th   the nearest to Jupiter was smallest the 4th was distant from the 3rd about double.

The spacing of the 3 to the west was no greater than the diameter of Jupiter and they were in a straight line.

long. 71°38' lat. 1°13'

Figure 10–16 The phases of Venus. Note that Venus is a crescent only when it is in a part of its orbit that is relatively close to the Earth, and so it looks larger at those times.

financial support in Denmark, and moved to Prague. Johannes Kepler came there as a young assistant to work with him. At Tycho's death, in 1601, Kepler was left to analyze all the observations that Tycho and his assistants had made.

Tycho's observational data showed that the tables then in use did not adequately predict planetary positions. Laboring for many years, without benefit of the modern computers that could now solve the problem in minutes, Kepler was finally able to make sense out of the observations of Mars that Tycho had made and clear up the discrepancy. In 1609 and 1618 Kepler presented three laws of motion that, to this day, we consider to be the basis of our understanding of the solar system.

Box 10.1   Kepler's Laws of Planetary Motion

(1) The planets orbit the Sun in ellipses, with the Sun at one focus.
(2) The line joining the Sun and a planet sweeps through equal areas in equal times.
(3) The period of revolution of a planet, and the semi-major axis of its orbit (half the longest dimension of the ellipse), are related to each other such that the square of the period is proportional to the cube of the semi-major axis of the orbit.

An ellipse *is a curve that has the property that the sum of the lengths of the lines drawn from any point on the ellipse to two points (each of which is called a focus) is constant.*

The major axis *of an ellipse is a line bounded by the ellipse and drawn through the two foci, or the length of that line; the semi-major axis is half that length. The* minor axis *is the perpendicular line bisecting the major axis, or the length of that part of the line included within the ellipse. In the special case when the major and minor axes are the same length, the ellipse is a circle.*

A sector *is the area bounded by two straight lines from a focus of an ellipse (or, for a circle, two radii) and the part of the ellipse joining their outer ends.*

Until Kepler worked out these laws, even the heliocentric calculations assumed that the planets followed "perfect" orbits, namely, circles. The realization that the orbits were in fact ellipses (Fig. 10–18), as stated in Kepler's first law, greatly improved the accuracy of the calculations.

The second law, also known as *the law of equal areas*, governs the speeds with which the planets travel in their orbits. When a planet is at its greatest distance from the Sun in its elliptical orbit, it must travel relatively slowly because the weaker gravity at the greater distance from the Sun accelerates the planet less. The line joining it with the Sun sweeps out a long skinny sector. This long skinny sector has the same area as the short fat sector formed when the planet is closer to the Sun, when the planet travels faster in its orbit (Fig. 10–19).

Figure 10–17 Tycho's observatory at Uraniborg, on the Island of Hveen in Denmark. In the illustration, Tycho is showing the mural quadrant that he used to measure the altitudes at which stars and planets crossed the meridian. The pre-telescopic observations made here enabled Kepler to discover his laws. The drawing dates from 1598.

To understand the third law (it is ordinarily referred to as *Kepler's third law* and is demonstrated in Figure 10–20), let us compare the orbit of, say, Jupiter with that of the Earth. We can choose to work in units that are convenient for the Earth. We call the average distance from the Sun to the Earth *1 Astronomical Unit*. Similarly, the unit of time that the Earth takes to revolve around the Sun is defined as *one year*. Using these values, if we know from observation that Jupiter's period of revolution around the sun is 11.86 years, we can use Kepler's third law. If P is the period of revolution of a planet and R is the radius of its orbit,

$$\frac{P \text{ (Jupiter)}^2}{P \text{ (Earth)}^2} = \frac{R \text{ (Jupiter)}^3}{R \text{ (Earth)}^3}.$$

Substituting,

$$\frac{(11.86)^2}{1} = \frac{R \text{ (Jupiter)}^3}{1}.$$

Now $(11.86)^2$ can be rounded off to 12^2, which is approximately 150. (Astronomers estimate calculations all the time, and it is important for students to realize that they need not do a calculation very exactly when only an approximate answer is required. At some other times, an exact answer may

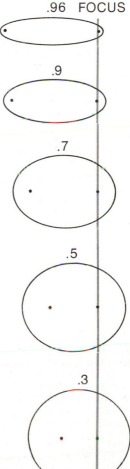

.96 FOCUS
.9
.7
.5
.3

Figure 10–18 A series of ellipses of the same major axis but different eccentricities. The *foci* are marked; these are the two points inside with the property that the sum of the distances from any point on the circumference to the foci is constant. As the eccentricity—distance between the foci divided by the major (longer) axis—approaches 1, the ellipse approaches a straight line. As the eccentricity approaches zero, the foci come closer and closer together. A circle is an ellipse of zero eccentricity.

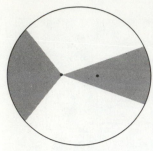

Figure 10–19 Kepler's second law states that the two shaded sectors, which represent the areas covered by a line drawn from a focus of the ellipse to an orbiting planet in a given length of time, are equal in area. The Sun is at this focus; nothing is at the other focus.

∝ means "is proportional to"; it means that the left hand side is equal to a constant times the right hand side. The constant, 6.67 × 10⁻⁸ when masses are in grams and distances are in centimeters, is called G or the universal gravitational constant. We can write, F = Gm₁m₂/d². This is known as the law of universal gravitation; it is a universal law in that it applies all over the Universe rather than being limited to local applicability.

be required.) So the radius of Jupiter's orbit around the Sun is approximately the cube root of 150, or approximately 5 A.U. The actual measured value is 5.2 A.U. — the small difference was almost entirely introduced in the rough estimation process. So it checks.

Kepler's third law applies not only to planets and the Sun, but also to all pairs of bodies in which one body containing relatively little mass is orbiting another, more massive, body. Planets and their moons are such cases. The form of the equation is the same; only the constant of proportionality between P^2 and R^3 is different. Since this constant depends on the mass of the central body, in this case the Sun, measurements of the period and size of the orbit give us the Sun's mass. Similarly, by studying the orbits of satellites around a planet, we can derive that planet's mass. (Actually, in determining the mass of the Sun, the planet's mass contributes slightly to the value of the proportionality constant, but the effect is too small to be detectable except for Jupiter and Saturn.)

It was not until the time of Isaac Newton, many years after these three laws were laboriously calculated and discovered by Kepler, that the laws were derived mathematically from basic principles. Newton invented and used the calculus to do this in his *Principia Mathematica*, published in 1687. He used the law of gravitation that he had discovered, for Newton — whether or not you believe that an apple fell on his head — was the first to realize the universality of gravity. He formulated the law that the force of gravity between two bodies varies directly with the product of the masses of the two bodies and inversely with the square of the distance between them: force of gravity $\propto m_1 m_2/d^2$.

10.4 THE REVOLUTION AND ROTATION OF THE PLANETS

The orbits of all the planets lie in approximately the same plane. They thus take up only a disk whose center is at the Sun, rather than a full sphere. Little is known of the parts of the solar system away from this disk, although the comets may originate in a spherical cloud around the Sun.

The plane of the Earth's orbit around the Sun is called the *ecliptic plane* (Fig. 10–22). Of course, since we are on the Earth rather than outside the solar system looking in, we actually see the Sun appear to move across the sky with respect to the stars. (Remember that on a given day of 24 hours, the Sun moves at almost the same rate as the stars as it rises, moves across the sky, and sets; one has to plot the position of the Sun with respect to the stars from day to day to see that it is in fact moving among them.) The path that the Sun takes among the stars, as seen from the Earth, is called *the ecliptic* (see also Section 3.11).

Figure 10–20 Kepler's third law relates the period of an orbiting body to the size of its orbit. The outer planets orbit at much slower velocities than the inner planets and also have a longer path to follow in order to complete one orbit. The distances the planets travel in their orbits in 1 year is shown here.

230° IN ONE HOUR

SATELLITE ORBIT AT 200 KM ABOVE SURFACE

EARTH ROTATION 15° PER HOUR

2 3 4 5 6
EARTH RADII

SATELLITE ORBIT AT 3¼ EARTH RADII

42°

SATELLITE ORBIT AT 6½ EARTH RADII

SATELLITE REVOLUTION 15° PER HOUR

Figure 10–21 Most orbiting satellites are only 200 km or so above the Earth's surface, and orbit the Earth in about 90 minutes. From Kepler's third law we can see that the velocity in orbit of a satellite decreases as the satellite gets higher and higher. At about 6½ Earth radii, the satellite orbits at the same velocity as that at which the Earth's surface rotates underneath. Thus the satellite is in *synchronous rotation*, and always remains over the same location on Earth. This property makes such synchronous satellites useful for relaying communications.

The fact that the planets all have orbits close to the ecliptic plane means that they are all in the part of our sky close to the ecliptic. Since the ecliptic is divided into 12 *signs of the zodiac* (see Fig. 19–10), each associated with a distinct constellation, the planets always fall in one of these 12 *houses*, and never stray into other constellations.

The *inclinations* of the orbits of the other planets with respect to the ecliptic are small, with the exception of Pluto (Fig. 10–23). Of the other planets, Mercury has an inclination of 7°; the remainder, except for Pluto, have inclinations of less than 4°. Pluto's much larger inclination of 17° is discrepant and is just one of the pieces of evidence suggesting that Pluto may not have formed in the same circumstances as the other planets in the solar system.

The inclination of the orbit of a planet is the angle that the plane in which its orbit lies makes with the plane in which the Earth's orbit lies.

The fact that the planets all orbit the Sun in essentially the same plane is one of the most important facts that we know about the solar system. Its explanation is at the base of most cosmogonical models, and central to that explanation is a property that astronomers and physicists use in analyzing spinning or revolving objects: *angular momentum*. The amount of angular momentum of a small body revolving around a large central body is (distance from the center) × (velocity) × (mass). The importance of angular momentum lies in the fact that it is conserved, that is, the total angular momentum of the system (the sum of the angular momenta of the different

URANUS
4.29°

NEPTUNE
2.18°

PLUTO
1.5°

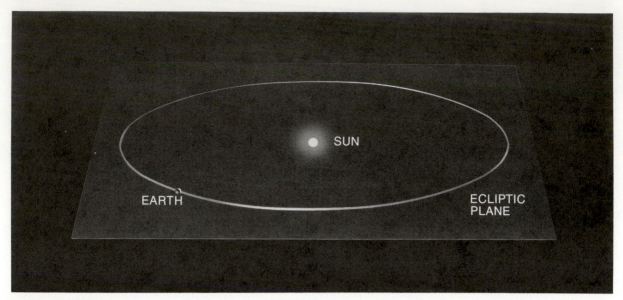

Figure 10–22 The ecliptic plane is the plane of the Earth's orbit around the Sun.

parts of the system) doesn't change even though the distribution of angular momentum among the parts may change. The total angular momentum of the solar system should thus be the same as it was in the past, unless some mechanism is carrying angular momentum away.

The most familiar example of angular momentum is an ice skater. She may start herself spinning by exerting force on the ice with her skates, thus giving herself a certain angular momentum. When she wants to spin faster, she draws her arms in closer to her (Fig. 10–25). This changes the distribution of her mass so that it is effectively closer to her center (the axis around which she is spinning), and to compensate for this she starts to rotate more quickly so that her angular momentum remains the same.

The motion of the planets around the Sun in their orbits is called revolution. *The spinning of a planet is called* rotation. *We must be careful and precise when using these words (Fig. 10–24).*

Thus from the fact that every planet is revolving in the same plane and in the same direction, we deduce that the solar system was formed of primordial material that was rotating in that direction. We would be very surprised to find a planet revolving around the Sun in a different direction—and we do not—or, to a lesser extent, to find a planet rotating in a direction opposite to that of its fellows. There are, however, two examples of such backwards rotation, which must each be carefully considered. Such rotation in the opposite sense is called *retrograde rotation*. Do not confuse it with *retrograde motion*, which has to do with the apparent motion of the position of a planet in the sky as seen from the Earth.

Figure 10–23 The orbits of the planets, with the exception of Pluto, have only small inclinations to the ecliptic plane.

PLUTO

MERCURY
VENUS
SATURN
MARS
NEPTUNE
JUPITER
URANUS
EARTH
ECLIPTIC

B.C.   by johnny hart

I WONDER WHAT THE BACKSIDE OF THE EARTH LOOKS LIKE.

THERE IS NO BACKSIDE, STUPID,... IT HAS A COMPLETE REVOLUTION EVERY 24 HOURS.

YOU MEAN ROTATION, DON'T YOU?

...THAT TOO.

Figure 10–24 It is all too easy to "misspeak" when discussing rotation, which refers to a body turning on its axis, and revolution, which refers to a body orbiting around another body.

10.5 BODE'S "LAW"

For over two hundred years, a numerical relation called "Bode's law" has been known to give the approximate distances of the planets from the Sun, though it has never been theoretically understood.

Bode's law says to write down first 0, then 3, and then keep doubling the previous number: 6, 12, 24, etc. Now add 4 to each number to get the series 4, 7, 10, 16, 28, and so on. Then divide each number by 10. The results: .4, .7, 1.0, etc., give the approximate radii of the planetary orbits in astronomical units (Table 10–1).

The relation was first published anonymously by Titius of Wittenberg in 1766 in a translation of a book by a Swiss scientist, but Bode's name became attached to the relation a half-dozen years later when Bode, director of the Berlin Observatory, popularized it. Sometimes we now call it the Titius-Bode law. The observational tests of Bode's law are connected with the outer planets and asteroids, which will be discussed in Chapters 17 and 18. When Uranus was discovered by William Herschel in England in 1781, its orbital distance fit Bode's law and the discovery spurred theoretical work to derive the "law." Bode's law also predicted a planet between Mars and Jupiter. The asteroids, some of which are at this distance, as predicted by Bode's law, began to be discovered twenty years after Uranus. The fact that there were many asteroids (minor planets), instead of one major planet, was disturbing to many. To have a perspective in time, let us realize that this active discussion of Bode's law was going on in Europe at the time of the American Revolution.

Figure 10–25 An ice skater draws her arms in, in order to redistribute her mass. Because her angular momentum is conserved, she begins to spin faster in order to compensate for the new mass distribution. Dorothy Hamill is shown here during her gold medal performance in the free style event at the Innsbruck Olympics in 1976.

TABLE 10–1 BODE'S LAW

Planet	Distance by Bode's Law (A.U.)	Actual Mean Distance (A.U.)
Mercury	0.4	0.39
Venus	0.7	0.72
Earth	1.0	1.00
Mars	1.6	1.52
Asteroids	2.8	–
Jupiter	5.2	5.20
Saturn	10.0	9.54
Uranus	19.6	19.2
Neptune	38.8	30.6
Pluto	77.2	39.4

But Neptune and Pluto, which were discovered later, do not fit Bode's law, and no mathematical derivation of Bode's law has ever been found. (Some numerical calculations with large computers do show planetary spacings that follow different rules.) Many astronomers now think that Bode's law is merely a numerical coincidence, devoid of deeper meaning.

It is no surprise that the planetary orbits have spaces between them such that we don't find two planets in orbits very close to each other. Most matter that would exist between the planets would be swept up by the gravitational forces of the planets as they go around the Sun.

10.6 THEORIES OF COSMOGONY (*Optional*)

The solar system exhibits many regularities, and theories of its formation must account for them. The orbits of the planets are almost, but not quite, circular, and all lie in essentially the same plane. All the planets revolve around the Sun in the same direction, which is the same direction in which the Sun rotates. Moreover, almost all the planets and planetary satellites rotate in that same direction. Some planets have families of satellites that revolve around them in a manner similar to the way that the planets revolve around the Sun. And cosmogonical theories must explain the spacing of the planetary orbits and the distribution of planetary sizes and compositions.

Note that "cosmogony" is pronounced with a hard "g."

René Descartes, the French philosopher, was one of the first to consider the origin of the solar system in what we would call a scientific manner. In his theory, proposed in 1644, circular eddies of all sizes called vortices were formed in a primordial gas at the beginning of the solar system and eventually settled down to become the various celestial bodies.

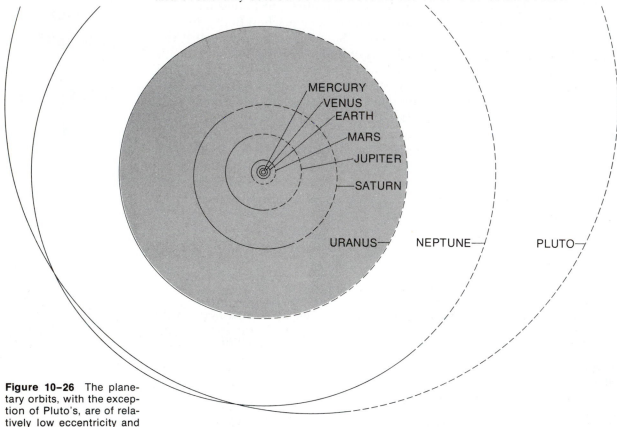

Figure 10–26 The planetary orbits, with the exception of Pluto's, are of relatively low eccentricity and appear at regular spacings. The portion of the orbits that are below the ecliptic plane are shown with dotted lines.

After Newton proved that Descartes's vortex theory was invalid, it was 60 years before the next major developments. The Comte de Buffon suggested in France in 1745 that the planets were formed by material ejected from the Sun when what he called a "comet" hit it. (At that time, the composition of comets was unknown, and it was thought that comets were objects as massive as the Sun itself.) Later versions of Buffon's theory, called *catastrophe theories*, followed a similar line of reasoning although they spoke explicitly of a collision with another star. The requirement of a collision could be loosened by postulating that the material for the planets was drawn out of the Sun by the gravitational attraction of a passing star. This latter possibility is called a *tidal theory*.

But catastrophe and tidal theories are currently out of fashion, for they predict that only very few planetary systems would exist, since calculations have been made that show that only very few stellar collisions or near-collisions would have taken place in the lifetime of the galaxy.

The theories of cosmogony that astronomers now tend to accept stem from another 18th century idea. Immanuel Kant, the noted German philosopher, suggested in 1755 that the Sun and the planets were formed in the same way as each other. In 1796, the Marquis Pierre Simon de Laplace, the French mathematician, independently advanced a similar kind of theory to Kant's when he postulated that the Sun and the planets all formed from a spinning cloud of gas called a *nebula*. Laplace called this the *nebular hypothesis*, using the word "hypothesis" because he had no proof that it was correct. The spinning gas supposedly threw off rings that eventually condensed to become the planets. But though these beginnings of modern cosmogony were laid down in the time of Benjamin Franklin and George Washington, the theory still has not been completely understood or quantified, for not all the stages that the primordial gas would have had to follow are understood.

Further, some of the details of Laplace's theory were later thought for a time to be impossible. For example, it was calculated in the last century

Figure 10–27 The leading model for the formation of the solar system has the protosolar nebula condensing and, between stages C and D, contracting to form the protosun and a large number of small bodies called planetesimals. The planetesimals clumped together to form protoplanets (D), which in turn contracted to become planets. Some of the planetesimals may have become moons or asteroids (which are discussed in Section 18.3).

that the planets would not have been able to condense out of the rings of gas, and also that they could not have been set to rotate as fast as they do.

So for a time, the nebular hypothesis was not accepted. But the current theories of cosmogony again follow Kant and Laplace (Fig. 10–27). In these *nebular theories* the Sun and the planets condensed out of what is called a *primeval solar nebula.* Some five billion years ago, billions of years after the galaxies began to form, smaller clouds of gas and dust began to contract out of interstellar space. Similar interstellar clouds of gas and dust can now be detected at many locations in our galaxy.

Because of random fluctuations in the gas and dust from which it formed, the primeval solar nebula probably would have had a small net spin from the beginning. As it contracted, it would have begun to spin faster because of the conservation of angular momentum (the same reason that the ice skater spins faster). Gravity would have contracted the spinning nebula into a disk, for in the directions perpendicular to the plane in which the nebula is rotating there was no force to oppose gravity's pull.

Perhaps rings of material were left behind by the solar nebula as it contracted towards its center. Perhaps there were additional agglomerations beside the *protosun,* the part of the solar nebula that collapsed to become the Sun itself. These additional agglomerations, which may have been formed from interstellar dust, would have grown larger and larger. Millimeter-size particles would have clumped into larger bodies and then larger still; the size of the clumps would have increased until the bodies were kilometers and then thousands of kilometers across. The intermediate bodies, hundreds of kilometers across, are called *planetesimals.*

The planetesimals combined under the force of gravity to form *protoplanets.* These protoplanets may have been larger than the planets that resulted from them because they had not yet contracted, though gravity would ultimately cause them to do so. Some of the larger planetesimals may themselves have become moons.

As the Sun condensed, the energy it gained from its contraction went to heat it until its center reached the temperature at which nuclear fusion reactions began taking place. The planets, on the other hand, were simply not massive enough to heat up sufficiently to have nuclear reactions start.

There are many possibilities that bedevil cosmogonical research. Some of the planets and moons may not have been formed in their present configurations. Perhaps the Earth captured another protoplanet, which became the Moon. Perhaps Pluto was ejected from an orbit around Neptune into an orbit as an independent planet. Perhaps gravitational encounters put some of the planets and moons into retrograde rotation.

The cosmogonical ideas we have discussed are all speculative, for the nature of the formation of the solar system is not definitively understood, nor are the details known. Professional astronomers join students of astronomy in wishing that the theory could be laid out neatly and conclusively for all to see, but we are not yet able to do that.

It is, of course, always best to test a scientific theory by seeing if it explains facts that were not known when the theory was advanced. With the recent exploration of the planets from space, we are finding tremendous amounts of basic information and checking to see if this information was predicted by any of the theories. As we proceed, now, to discuss the planets individually and in detail, we must always be aware when a particular set of observations that we discuss is of special significance in the understanding of our origins.

We may even have an example of a planetary system in formation to study. It has been realized that a young main-sequence star in the constellation Cygnus is 10 times brighter than it should be for its spectral type, which indicates that something else besides the star is contributing light. (The star is called MWC 349, after its number in the Mount Wilson Catalogue.) The star has been observed not only in the visible from the ground, but also

The New York Times

Again

There it is, going on 10,000 years ago, right in front of our eyes, and there's nothing we can do about it. Even our highest-pitched, deepest-probing laser shouts would not be heard. If they were, they'd arrive too late to do anyone any good. Never was it more plain that what's past is present and what's present is past.

We speak of the glow in the constellation Cygnus that astronomers believe may be a rerun of the process whereby our own solar system was formed 4.6 billion years ago. Unfortunately, we shall never know whether events evolved out there as they have in the heavens hereabouts. But it is nonetheless intriguing to ponder a Creation elsewhere in the universe, to wonder what form things will take, or have taken. Will it be Eden before the snake or the other way around?

And what name has been given to this new Genesis, this thrilling solar rebirth, this evocation of the cosmic conundrum whence we came? MWC 349. Identified by numbers, even *before* birth. What is the universe coming to?

Figure 10–28   (A) An artist's conception of MWC 349, showing the disk which is believed to be present and from which planets may be forming. (B) An editorial from *The New York Times*, June 23, 1977.

in the infrared from an airborne observatory (actually an instrumented airplane). Recent theoretical analysis of the results, particularly of the spectra, suggests that we are observing a hot disk of gas and dust, just like a protostar before the protoplanets form, seen face on (Fig. 10–28). Each month, the brightness declines by about 1 per cent, and has been doing so for 40 years, just as we would expect if the disk is condensing into planets. Since MWC 349 is a star of spectral type O6, it should evolve substantially more rapidly than did the Sun. Two other cases with similar pre-planetary disks are also under study.

SUMMARY AND OUTLINE OF CHAPTER 10

Aims: To discuss the scale of the solar system, the history of our understanding of the heliocentric theory, Kepler's laws, and the origin of the solar system

The scale of the solar system and the planets themselves
The phases of the moon and lunar eclipses (Section 10.1)
The apparent motions of the planets; retrograde motion (Section 10.2)
The development of our knowledge of the solar system (Section 10.2)
    Earth-centered system: Aristotle and Ptolemy
        Epicycles and deferents
    Heliocentric system: Copernicus
        De Revolutionibus, 1543
        Retrograde motion explained as a projection effect
    First telescopic observations: Galileo
    Improved positional observations and deductions from them: Tycho
      and Kepler
Mathematical and physical deductions
    Kepler's three laws and how to use them
        (1) Planets orbit the Sun in ellipses
        (2) Law of equal areas
        (3) Period^2 proportional to distance^3
           (Period is a time interval; think of Times Square, New York)
    Newton and the universality of gravity
The ecliptic; inclinations of the orbits of the planets (Section 10.3)
    Pluto has largest inclination by far
    Other planets have only slight inclinations
Conservation of angular momentum (Section 10.3)
    All planets revolve in same direction
    Most planets rotate in the same sense
Bode's "law" (Section 10.4)
    0, 3, 6, 12, and so on
    Add 4 and divide by 10
    Not known if the "law" has basic significance
Theories of cosmogony (Section 10.5)
    Catastrophe theories
    Nebular theories
        Kant and Laplace
        Newer versions: protosun, planetesimals, and protoplanets

QUESTIONS

1. What are the features that distinguish the terrestrial from the giant planets? What features does Pluto have in common with either group?

2. Suppose that you live on the Moon. Sketch the phases of the Earth that you would observe for various times during the Earth's month.

3. If you lived on the Moon, would the motion of the planets appear any different than from Earth?

4. If you lived on the Moon, how would the position of the Earth change in your sky over time?

5. If you lived on the Moon, what would you observe during an eclipse of the Moon. How would an eclipse of the Sun by the Earth differ from an eclipse of the Sun by the Moon that we observe from Earth?

6. If you lived on Saturn, which planets would appear to go through phases? Which planets would exhibit retrograde motion?

7. Discuss the following statement: "With the addition of epicycles, the geocentric theory of the solar system could be made to agree with observations. Since it was around first, and therefore better known, it should have been kept."

8. Do Kepler's laws permit circular orbits?

9. At what point in its orbit is the Earth moving fastest?

10. Use Kepler's 3rd law and the fact that Mercury's orbit has a semimajor axis of 0.4 A.U. to deduce the period of Mercury. Show your work.

11. Use the data in Appendix 5 to show for which of the inner six planets the square of the period does not equal the cube of the semi-major axis to the accuracy given.

12. The occupants of Planet X note that they are 1 greel from their sun (a greel is the Planet X unit of length), and they orbit with a period of 1 fleel. They observe Planet Y which orbits in 8 fleels. How far is Planet Y from their sun (in greels)?

13. Would an observer on the Sun see the Earth moving along the ecliptic? Would the apparent path of the Earth across the sky be east to west, or west to east? Explain.

14. Explain how conservation of angular momentum applies to a diver doing somersaults or twists. What can divers do to make sure they are vertical when they hit the water?

15. If two planets are of the same mass but different distances from the Sun, which will have the higher angular momentum around the Sun? (Hint: Use Kepler's third law.)

16. If the planets condensed out of the same primeval nebula as the Sun, why didn't they become stars?

11

The Moon

The Earth's nearest celestial neighbor—the Moon—is only 380,000 km (238,000 miles) away from us on the average, close enough that it appears sufficiently large and bright to dominate our nighttime sky. The Moon's stark beauty has called attention to it since the beginning of history, and studies of the Moon's position and motion led to the earliest consideration of the solar system, to the prediction of tides, and to the establishment of the calendar.

11.1 THE APPEARANCE OF THE MOON

The fact that the Moon's surface has different kinds of areas on it is obvious to the naked eye. Even a small telescope—Galileo's, for example—reveals a surface pockmarked with craters. The *highlands* are heavily cratered; other areas, called *mare* (pronounced mar′-ā; plural, *maria*, pronounced mar′ē-a), are relatively smooth, and indeed the name comes from the Latin word for sea (Fig. 11–2). But there are no ships sailing on the lunar seas and no water in them; the Moon is a dry, airless, barren place. The Moon's mass is only 1/81 that of the Earth, and the gravity at its surface is only 1/6 that of the Earth. Any atmosphere and any water that may once have been present would long since have escaped into space. The Moon is 3476 km (2160 miles) in diameter, about one-fourth the diameter of the Earth. No other moon in the solar system is such a large fraction of the size of its planet.

Many types of structure are visible on the Moon besides the smooth maria and the cratered highlands. In the highlands, there are *mountain ranges* and *valleys* (Fig. 11–2A). Lunar *rilles* are clefts that can extend for hundreds of kilometers along the surface (Fig. 11–2B). Some are relatively straight while others are sinuous. On the other hand, raised *ridges* also occur. The craters themselves come in all sizes, ranging from as much as 295 km across for one called Bailly down to tiny fractions of a millimeter. Crater *rims* can be as much as several kilometers high, much higher than the rim of the Grand Canyon stands above the Colorado River. Examples of all these features will be seen in the photographs that illustrate this chapter.

Figure 11–1 The crescent phase of the Moon 3 days after the new moon.

Scientist-astronaut Harrison Schmitt collecting small rocks and rock chips with a lunar rake during the Apollo 17 mission to the Taurus-Littrow region of the Moon.

Figure 11–2 Ground based views of the Moon. (*A*) The northern region, with Mare Serenitatis at center, and including the Caucasus Mountains and the Alpine Valley. (*B*) Ridges and the Ariadaeus Rille. (*C*) In order to show the whole Moon, but still show the detail that does not show up well at full moon, the Lick Observatory has put together this composite of first and third quarters. Note the dark maria and the lighter, heavily cratered highlands. Two young craters, Copernicus and Kepler, can be seen to have rays of light (that is lighter than the background) material emanating from them.

A

B

MARE NUBIUM

MARE HUMORUM

MARE NECTARIS

MARE FOECUNDITATIS

MARE TRANQUILLITATIS

Copernicus

Kepler

MARE VAPORUM

OCEANUS PROCELLARUM

MARE CRISIUM

MARE SERENITATIS

MARE IMBRIUM

MARE FRIGORIS

Plato

C

When the Moon is full, it is especially bright in the nighttime sky — bright enough, in fact, to cast shadows or even to read by. But full moon is a bad time to try to observe surface structure, for we are looking along the same line that is being traveled by the sunlight, and any shadows we see are short. When the Moon is in a crescent phase or even a half-moon, however, the part of the lunar surface where the shadows are long faces us. At those times, the lunar features stand out in bold relief (Fig. 11–2).

Shadows are longest near the *terminator*, the line separating day from night. Since the Moon goes around the Earth and returns to the same position with respect to the Earth and Sun once every 29 1/2 days (this is the

---

Box 11.1   Sidereal and Synodic Periods

Sidereal: with respect to the stars. For example, the Moon revolves around the Earth in about 27 1/3 days with respect to the stars.

Synodic: with respect to some other body, usually the Earth. For example, by the time that the Moon has made one sidereal rotation around the Earth, the Earth has moved about 1/13 of the way around the Sun (about 27°). Thus the Moon must travel a little farther to catch up (Fig. 11–3). The synodic period of the Moon, the synodic month, is about 29 1/2 days.

Another example concerns revolution rather than rotation. Consider Mars, which has a sidereal period of revolution around the Sun of 687 days. However, only every 780 days does the Earth, which is traveling in orbit faster than Mars, catch up with Mars. (This is equivalent to stating that it is 780 days from opposition to opposition.) The synodic period of Mars is thus 780 days.

---

synodic revolution period of the Moon), the terminator moves around the Moon in that time. The phases repeat with this period, which is called the *synodic month*. Most locations on the Moon are thus in about 15 days of sunlight, during which time they become very hot, and then 15 days of darkness, during which time their temperature drops to as low as 160 K.

The Moon rotates on its axis at the same rate as it revolves around the Earth, always keeping the same face in our direction. The Earth's gravity has locked the Moon in this pattern, interacting with a bulge in the distribution of the lunar mass to prevent the Moon from rotating freely.

As a result of this interlock, we always see the same side of the Moon from our vantage point on Earth. Actually, we can see a bit more than half. For one thing, the Moon's speed in its orbit varies because of the orbit's eccentricity and Kepler's second law, but its rate of spin is constant, so sometimes we can see around one side of the Moon or the other. Also, the Moon's equator is inclined by about 1½° to the plane of its orbit, which is inclined about 5° to the ecliptic, so we can alternately see over the lunar north pole and south pole as the Moon moves around the Earth. These motions are called *librations* (Fig. 11–4); as a result we can see 59 per cent of the lunar surface. Another contribution to libration comes from the fact that because of the Earth's diameter we look from different vantage points in the course of each day. An actual small variation of the Moon's period of rotation is a fourth, minor contribution to libration.

Figure 11–3 After the Moon has completed one rotation on its axis with respect to the stars, which it does in 27⅓ days, it has moved from A to B. A given feature has still not swung far enough around to again be in the same position with respect to the Sun because the Earth has revolved one month's worth around the Sun. It takes about an extra two days for the Moon to complete its rotation with respect to the Earth, which gives a synodic rotation period of about 29½ days. The extra angle it must cover is shaded. In the extra two days, both Earth and Moon have continued to move around the Sun (toward the left in the diagram).

Figure 11–4 Libration allows us to see part way around the far side of the Moon. From the Earth, we can see 59 per cent at one time or another. Note, for example, the position of Mare Crisium on the lower left of both photographs.

11.2 LUNAR EXPLORATION

The space age began on October 4, 1957, when the U.S.S.R. launched its first Sputnik (the Russian word for *satellite*) into orbit around the Earth. The shock galvanized the American space program into action and within months American spacecraft were also in Earth orbit. We have seen in our discussions of stars, and will continue to see in our discussions of the Milky Way Galaxy and of other galaxies, how the ability to observe from space has benefited astronomy.

In the case of the Moon and planets, the ability to travel into space did not merely free us from the obscuration of the Earth's atmosphere but also allowed us to explore the planets directly. The Moon, as the closest celestial body, was obviously the place to begin.

In 1959, the Soviet Union sent its Luna 3 spacecraft around the Moon; it radioed back the first murky photographs of the Moon's far side. Now that we have high-resolution maps of most of the lunar surface, it is easy to forget how big an advance that was.

In 1961, President John F. Kennedy announced that it would be a national goal of the United States to put a man on the Moon, and bring him safely back to Earth, by 1970. This grandiose goal led to the largest coordinated program in the history of the world, and had an outstandingly successful conclusion.

The American lunar program, under the direction of the National Aeronautics and Space Administration (NASA), proceeded in gentle stages. On one hand, the ability to carry out manned space flight was developed with single-astronaut sub-orbital and orbital capsules, called Project Mercury, and then with two-astronaut orbital spacecraft, called Project Gemini. Simultaneously, on the other hand, a series of unmanned spacecraft were sent to the Moon.

The manned and unmanned trains of development came together with Apollo 8, which circled the Moon on Christmas Eve, 1968, and returned to Earth. The next year, Apollo 11 brought humans to land on the Moon for the first time. It went into orbit around the Moon after a three-day journey from Earth, and a small spacecraft called the Lunar Module (LM) separated from the larger Command Module. On July 20, 1969 — a date that from the long-range standard of history may be the most significant of the last millennium — Neil Armstrong and Buzz Aldrin left Michael Collins orbiting in the Command Module and landed on the Moon (Fig. 11–5). In the preceding days there had been much discussion of what Armstrong's historic first words should be, and millions listened as he said "One small step for a man, one giant leap for mankind."

The Lunar Module carried a variety of experiments, including devices to test the soil, a camera to secure close-up stereo photos of the lunar soil, a sheet of aluminum with which to capture particles from the solar wind, and a seismometer. Later Lunar Modules carried additional experiments, some even including a vehicle (Fig. 11–6). In all, six Apollo missions carried people to the Moon. Unfortunately, the missions seemed to get confused in the popular mind with just one of the experiments — the collection of rocks. These rocks and dust, returned to Earth for detailed analysis, were indeed important, but represented only a fraction of the missions.

At present, the United States has no definite plans of any kind for sending manned or even unmanned spacecraft to the Moon, though there is a project being discussed to send an unmanned vehicle into a polar orbit around the Moon in order to map regions at high lunar latitudes.

The Soviet Union, which has sent three unmanned spacecraft to land on the lunar surface, collect lunar soil, and return it to Earth, seems to be continuing lunar work. Its first two round-trip spacecraft each collected a few grams of lunar soil. Luna 24, which went to the Moon in August 1976, drilled to a depth of 1/2 meter (2 feet) below the lunar surface and brought this long, thin cylinder of material back to Earth. In 1970 and 1973, two

Figure 11–5 Neil Armstrong, the first person to set foot on the Moon, took this photograph of his fellow astronaut Buzz Aldrin climbing down from the Lunar Module of Apollo 11 on July 20, 1969. The site is called Tranquility Base, as it is in the Sea of Tranquility (Mare Tranquillitatis).

Figure 11–6 *(A)* Eugene Cernan riding on the Lunar Rover during the Apollo 17 mission. The mountain in the background is the east end of the South Massif. *(B)* Drawing by Alan Dunn; © 1971 *The New Yorker Magazine, Inc.*

A

B

TABLE 11-1 MISSIONS TO THE LUNAR SURFACE AND BACK TO EARTH

Apollo 11	U.S.A.	1969	manned
Apollo 12	U.S.A.	1969	manned
Luna 16	U.S.S.R.	1970	unmanned
Apollo 14	U.S.A.	1971	manned
Apollo 15	U.S.A.	1971	manned
Luna 20	U.S.S.R.	1972	unmanned
Apollo 16	U.S.A.	1972	manned
Apollo 17	U.S.A.	1972	manned
Luna 24	U.S.S.R.	1976	unmanned

other Soviet spacecraft had carried remote-controlled rovers, Lunokhods 1 and 2, that traveled over 10 or so kilometers of the lunar surface over a period of many months each.

11.3 THE RESULTS FROM APOLLO

The kilometers of film exposed by the astronauts, the 382 kilograms (843 pounds) of rock brought back to Earth, the rolls and rolls of magnetic tape recording the results from lunar seismographs, and other data, all studied by hundreds and hundreds of scientists from countries all over the Earth, have led to new views of several basic questions. They have raised many new questions about the Moon and the solar system as a whole. Let us consider (a) the composition of the lunar surface, (b) the chronology of the Moon, (c) the origin of the lunar craters, (d) the structure of the lunar interior and whether it is hot or cold, and (e) how and where the Moon was formed.

11.3a The Composition of the Lunar Surface

The types of rocks that were encountered on the Moon arc types that are familiar to terrestrial geologists. In the maria, the rocks are mainly *basalts* (a type of rock that results from the cooling of molten material like lava). The highland rocks are *anorthosites*. (Anorthosites—defined by their particular combination of minerals—are rare on Earth, though the Adirondack Mountains are made of them. Anorthosites, though they have also cooled from molten material, have done so under different conditions than basalts and have taken longer to cool.)

Some of the rocks in each type of location are *breccias*, mixtures of fragments of several different types of rock that have been compacted and welded together. The ratio of breccias to crystalline rocks is much higher in the highlands than in the maria because of the much greater number of impacts in the highlands.

The astronauts also collected some *lunar soils*, bits of dust plus larger fragments from the Moon's surface. Some of these proved, on examination through the microscope, to contain a type of structure that is not common on Earth. Small glassy globules (Fig. 11–7) are mixed in. In some of the soils (Color Plate 28) glass coatings and glass globules were present; the bubbly globules in these cases undoubtedly resulted from the melting of rock during its ejection from the site of a meteorite impact and the subsequent cooling of the molten material. In some cases, where most of the soil consisted of glass beads, a volcanic origin is probable.

Figure 11–7 An enlargement of a glassy spherule from the lunar dust collected by the Apollo 11 mission. The shape and composition of the spherule indicate that it was created when a meteorite crashed into the Moon, melting lunar material and splashing it long distances. The glassy bead is enlarged 3200 times in this photograph taken with a scanning electron microscope.

Almost all of these rocks and soils have lower proportions of elements with low melting points (*volatile* elements) than does the Earth. On the other hand, there are relatively high proportions of elements with high melting points (*refractory* elements) like calcium, aluminum, and titanium compared to terrestrial abundances. Titanium, which is only a minor constituent of Earth rocks, amounted to 10 per cent of some lunar samples. Of elements that are even rarer on Earth, the abundances of uranium, thorium, and the rare-earth elements are also increased. On the other hand, analyses of basalts from the maria, rocks that may have been formed in the lunar interior instead of on the surface, are not so very rich in such refractory elements as aluminum and calcium. In sum, the Moon and the Earth now seem to be more similar than had been thought for the last few years, though some differences are real. Since the comparison of abundances on the Earth and the Moon has changed drastically over the last few years, it seems wise not to be too confident that the accepted abundance values won't change again.

Something that seems fairly definite is that none of the lunar rocks contain any trace of water bound inside their minerals. This ends all hope that water existed on the Moon at any time in the past, and so seems to eliminate the possibility that life evolved there.

Elements 57 to 71 (Appendix 4) are known as rare earths.

Figure 11–8 This view from the orbiting Apollo 15 Command Module shows a smaller crater, Krieger B, superimposed on a larger crater, Krieger. Obviously, the smaller crater is younger than the larger one. Several *rilles* (clefts along the lunar surface that can be hundreds of kilometers in length) and *ridges* are also visible. Sometimes the areas in the centers of craters, the *floors*, are smooth but sometimes craters have *central peaks*.

11.3b Lunar Chronology

One way of dating the structures on the surface of a moon or planet is to observe the number of craters on them. If we assume that the events that cause craters—whether they are impacts of meteors or the eruptions of volcanoes—continue over a long period of time and that there is no strong erosion, surely those locations with the greatest number of craters must be the oldest. Relatively smooth areas—like the maria—must have been covered over with volcanic material at some relatively recent time (which is still billions of years ago). Even from the Earth we can count the larger craters. And when one crater is superimposed on another (Fig. 11–8), we can be certain that the superimposed crater is the younger one.

A few craters on the Moon, notably Copernicus (Figs. 11–2 and 11–9), have thrown out obvious rays of lighter-colored matter. Since these rays extend over other craters, they are younger than these other craters. The youngest rayed craters may be very young indeed—perhaps only a few hundred million years. The rays darken with time, so rays that may have once existed near other craters are now indistinguishable from the rest of the surface. Rays are visible for about 3 billion years.

The oldest rocks that were found at the locations sampled on the Moon were formed 4.42 billion years ago. The youngest rocks were formed 3.1 billion years ago.

There are overall differences between the ages of highland and maria rocks. The highland rocks were formed between about 3.9 and 4.4 billion years ago, and the maria between 3.1 and 3.8 billion years ago. One highland rock even showed signs of having been melted and re-formed twice— mainly at 3.93 billion years but with some crystals that were 4.4 billion years old included. Several highland rocks are exactly the same age, all 4.42 billion years old. So 4.42 billion years ago may therefore have been the origin of the lunar surface material, that is, the time when it last cooled.

One can use information like the above to build a model of what happened on the Moon. For example, potentially the lunar surface either could have been melted by the heat of impact of meteorites or could have been formed by lava from below the surface, but it would have taken at least

Meteorites hit the Moon with such high velocities that huge amounts of energy are released at the impact. The effect is that of an explosion, as though it had been TNT or an H-bomb exploding.

Figure 11–9 The crater Copernicus, seen in this ground-based photograph, has rays of material that is relatively light in shade emanating from it. This light-colored material was thrown out radially when the meteorite that formed the crater impacted.

half a billion years for the internal heat to generate lava, and the highland rocks do not seem to be formed from cooled lava. So any features that were formed before the origin of the Moon plus half a billion years would have had to have been heated by meteoritic impact or been molten because of the heat that resulted when the Moon was formed.

All the observations can be explained on the basis of the following general picture. The Moon formed 4.6 billion years ago. We know that the top 100 km or so of the surface was molten after about 200 million years. This could have been caused by the original heat or by an intense bombardment of meteorites (or debris from the period of formation of the Moon and the Earth) that melted the surface entirely. The presence of radioactive aluminum (Al^{26}) also provided heat. Then the surface cooled definitively. From 4.2 to 3.9 billion years ago, bombardment (perhaps by planetesimals) caused most of the craters we see today. About 3.8 billion years ago, the interior of the Moon heated up sufficiently (from radioactive elements inside) that vulcanism began; lava flowed on the lunar surface and filled the largest basins that resulted from the earlier bombardment, thus forming the maria. By 3.1 billion years ago, the era of vulcanism was over, and the Moon has been geologically pretty quiet since then.

Up to this time, the Earth and the Moon shared similar histories. But active lunar history stops about 3 billion years ago, while the Earth continued to be geologically active. Further, because the Earth's interior continued to send gas into the atmosphere and because the Earth's higher gravity retained that atmosphere, the Earth developed conditions in which life evolved. The Moon, because it is smaller than the Earth, presumably lost its heat more quickly and also generated a thicker crust.

Almost all the rocks on the Earth are younger than 3 billion years of age; erosion and the remolding of the continents as they move slowly over the Earth's surface, according to the theory of *plate tectonics* (known colloquially as *continental drift*), have taken their toll. The oldest single rock ever discovered on Earth has an age of 3.7 billion years. So we must look to extraterrestrial bodies—the Moon or meteorites—that have not suffered the effects of plate tectonics or erosion by such processes as occur in the presence of water or an atmosphere. Only in locations that did not suffer the effects of major erosive processes can we study the first billion years or more of the solar system.

11.3c The Origin of the Craters

The debate over whether the craters were formed by meteoritic impact or by volcanic action began in pre-Apollo times. The results from Apollo indicate that most craters resulted from meteoritic impact, even though some very small fraction may represent the effects of vulcanism. It is almost as though the wrong question was being asked all along. The question should not have been, "Are the craters volcanic?" but rather, "Is there evidence of vulcanism?" The answer to the latter question is "yes." Only a very few craters result from vulcanism, but there are many other signs of volcanic activity, including the lava flows that filled the maria and modifications of many impact craters by subsequent vulcanism. In any case, over 99 per cent of the lunar craters that we see with our telescopes were caused by the impacts of meteorites.

The photographs of the far side of the Moon have shown us that the near and far hemispheres are quite different in overall appearance. The maria that are so conspicuous on the near side are almost absent from the

Figure 11–10 The far side of the Moon looks very different from the near side in that there are few maria (compare with Fig. 11–2). This photograph was taken from Apollo 16, and shows some of the near-side maria at upper left.

far side, which is cratered all over (Fig. 11–10). Since the many ways that the Earth affects the Moon extend even to the distribution of mass in the outer lunar layers, the asymmetry in the distribution of maria may arise inside the Moon itself. Indeed, once any asymmetry is set up, then the Earth's gravity would lock one side toward us. It is interesting to note that not only the Moon but also the Earth, Mars, and probably Mercury have asymmetric hemispheres. The cause of this is unknown.

11.3d   The Lunar Interior

Before the Moon landings, it was widely thought that the Moon was a simple body, with the same composition throughout. But we now know it to be a *differentiated* body like the planets, with a metallic *core* at its center, a silica-rich *mantle* making up the rest of the interior, and a *crust* at the surface of lighter material. The crust is perhaps 65 kilometers thick on the near side and twice as thick on the far side; this asymmetry may explain the different appearances of the near and far sides.

The interior of a moon or planet cannot be observed directly, though there are many indirect methods of deduction that can be brought to bear. One of the best methods has scientists using the seismograph, the device employed on Earth to detect earthquakes. The speeds at which different

Differentiated bodies have layers of materials in different states and of different compositions, as opposed to bodies that are homogeneous throughout.

types of waves are transmitted through the lunar interior tell us of the condition of the interior.

The instruments that were left on the Moon were shut off by NASA as an economy measure in 1977.

Working seismometers were left by Apollo astronauts at four widely spaced locations on the Moon. The seismometers enabled us to locate the origins of the thousands of weak moonquakes that occur each year—almost all of which are magnitudes 1 to 2 on the Richter scale used on Earth. Perhaps three moonquakes per year reach Richter 4, a strength which on Earth could be felt but which would not cause damage. On one occasion (July 21, 1972) a meteorite hit the far side of the Moon and generated seismic waves strong enough that we would have expected them to travel through to the near side, where the seismometers are all located. From the fact that one type of seismic wave—so-called *shear waves*—did not travel through the core while the other types did, most researchers deduce that the core is molten or at least plastic in consistency.

Another observation that can be brought to bear here is the fact that the Moon has no general magnetic field. If the interior of the Moon were molten, as is the Earth's interior, it might be expected that a more intense magnetic field than is detected would be set up. The weak magnetic field we do detect could be left over from a core that could have been molten long ago but has since cooled. We shall see in subsequent chapters that the unexpected detection of a magnetic field on Mercury has shown us that we do not understand as well as we had thought how planetary magnetic fields are formed.

Tracking the orbits of the Apollo Command Modules and other satellites that orbited the Moon also told us about the lunar interior. If the Moon were a perfect, uniform sphere of mass, the spacecraft orbits would have been perfect ellipses. Any deviation of the orbit from an ellipse can be interpreted as an effect of an asymmetric distribution of lunar mass. (The effects of the Earth, Sun, or other planets are relatively small.)

It might be noted that although the finding of mascons on the Moon was unexpected, there are two dozen locations on Earth where terrestrial "mascons" have even larger gravitational effects than does the largest lunar mascon.

One of the major surprises of the lunar missions was the discovery of *mascons*, regions of **mass con**centrations near and under most maria that lead to anomalies in the gravitational field, that is, deviations of the gravitational field from being spherical. The mascons may be lava that is denser than the surrounding matter. The existence of the mascons is evidence that the whole lunar interior is not molten, for if it were, then these mascons could not remain near the surface. However, the mascons could be supported by a crust of sufficient thickness.

In sum, we have contradictory evidence as to whether the lunar interior is hot or cold. Still, as a result of the meteorite impact detected with the seismic experiment, most scientists believe that the Moon's core is molten. If such a small body as the Moon formed with a hot interior, then we know that the planetesimals (Section 10.6) would have had to coalesce very rapidly into planets, since the heat did not have time to be radiated away. The study of the Moon gives us insight into the formation of the whole solar system.

11.3e The Origin of the Moon

Among the models that have been considered in recent years for the origin of the Moon are the following:

    (1) *Condensation*: the Moon was formed near to and simultaneously with the Earth in the solar system;

    (2) *Fission*: the Moon was separated from the material that formed the Earth; and

(3) *Capture*: the Moon was formed far from the Earth in another part of the solar system, and was later captured by the Earth's gravity.

Studies of the chemical composition of the lunar surface, and comparison with the composition of the terrestrial surface, have been important in narrowing down the possibilities. The mean lunar density of 3.3 grams/cm³ is close to the average density of the Earth's mantle, which had led to some belief in the fission hypothesis. However, detailed examination of the lunar rocks and soils indicates that the abundances of elements are sufficiently different to rule out the possibility that the Moon formed directly from the Earth. (Some minerals (Fig. 11–11) that do not exist on Earth have been discovered on the Moon, though this results from different conditions of formation rather than from abundance differences.)

Figure 11–11 A crystal of armalcolite (**Arm**strong-**Al**drin-**Col**lins, the crew of Apollo 11), examined under a polarization microscope. This mineral has been found only on the Moon.

Still, the fission hypothesis in the version that the Moon separated from the Earth very early on cannot be excluded. The length of the interval in which this could have happened, however—given the extreme ages of lunar soil and some of the rocks—was only a short one.

The third model, the capture model, is also currently thought by most astronomers to be unlikely. The conditions necessary for the Earth to capture such a massive body by gravity seem too unlikely for this to have occurred. However, the evidence against the model is not conclusive—one can't apply statistical methods to one example—and several possible ways have been suggested in which the Moon could have been captured by the Earth.

Although there are chemical differences between the Moon and the Earth, they are not so overwhelming as to exclude the condensation possibility, and most astronomers currently believe in some version of this model. The idea that the Moon has an iron core adds backing to this theory. In sum, it seems most probable that the Earth and the Moon formed near each other as a double planet, probably by accretion of planetesimals. The condensation model thus closely connects the question of the origin of the Moon to the larger question of the origin of the solar system, which we also think involved the accretion of planetesimals.

It had been hoped that landing on the Moon would enable us to clear up the problem of the lunar origin. But though the lunar programs have led to modifications and updating of the models described above, none of these models has been entirely ruled out. Nobody can say definitely which, if any, of the three is correct.

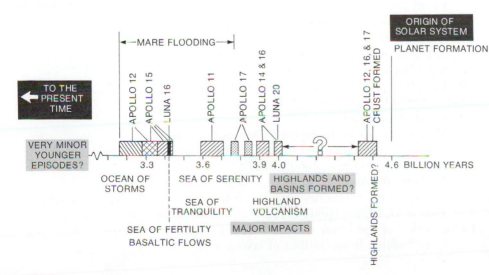

Figure 11–12 The chronology of the lunar surface, based on work carried out at the Lunatic Asylum, as the Caltech laboratory of Gerald Wasserburg is called.

11.4   THE VALUE OF MANNED LUNAR RESEARCH

Color Plates 26 to 31 are devoted to the Moon and lunar exploration.

The issue of manned versus unmanned space research has long been controversial among astronomers, who tend to favor unmanned research because its substantially lower cost would allow more work to be carried out. The success of the Apollo missions plus the ability of the Skylab astronauts to rescue a damaged spacecraft and operate complex equipment requiring on-the-spot judgment has tempered this opinion for many.

Further, it must be recalled that the decision to start the program of manned lunar exploration was largely the result of political factors, and the sending of a person to the Moon was a necessary part of the project. While it is true that much of the cost of the project was related to the cost of keeping humans alive and safe, it is also true that this part of the cost was justified on this ground and the equivalent amount of money would not have been put into straight astronomical research.

We have also succeeded through the Apollo program (and in the succeeding Skylab program, which used rockets and spacecraft developed and built for Apollo) in showing that humans can function in space and carry out useful work over long periods of time. Weightlessness and other effects of spaceflight (such as calcium depletion in bones) do not seem to be major barriers to space exploration. Perhaps this knowledge of our human capabilities will be the longest-lasting benefit of the Apollo and Skylab programs.

For those of us who remember the shock of Sputnik, and followed each small step farther and farther off the Earth's surface and into space, it is hard to believe that the era of manned lunar exploration not only has begun but also has already ended. At present we have no plans to send more people to the Moon. Perhaps by the turn of the 21st century, manned lunar exploration will resume. Twenty or thirty years from now, we may each be able to visit the Moon as researchers or even as tourists.

SUMMARY AND OUTLINE OF CHAPTER 11

Aims: To see how direct exploration of the Moon has increased our knowledge manyfold, but how such fundamental questions as how the Moon was formed remain unanswered

Lunar features (Section 11.1)
 Fixed: mare, highlands, mountains
 Dependent on sun angle: terminator
Revolution and rotation (Box 11.1)
 Sidereal: with respect to the stars
 Synodic: with respect to another body
 Libration allows us to see 59% of the surface from the Earth
Lunar exploration (Section 11.2)
 Unmanned series
 Manned series in Earth orbit: Mercury, Gemini
 Manned and unmanned lines of development merge in Apollo
 Six Apollo landings: 1969 to 1972
 No more manned landings planned
 Three Soviet unmanned spacecraft brought samples back to Earth
 Two Soviet Lunokhods roved many kilometers over the lunar surface
Composition of the lunar surface (Section 11.3a)
 Mare basalts: from cooling of lava

Highland anorthosites
More breccias—broken up and re-formed rock—in highlands
Soils

Chronology (Section 11.3b)
Dating by crater counting
Oldest soils: 4.6 billion years; oldest rocks: 4.2 billion years, almost back to origin of the solar system
Highland rocks (3.9–4.4 billion years) older than maria (3.1–3.8 billion years)
Probable model: Moon formed 4.6 b.y. At 4.4 b.y., the surface, which had been molten, cooled. Meteorite bombardment from 4.2 to 3.9 b.y. made most of the craters. Interior heated up from radioactive elements; the resulting vulcanism from 3.8 to 3.1 b.y. caused lava flows that formed the maria. Then lunar activity stopped.

Craters (Section 11.3c)
Almost all formed in meteoritic impact
Signs of vulcanism also present on surface, though only very few craters are volcanic
Far side has no maria, quite different in appearance from near side

Interior (Section 11.3d)
Moon is differentiated into a core, a mantle, and a crust
Seismographs left by Apollo missions reveal that interior is molten
Lack of lunar magnetic field, on the other hand, is an indication that interior is not molten, though seismic evidence seems definitive
Weak moonquakes occur regularly
Mascons discovered from their gravitational effects

Origin (Section 11.3e)
Condensation, fission, and capture theories
All still viable, with most astronomers tending to believe in a condensation theory

QUESTIONS

1. Compare the lengths of sidereal and synodic months. Which is longer? Why?

2. If the mass of the Moon is 1/81 that of the Earth, why is the gravity at the Moon's surface as great as 1/6 that at the Earth's surface?

3. To what location on Earth does the terminator on the Moon correspond?

4. Why is the heat flow rate related to the radioactive material content in the lunar surface?

5. What does cratering tell you about the age of the surface of the moon?

6. Why is it not surprising that the rocks in the lunar highlands are older than those in the maria?

7. Why are we more likely to learn about the early history of the Earth by studying the rocks from the Moon than those on the Earth?

8. What do the mascons tell us about the interior of the Moon?

9. Choose one of the proposed theories to describe the origin of the Moon and discuss the evidence pro and con.

10. (a) Describe the American and Soviet lunar explorations.
    (b) What can you say about future plans?

TOPIC FOR DISCUSSION

Discuss the scientific, political, and financial arguments for resuming manned exploration of the Moon.

Mercury is the innermost planet, and until very recently has been one of the least understood. Except for distant Pluto, its orbit around the Sun is the most elliptical (the difference between the maximum and minimum distances of Mercury from the Sun is as much as 40 per cent of the average distance, compared with less than 4 per cent for the Earth). Its average distance from the Sun is 58 million kilometers (36 million miles), which is 4/10 of the Earth's average distance. In order to use convenient numbers, we have defined the Earth's average distance from the Sun to be 1 Astronomical Unit. Thus Mercury is 0.4 A.U. from the Sun.

Since we on the Earth are outside Mercury's orbit looking in at it, Mercury always appears close to the Sun in the sky (Fig. 12–1). At times it rises just before sunrise, and at times it sets just after sunset, but it is never up when the sky is really dark. The maximum angle from the Sun at which we can see it is 28°, which means that the Sun always rises or sets within about two hours of Mercury's rising or setting. Of course, the difference in

In order to know when Mercury will be favorably located in the sky for you to observe it, you should watch the sky descriptions published in monthly magazines like Sky & Telescope *or* Astronomy *and in some local newspapers.*

Figure 12–1 Since Mercury's orbit is inside that of the Earth, Mercury is never seen against a really dark sky. A view from the Earth appears at right, showing Mercury and Venus at their greatest respective distances from the Sun. The open view of their orbits is an exaggeration; we actually see the orbits nearly edge on.

Mercury, photographed at a distance of 200,000 km from Mariner 10, revealed a cratered surface. The symbol for Mercury appears at the top of this page.

Figure 12-2 From the Earth, we cannot see much surface detail on Mercury. These views, among the best ever taken on Earth, were made by the New Mexico State University Observatory. Mercury was only 7.1 (*top*) and 5.1 (*bottom*) seconds of arc across.

time is usually even less than this maximum. Because of Mercury's closeness to the Sun in the sky, astronomers have never gotten a really good view of Mercury from the Earth, even with the largest telescopes. Many people have never seen it at all. (Copernicus's deathbed regret was that he had never seen Mercury.) The best photographs taken from the Earth show Mercury as only a fuzzy ball with faint, indistinct markings (Fig. 12-2). These photographs could be taken only when Mercury was low in the sky near sunrise and sunset. Consequently the light from Mercury had to pass obliquely through the Earth's atmosphere, making a long path through turbulent air. Hence the photographs are blurred.

12.1 THE ROTATION OF MERCURY (*Optional*)

Astronomers, from studies of drawings and photographs, did as well as they could to describe Mercury's surface. A few features could barely be distinguished, and the astronomers watched to see how long those features took to rotate around the planet. From these observations they decided that Mercury rotated in the same amount of time that it took to revolve around the Sun. It seemed reasonable that there could be a bulge in the distribution of mass of Mercury. The side that was bulging would be attracted to the Sun by gravity, locking the rotation to the revolution, just as the Moon is locked to the Earth. This is called *synchronous rotation*. An 88 day period of rotation for Mercury, matching its known 88 day period of revolution around the Sun, thus appeared in all the reference works and textbooks.

In the 1960's, a new method of studying some of the planets—radar—advanced to the point where it could be used to study Mercury. Generally, radio telescopes just sit and listen; that is, they receive signals from the planets and outer space. The intensity of the signal received in this passive manner indicated that the dark side of Mercury was hotter than had been thought when it had been assumed that one half of Mercury's surface was never struck by light from the Sun.

Radio signals can also be used in an active, radar mode, sending out signals and receiving their echoes. This is the same method that is routinely used to track airplanes. (*Radar* is an acronym for "**ra**dio **d**etection **a**nd **r**anging.") When radio radiation is sent out of a radio telescope, it travels across space in a widening cone. Sharp bursts of very intense radiation are sent out from a large radio telescope on Earth (the radiation travels, of course, at the speed of light), a small fraction hits the planet being observed and is bounced back toward Earth, and an even smaller fraction of the original burst is received by the same radio telescope on Earth, minutes later. Through complex computer studies of the returned signals, astronomers can make maps of Mercury's surface showing the height and roughness of each point.

There is even one more important item that can be told from this radar technique. Since Mercury is rotating, one side of the planet is always receding relative to the other. Studies of how the Doppler effect changes the radar signals as they bounce off the planet tell us the speed of the planet's rotation.

The results of the radar experiment were a surprise: scientists had been wrong about Mercury's rotation. It actually rotates in 59 days. This is about ⅔ of the period of its revolution, so the planet rotates three times for each two times it revolves around the Sun.

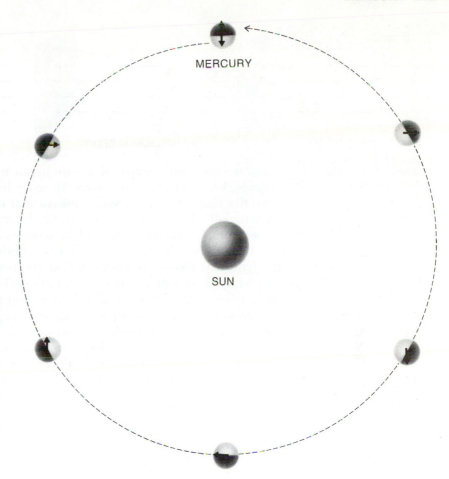

Figure 12–3 Follow the arrow that starts facing down toward the Sun in the image of Mercury at the top of the figure, as Mercury revolves along the dotted line. Mercury, and thus the arrow, rotate once with respect to the stars in 59 days, when Mercury has moved only 2/3 of the way around the Sun. (Our view is as though we were watching from a distant star.) Note that after one full revolution of Mercury around the Sun, the arrow is facing away from the Sun. It takes another full revolution, a second 88 days, for the arrow to be again facing the Sun. Thus the rotation period with respect to the Sun is twice 88, or 176, days.

Because rotation is always measured from the time any given star is directly overhead of a position on Mercury until it is again at the same position in the sky (a sidereal day), Mercury's rotation and revolution combine to give a value for the rotation of Mercury relative to the Sun (that is, a Mercurian solar day) that is neither the 59 day *sidereal rotation period* nor the 88 day period of revolution (Fig. 12-3). Actually, if we lived on Mercury we would measure each day and each night to be 88 Earth days long; we would alternately be fried and frozen for 88 Earth days at a time. For Mercury, the *solar rotation period* (not to be confused with the rotation period of the sun) is thus 176 days long.

No harm was done by the scientists' misconception of Mercury's rotational period for all those years, but the story teaches all of us a lesson: we should not be too sure of so-called facts, even when they are stated in all the textbooks. Don't you believe everything you read here, either.

12.2 OTHER KNOWLEDGE FROM GROUND-BASED OBSERVATIONS

Even though the details of the surface of Mercury can't be studied very well from the Earth, there are other properties of the planet that can be better studied. For example, we can measure Mercury's *albedo*, the fraction of sunlight hitting Mercury that is reflected from it (Fig. 12–4). We can measure this because we know how much sunlight hits Mercury (we know how bright the Sun is and how far away Mercury is from it).

The albedo *(from the Latin for whiteness) is the ratio of light reflected from a body to light received by it.*

LIGHT RAYS

HIGH ALBEDO   MEDIUM ALBEDO   LOW ALBEDO

Figure 12–4 *Albedo* is the fraction of radiation reflected. A surface of low albedo looks dark.

Then we can easily calculate at any given time how much light Mercury reflects, from both (1) how bright Mercury looks to us and (2) its distance from the Earth. Once we have a measure of the albedo, we can compare it with the albedo of materials on the Earth and on the Moon and thus learn something of what the surface of Mercury is like.

Let us consider some examples of albedo. An ideal mirror reflects all the light that hits it; its albedo is thus 100 per cent (the very best real mirrors have albedos of as much as 96 per cent). A black cloth reflects essentially none of the light; its albedo is almost 0 per cent. Mercury's overall albedo is only about 6 per cent. Its surface, therefore, must be made of a dark—that is, poorly reflecting—material. The albedo of the Moon is similar. In fact, Mercury (or the Moon) appears bright to us only because it is contrasted against a relatively dark sky; if it were silhouetted against a bedsheet, it would look relatively dark, as if it had been washed in Brand X instead of Tide.

12.3 MARINER 10

The tenth in the series of Mariner spacecraft launched by the United States went to Mercury. First it passed by Venus, as will be described in the next chapter, and then had its orbit changed by Venus's gravity to direct it to Mercury (Fig. 12–5). Tracking the orbits improved our measurements of the gravity of these planets and thus of their masses. Further, the 475-kilogram (1042-pound) spacecraft had a variety of instruments on board, including devices to measure the magnetic fields in space and near the two planets, and devices to measure the infrared emission of the planets and thus their temperatures. Two of these instruments—a pair of television cameras—provided not only the greatest popular interest but also many important data.

Mariner 10 flew by Mercury the first time (yes, it went back again) from March 29th to April 3rd, 1974, and took 1800 photographs that were transmitted to Earth. It came as close as 750 km from Mercury's surface.

The most striking overall impression is that Mercury is heavily cratered (see the photograph that opened this chapter). At first glance, it looks like the Moon! But there are several basic differences between the features on the surface of Mercury and those on the lunar surface. We can compare how the mass and location in the solar system of these two bodies affected the evolution of their surfaces.

Mercury's craters (Figs. 12–6 and 12–7) may have been eroded by any of a number of methods, including the impacts of meteorites or micrometeorites (large or small bits of interplanetary rock); the solar wind (the expansion of the particles in the solar corona into interplanetary space);

Figure 12–5 Mariner 10 flew by Mercury on three separate occasions. The position of the Earth is shown at the time of launch, and at the time of two of the flybys.

or, perhaps, something that occurred during a much earlier period when Mercury may have had an atmosphere, or internal activity, or been flooded by lava.

Most of the craters themselves seem to have been formed by impacts of meteorites. In many areas they appear superimposed on relatively smooth plains. There is a class of smaller, brighter craters that are sometimes in turn superimposed on the larger craters and thus must have been made afterwards (Figs. 12–7 and 12–8). The rate at which objects hit and formed craters is probably about the same for all the terrestrial planets.

Some craters have rays of higher albedo emanating from them (Fig. 12–9), just as some lunar craters do. The ray material represents relatively recent crater formation (that is, within the last hundred million years). The ray material must have been tossed out in the impact that formed the crater.

There are some maria and craters flooded with some material — presumably lava. This observation also indicates that there may well have been a period of internal heating and vulcanism — although it couldn't have been too extensive or too recent, or else the many craters that now exist would have been obliterated. But lava could also have been formed by the heat generated in impacts of large meteorites, so the presence of lava does not prove the presence of vulcanism.

One interesting kind of feature that is visible on Mercury is a line of cliffs hundreds of miles long; on Mercury as on Earth such lines of cliffs are called *scarps*. The scarps are particularly apparent in the region of Mercury's south pole (Fig. 12–10). Unlike fault lines we know on the Earth, such as the San Andreas Fault in California, there are no signs of geologic tensions like rifts or fissures nearby. The more that Mercury has been photographed and the photographs analyzed, the more we realize that these cliffs or scarps are global in scale, and not just isolated occurrences.

It seems that these scarps may actually be wrinkles in the crust of the planet. Mercury's core, judging by the fact that its density is about the same as the Earth's, is probably iron and takes up much of the central volume, perhaps 50 per cent of the volume or 70 per cent of the mass. At one time, perhaps the core was molten, and shrank by 1 or 2 km as it cooled. The

Figure 12–6 This 150 × 300 km area of Mercury, photographed from a distance of 35,000 km as Mariner 10 approached the planet for the first time, shows a heavily cratered surface with many low hills. The valley at bottom is 7 km wide and over 100 km long. The large flat-floored crater is about 80 km in diameter.

Figure 12–7 A fresh new crater, about 12 km across, in the center of an older crater basin.

Figure 12–8 Another fresh crater on Mercury, about 120 km across. Because Mercury's gravitational field is higher than that of the Moon, material ejected by an impact on Mercury does not travel as far across the surface.

Figure 12–9 (*A*) Many rayed craters can be seen on this photograph of Mercury, taken six hours after Mariner 10's closest approach on its first pass. The north pole is at the top and the equator extends from left to right about 2/3 of the way down from the top. (*B*) A field of light (that is, not dark) rays radiating from a crater off to the top left, photographed on the second pass of Mariner 10. The crater at top is 100 km in diameter.

Figure 12–7

Figure 12–8

A

B

Figure 12–10 This first-pass view of Mercury's northern limb shows a prominent scarp extending from the limb near the middle of the photograph. The photograph shows an area 580 km from side to side.

crust would have settled down with it, making the scarps in the quantity that is in fact observed. Alternatively, it seems possible that Mercury formerly rotated more rapidly. As the rotation slowed down, the shape of the planet would have become less oblate (that is, closer to round instead of bulging at the equator). The crust might have cracked and wrinkled as it tried to match the new shape of the interior.

The Mariner 10 mission was a navigational coup not only because it used the gravity of Venus to get the spacecraft to Mercury, but also because scientists and engineers were able to find an orbit around the Sun that brought the spacecraft back to Mercury several times over. Every six months the Mariner 10 and Mercury returned to the same place at the same

Box 12.1 Naming the Features of Mercury

The mapping of the surface of Mercury leads to a need for names. The scarps are being named for historical ships of discovery and exploration, such as Endeavour (Captain Cook's ship), Santa Maria (Columbus's ship), and Vostok (a Soviet spacecraft). Plains are being given the name of Mercury in many different languages, such as Tir, Odin, and Suisei. Craters are being named for non-scientific authors, composers, and artists, in order to complement the lunar naming system, which honors scientists.

Figure 12–11 A high resolution view of the fractured and ridged plains of the Caloris Basin, photographed from a distance of 19,000 km on Mariner 10's third pass.

By "high resolution," we mean that smaller details are visible than would be visible under "low resolution."

The Van Allen belts are belts of charged particles surrounding the Earth. They were perhaps the first major discovery of orbiting Earth satellites, the earliest of which was Sputnik, launched on October 4, 1957. American satellites soon followed; James Van Allen of the University of Iowa analyzed data from them and discovered that the satellites regularly passed through belts of particles around the Earth.

time. As long as the gas jets for adjusting and positioning Mariner functioned, it was able to make additional measurements and to send back additional pictures in order to increase the photographic coverage. On its second visit, for example, in September 1974, Mariner 10 was able to study the south pole and the region around it for the first time. This pass was devoted to photographic studies. The spacecraft came within 48,000 kilometers of Mercury, farther away than the 750 kilometer minimum of the first pass, but in order to photograph large areas and acquire contiguous coverage of the south polar area. On its third visit, in March 1975, it had the closest encounter ever—only 300 km above the surface (Fig. 12–11). Thus it was able to photograph part of the surface with a high resolution of only 50 meters. Then the spacecraft ran out of gas for the small jets that control its pointing, so even though it still passes close to Mercury every few months, it can no longer record data or send them back to Earth.

The cameras were not the only instruments on board Mariner. The infrared radiometer, for example, gave data that indicate that the surface of Mercury is covered with fine dust, as is the surface of the Moon, to a depth of at least several centimeters. Astronauts sent to Mercury, whenever they go, will leave footprints behind them.

The Mariner 10 mission gave more accurate measurements of the temperature changes across Mercury than had been determined from the Earth. In a few hundred kilometers at the terminator, the line between Mercury's day and night, the temperature falls from about 775 K (500°C) to about 425 K (150°C), and then drops even lower farther across into the dark side of the planet.

Mariner 10 detected lots of electrons near Mercury. Perhaps they are trapped in some sort of belt by the magnetic field, similar to the Van Allen belts around the Earth. But perhaps they are bound to Mercury for shorter times than electrons trapped by the Earth's magnetic field.

One more surprise—perhaps the biggest of the mission—was that a magnetic field was detected in space near Mercury. It was discovered on Mariner 10's first pass, and then confirmed on the third pass. The field is weak; extrapolated down to the surface it is about 1 per cent of the Earth's. It had been thought that magnetic fields were generated by the rapid rotation of molten iron cores in planets, but Mercury doesn't rotate fast enough to cause this effect. Perhaps Mercury rotated faster once upon a time, and the magnetic field generated has been frozen into Mercury since then.

Spacecraft encounters like that of Mariner 10 have revolutionized our knowledge of Mercury. The kinds of information that we can bring to bear on the basic questions of the formation of the solar system and of the evolution of the planets are much more varied now than they were just a short time ago.

SUMMARY AND OUTLINE OF CHAPTER 12

Aims: To discuss the difficulties in studying Mercury from the Earth, and what space observations have told us

Difficulty in observing from the Earth
Never far from the Sun in the sky
Radar astronomy (Section 12.1)
    The rotation period of Mercury is 2/3 its orbital period
    Radar can map the surface of a planet
    Sidereal and solar rotation periods and their relationship
Albedo (Section 12.2)
  Low albedo means a poor reflector
  Mercury has a low albedo
Mariner 10 observations (Section 12.3)
  Photographic results
    Types of objects
      Craters
      Maria
      Scarps
    Mechanisms
      Impact
      Vulcanism
      Shrinkage of the crust
  Results from infrared observations
    Dust on the surface
    Temperature measurements
  Results from other types of observations
    A big surprise: a magnetic field

QUESTIONS

1. Assume that on a given day, Mercury sets after the Sun. Draw a diagram, or a few diagrams, to show that the height of Mercury above the horizon depends on the angle that the Sun's path in the sky makes with the horizon. Discuss how this depends on the latitude or longitude of the observer.

2. If Mercury did always keep the same side towards the Sun, does that mean that the night side would always face the same stars? Draw a diagram to illustrate your answer.

3. Explain why a day on Mercury is 176 Earth days long.

4. Estimate the angular size of Mercury as viewed from Earth (at its closest approach). In a radar ranging experiment, the radar pulse is spread out over a cone whose point (technically, apex) has an angle of one arc minute. What fraction of this radiation strikes Mercury?

5. If you increased the albedo of Mercury, would its temperature increase or decrease? Explain.

6. List those properties of Mercury that could best be measured by spacecraft observations.

7. Think of how you would design a system that transmits still photographs taken by a spacecraft back to Earth and then produces the picture. The system can be simpler than your TV since still pictures are involved.

8. How would you distinguish an old crater from a new one?

9. What evidence is there for erosion on Mercury? Does this mean there must have been water on the surface?

10. List three major findings of Mariner 10.

Venus and the Earth are sister planets; their sizes, masses and densities are about the same. But they are as different from each other as the wicked stepsisters were from Cinderella. The Earth is lush, has oceans and rainstorms of water, an atmosphere containing oxygen, and creatures swimming in the sea, flying in the air, and walking on the ground. On the other hand, Venus is a hot, foreboding planet with temperatures constantly over 750 K, a planet on which life seems unlikely to develop. Why is Venus like that? How did these harsh conditions come about? Can it happen to us here on Earth?

Venus orbits the Sun at a distance of 0.7 A.U. Although it comes closer to us than any other planet—it can approach as close as 45 million kilometers (30 million miles)—we still do not know much about it because it is always shrouded in heavy clouds (Fig. 13–1). Observers in the past saw faint hints of structure in the clouds, which seemed to indicate that these clouds might circle the planet in about 4 days, revolving in the opposite sense from Venus's orbital revolution, but the clouds never parted sufficiently to allow us to see the surface.

Figure 13–1 A crescent Venus, observed with the 5-m Hale telescope in blue light. We see only a layer of clouds.

A sidereal day on Venus is 243 Earth days long.

13.1 THE ROTATION OF VENUS

In 1961, radar astronomy penetrated the clouds and provided an accurate rotation period for the surface of Venus. Venus, because of its relative proximity to Earth, is an easier target for radar than Mercury. Venus rotates in 243 days with respect to the stars in the direction opposite from the other planets; this backward motion is called retrograde rotation, to distinguish it from forward (direct) rotation.

The notion that Venus is in retrograde rotation seems very strange to astronomers, since the other known planets revolve around the Sun in the same direction, and almost all the planets and satellites also rotate in that same direction. Because of the conservation of angular momentum (which was described in Section 10.4) and since the original material from which

A mosaic of photographs of Venus, made in ultraviolet light from the Mariner 10 spacecraft in 1974. The symbol for Venus appears at the top of this page.

the planets coalesced was undoubtedly rotating, we expect that its original angular momentum should now be divided up among the Sun and the planets. We thus expect all the planets to revolve and rotate in the same sense.

It is thus a problem to explain why Venus rotates "the wrong way." (The rotation of Uranus, as we shall see in Section 17.2, is also a problem.) Nobody knows the answer. One possibility depends on the idea that when Venus was in the process of forming, the planetesimals formed clumps of different sizes. Perhaps the second largest clump struck the largest clump at such an angle as to cause the result to rotate backwards. Scientists do not like *ad hoc* ("for this special purpose") explanations like this one, because they are constructed to explain specific situations and do not permit generalization. Nevertheless, that's all we can do in this case.

The slow rotation of the solid surface contrasts with the rapid rotation of the clouds. The clouds rotate in the same sense as the surface of Venus but much more rapidly, once every 4 days. The tops of the clouds have been studied from the Earth, and the composition of the clouds—a mixture of droplets of water and of sulfuric acid—was deduced. Sulfuric acid takes up water very efficiently, so there is little water vapor above the clouds. It took especially careful work from high-altitude sites on the Earth, and from balloons, to detect the presence of the small amount of Cytherean water vapor above the clouds. This observation is difficult, because the spectral lines of water vapor from Venus were masked by the spectral water vapor lines that arose from the Earth's own atmosphere.

Astronomers often use the adjective Cytherean to describe Venus. Cythera was the island home of Aphrodite, the Greek goddess who corresponded to the Roman goddess Venus.

13.2 THE SURFACE OF VENUS

The surface of Venus can be detected from Earth with radio telescopes. Radio waves emitted by the surface penetrate the clouds to give us information about the surface of this planet. Just as we can determine the surface temperature of a star from studying the distribution of its radiation over optical wavelengths and by the strength of the emission, so can we also determine the temperature of the Cytherean surface by studying the radio emission (the optical radiation doesn't pass through the clouds). The surface is very hot, about 750 K (475°C).

In addition to directly measuring the temperature on Venus, we can calculate theoretically approximately what it should be if the Cytherean atmosphere did not impede the radiation. This value—less than 375 K (100°C)—is much lower than the measured values. The high temperatures derived from radio measurements indicate that energy does not leave Venus as readily as it leaves the Earth.

Since we are speaking in approximate values, numbers are rounded off to the nearest multiple of 5.

What accounts for this difference? What is happening on Venus is similar to the process that is generally—though incorrectly—thought to occur in

Figure 13-2 Venus receives energy from only one direction, and heats up and radiates energy (mostly in the infrared) in all directions. From balancing the energy input and output, astronomers can calculate what Venus's temperature would be if Venus had no atmosphere. This type of calculation is typical of those often made by astronomers.

Figure 13–3 Sunlight can penetrate Venus's clouds, so the surface is illuminated with radiation in the visible part of the spectrum. Venus's own radiation is mostly in the infrared, and the presence of carbon dioxide in the atmosphere blocks the transmission of most of this radiation. This is called the greenhouse effect. Water vapor and sulfuric acid particles block the rest of the infrared. The Earth's atmosphere is not as opaque as is Venus's in the infrared, so the Earth's greenhouse effect is much smaller, about 35 K of heating.

greenhouses here on the Earth. The process is thus called the *greenhouse effect* (Fig. 13–3). Sunlight passes through the Cytherean atmosphere in the form of radiation in the visible part of the spectrum. The sunlight is absorbed by, and so heats up, the surface of Venus. At the temperatures that result, the radiation given off is mostly in the infrared. But the carbon dioxide in Venus's atmosphere is opaque to infrared radiation, and the energy is trapped. Thus Venus heats up above the temperature it would have if the atmosphere were transparent. (Greenhouses on Earth don't work quite this way. The closed glass of greenhouses on Earth prevents convection and the mixing in of cold outside air. The trapping of energy by the "greenhouse effect" is a less important process in an actual greenhouse. Try not to be bothered by the fact that the greenhouse in your backyard is not heated by the "greenhouse effect.") Understanding such processes involving the transfer of energy is but one of the practical results from the study of astronomy.

Some additional information we have about the surface of Venus comes

Figure 13–4 Radar maps of Venus. The area circled in the large-scale map (*B*) is enlarged at left (*A*). An area that could not be well observed is shown as black. Radar results only show the radar reflectivity of the surface, but the results can be interpreted as showing shallow craters as big as 160 km across. The region marked β (beta) seems to be a huge volcano, 700 km across. A clearer view of *B*, from a more recent radar map, is shown at right (*C*).

A   B   C

from radar observations. Maps of the surface have painstakingly been built up at several of the huge radio telescopes capable of radar work. The radar maps of Venus (Fig. 13–4) show large-scale surface features, some very rough and others relatively smooth, just as surfaces on the Moon. Many craters have been found, all relatively shallow, but ranging up to 1000 km across. Most have probably been formed by meteoric impact.

From Venus's size and from the fact that its mean density is similar to that of the Earth, we conclude that its interior is also probably similar to that of the Earth. This means that we might expect to find volcanoes and mountains on Venus, and that venusquakes probably occur, too.

13.3 SPACE OBSERVATIONS

Venus was an early target of both American and Soviet space missions (Table 13–1). The American craft were in the Mariner series, a heavier and more elaborately instrumented series than the Pioneers. The Soviet series was called Venera.

In 1970, the Venera 7 spacecraft radioed 23 minutes of data back from the surface of Venus. It confirmed the high temperatures of 750 K and high pressures. Subsequently, in 1972, the lander from Venera 8 survived on the surface of Venus for 50 minutes and confirmed the findings of Venera 7. The Cytherian atmosphere (Fig. 13–5) is composed of 90 or more per cent carbon dioxide (CO_2). The pressure is 100 times higher than the pressure of Earth's atmosphere, because of the high amount of carbon dioxide in the atmosphere of Venus; carbon dioxide makes up less than 1 per cent of the terrestrial atmosphere. Carbon dioxide on Earth dissolved in sea water and eventually formed our terrestrial rocks in the form of carbonates. If this carbon dioxide were to be released from the Earth's rocks, along with other carbon dioxide trapped in sea water, our atmosphere would become as dense and have as high a pressure as that of Venus. Venus, of course, has no oceans in which the carbon dioxide can dissolve.

Figure 13–5 The composition of Venus's atmosphere.

TABLE 13–1   UNMANNED PROBES LAUNCHED TO VENUS

Spacecraft	Launched by	Arrival Date	Comments
Venera 1	USSR	1961	Failed en route
Mariner 2	USA	1962	Flyby
Venera 2	USSR	1966	Flyby
Venera 3	USSR	1966	Crash-landed
Venera 4	USSR	1967	Probed atmosphere
Mariner 5	USA	1967	Flyby
Venera 5 and 6	USSR	1969	Probed atmosphere
Venera 7	USSR	1970	23 minutes of operation on surface
Venera 8	USSR	1972	50 minutes of operation on surface
Mariner 10	USA	1974	Flyby
Venera 9	USSR	1975	53 minutes of operation on surface; photograph
Venera 10	USSR	1975	65 minutes of operation on surface; photograph

Figure 13–6   The Mariner 10 spacecraft.

The spacecraft observations have improved our understanding of the clouds that surround Venus, high above its surface. Terrestrial clouds rarely go above 10 km but the denser Cytherean clouds may not start until 50 km above the surface and then thinner clouds extend upward for an additional 30 km or so. These clouds may be composed of liquid droplets of sulfuric acid.

The United States spacecraft Mariner 10 (Fig. 13–6) took thousands of photographs of Venus in 1974 as it passed by en route to Mercury. It went sufficiently close and its imaging optics were good enough that it was able to study the structure in the clouds with resolution of up to 200 meters, about an eighth of a mile (Fig. 13–7). The structure shows only when viewed in ultraviolet light (see the figure opening this chapter). We could observe much finer details from space than we can see from the Earth (Fig. 10–16).

Studies of Venus like these have great practical value. The better we understand the interaction of solar heating, planetary rotation, and chemical composition in setting up an atmospheric circulation, the better we will understand our Earth's atmosphere. We then may be better able to predict the weather and discover jet routes that would aid air travel, for example. The potential saving of large amounts of money from this knowledge is enormous: it would be many times the investment we have made in planetary exploration.

In 1975 Soviet scientists succeeded in landing a spacecraft on Venus and having it survive long enough to send back photographs. Venera 9 and Venera 10 traveled over four months from Earth to reach Venus, and arrived in October 1975. Each sent down a lander (Fig. 13–8). The single photograph that the Venera 9 lander took in the 53 minutes before it succumbed

Figure 13–7   Several layers of haze at the limb of Venus, photographed by Mariner 10.

Figure 13–8 A model of the Soviet Venera 9 lander.

to the tremendous temperature and pressure showed a clear image of sharp-edged, angular rocks (Fig. 13–9). This came as a surprise to some scientists, who had thought that erosion would be rapid in the dense Cytherean atmosphere, and that the rocks should therefore have become smooth or disintegrated into sand.

Three days later, the Venera 10 lander reached the surface and transmitted data for 65 minutes, including a photograph (Fig. 13–10). The landing site was 2000 kilometers distant on the surface of Venus from the Venera 9 site. The rocks in the Venera 10 site were not sharp; they resembled huge pancakes, and between them were sections of cooled lava or debris of weathered rock.

Thus we have two photographs of the surface of Venus: mankind's first photographs from the surface of another planet. The one from Venera 9 looked typical of a "young mountainscape," with the sharp rocks possibly ejected from volcanoes. On the other hand, the one from Venera 10 showed "a landscape typical of old mountain formations," in the words of Soviet scientists.

The Venera landers also made measurements of the soil, determining that its chemical composition and density correspond to that of basalt, in common with the Earth, the Moon, and Mars. They measured temperatures of 760 K and 740 K at the two sites, and a pressure over 90 times that of the Earth's atmosphere.

13.4 OF WATER AND LIFE

Figure 13–9 The surface of Venus photographed from Venera 9. The photograph is reproduced in a flat strip, but was actually a scan from upper left, down toward the center, and up toward the right. Think of looking to your left at the horizon, then down at your feet, and then up to the horizon at the right. That is why the horizon is visible as tilted lines at upper left and at upper right.

Why is it that there is so much water on Earth, in our atmosphere, and in our oceans, and so little liquid water on the surface of Venus? Is it there, and just trapped inside the rocks, or is it trapped under the surface? What features that have made Earth a hospitable planet for life as we know it have made Venus so unattractive? Can we make sure we are not changing the Earth's atmosphere so that we will wind up with an atmosphere like that of Venus?

On the Earth, the origin of life changed the future course of our planet. The atmosphere of Earth in its primitive state may well have been of carbon dioxide, methane, ammonia, and other gases we now find distasteful or even fatal, but life forms produced much of the oxygen that we now have. One form of an oxygen molecule, ozone, consists of three oxygen atoms bound together (O_3). In the presence of ozone, water (H_2O) breaks down into oxygen and hydrogen. The hydrogen is so light that it has long

Figure 13-10 The surface of Venus photographed from Venera 10.

since escaped from Earth. That's fine with us, as the oxygen is left behind for us to breathe. Perhaps if life had arisen on Venus at the same time that it arose on Earth, it would have modified the Cytherean atmosphere as it did the terrestrial one.

The water on the Earth has acted as a catalyst to spur the fixation of carbon dioxide into certain types of rocks. (A catalyst is a substance that causes or influences a chemical change without itself being affected by it.) On Venus the carbon dioxide is still in the atmosphere—in fact, makes up 95 per cent of the atmosphere. The carbon dioxide causes the greenhouse effect to occur, thus keeping the surface of Venus very hot. Apparently water and plant life have saved the Earth from suffering the same fate as Venus.

The ozone plays another vital role on Earth: it keeps out ultraviolet radiation from the Sun. The part of the ultraviolet that has wavelengths slightly longer than 3000 Å, where the Earth's atmosphere becomes transparent, causes suntanning at the beach. None of the wavelengths shorter than 3000 Å pass through the ozone. If they did, they would kill us. Most scientists think that amounts of ultraviolet radiation even slightly increased over the amount we already receive, whether longer or shorter than 3000 Å, would increase the incidence of skin cancer. Scientists have recently been concerned about the way that man's technology may threaten the ozone layer of our Earth.

Figure 13-11 The layers of the Earth's atmosphere showing the location of the part of the stratosphere where ozone is formed. Once formed, the ozone is transported to lower stratospheric layers.

An official report of the National Research Council, released in 1976, says that the fluorocarbon problem is a definite hazard to the Earth's ozone layer. Plans to restrict the use of fluorocarbon propellants are under way. The use of Freons in aerosol cans seems non-essential and will be limited. Essential uses, such as refrigeration, will still be permitted.

A major worry that surfaced recently is the possibility that the gases that are used as refrigerants in refrigerators and air conditioners and as propellants in some aerosol cans are accumulating in the stratosphere and depleting the ozone. These gases were originally chosen for aerosol use because they were thought to be inert, that is, non-interacting. However, it now seems that they are not as inert as they originally had been thought to be. Once they are released into the atmosphere, the propellant gases (such as some types of Freon, a commonly used trade name for chloro-fluorocarbons including $CFCl_3$ and CF_2Cl_2), eventually break down, and the individual chlorine atoms can transform the ozone into other molecules. The chlorine returns to its atomic state after breaking down the ozone (chemically, it acts as a catalyst), and so each chlorine atom can wreak havoc on the ozone for years. We may already have put enough gases into the atmosphere to make a permanent change in the ozone layer.

One lesson to learn from all these considerations is that we can be glad that the basic scientific research is being done, because it exposes the possible side effects of ordinary things we otherwise take for granted: for example, cans of deodorant. Studies of the Cytherean atmosphere make us appreciate the fundamental benefits of our own atmosphere. With the new technology continuing to be developed, we must make certain that basic scientific research goes forward on all fronts so that we will be prepared in the future for whatever problems might arise. Research limited to specific problems won't be enough.

13.5 FUTURE EXPLORATION

NASA is planning an elaborate mission to Venus in 1978 that will send four probes through the atmosphere of Venus. The probes will be launched from a Pioneer spacecraft and will radio back signals via a second Pioneer that will orbit Venus. The probes will send back data as they pass through the Cytherean atmosphere but are not being designed to survive the impact of the surface. The orbiter will contain a radar for high resolution mapping of the surface.

Soviet plans for space exploration are ordinarily not disclosed in advance, and the fact of a mission is disclosed only when the rocket is already en route to a planet. But on the basis of their past record, we can assume that the Soviets will also continue their launches of spacecraft to Venus.

SUMMARY AND OUTLINE OF CHAPTER 13

Aims: To study Venus's clouds and surface, and to understand why and how Venus became inhospitable to life as we know it

The rotation period (Section 13.1)
The surface of Venus (Section 13.2)
    The high temperature: 750 K
    The greenhouse effect
    Surface maps from radar observations
Space observations (Sections 13.3 and 13.5)
    U.S. Mariner 10 observations of clouds and atmosphere

Soviet Venera 9 and 10 landers: temperature measurements and photo-
graphs of the surface
Why Venus and the Earth evolved differently (Section 13.4)
Danger to the Earth's ozone layer

QUESTIONS

1. Make a table displaying the major similarities and differences between the Earth and Venus.

2. Two main factors determine how bright Venus appears from the Earth: how much of the lighted side faces the Earth and how far Venus is from Earth. What are the relative positions of the Earth, Sun, and Venus when Venus appears brightest? You may wish to draw diagrams to demonstrate this.

3. Why do we think that there have been significant external effects on the rotation of Venus?

4. If observers from another planet tried to gauge the rotation of the Earth by watching the clouds, what would they find?

5. Suppose a planet had an atmosphere that was opaque in the visible but transparent in the infrared. Describe how the effect of this type of atmosphere on the planet's temperature differs from the greenhouse effect.

6. Why do radar observations of Venus provide more data about the surface structure than a Mariner flyby?

7. If one removed all the CO_2 from the atmosphere of Venus, the pressure of the remaining constituents would be how many times the pressure of the Earth's atmosphere?

8. The Earth's magnetic field protects us from the solar wind. What does this tell you about the charge (or lack of it) of the solar wind particles?

9. Some scientists have argued that if the Earth had been slightly closer to the Sun, it would have turned out like Venus; outline the logic behind this conclusion.

10. Outline what you think would happen if we sent an expedition to Venus that resulted in the loss of almost all of the CO_2 from the Cytherean atmosphere.

TOPIC FOR DISCUSSION

1. Evaluate the consequences of continued use of chloro-fluorocarbons in aerosol cans. Distin-guish essential from non-essential uses of chloro-fluorocarbons.

14

The Earth

Since it doesn't seem fair to know so much more about the Earth than the other planets simply because we live on it, let us consider our own planet as though it were being described by an inhabitant of Mars. Other material about the Earth—on continental drift and the composition of the atmosphere, for example—is covered for purposes of comparison in other chapters on the planets.

REPORT OF THE MARTIAN ACADEMY OF SCIENCE

For centuries we have known Earth as an interesting object in our sky; sometimes it is the morning star and sometimes it is the evening star. Since it is an inner planet for us, Earth is always within a few hours of rising or setting, so we can never observe it high overhead in a dark sky and through the minimum amount of our atmosphere.

The major problem we have in observing Earth is not our atmosphere but its own. Often major portions of Earth are covered with white clouds that prevent us from seeing the surface.

When we can see through the clouds, we see that Earth is mostly covered with a blue-greenish dark substance of much lower albedo than the clouds (as shown on the photograph on the facing page and on Color Plate 32). A smaller fraction of the surface is covered with lighter-colored material, and this lighter material changes in color somewhat with the seasons. As springtime comes to each hemisphere of Earth, the lighter material becomes greenish.

There is much less of the life-giving carbon dioxide on Earth than there is in our own atmosphere. But oxygen, a gas that is deadly to us, does exist by itself in Earth's atmosphere, both as O_2 and O_3. Earth has polar caps, one in the north and another in the south, and their sizes change with the season. Perhaps they are tremendous reservoirs of carbon dioxide in the form of ice (frozen CO_2), as are our polar caps. Some Martian scientists, but a minority, think that the polar caps on Earth might partially be made of frozen dihydrogen oxide (H_2O), which we call "wet ice," and are planning to take better spectra to study this possibility. They claim that their infrared measurements of Earth's temperature indicate that the planet is too warm for ordinary carbon dioxide ice.

Figure 14–1 A negative print of Hawaii, showing the shield volcano Mauna Kea. Measured from the ocean floor, it is the highest mountain on Earth. (Everest rises from a high plain.) The largest canyon-like feature on Earth is the Great Rift Valley in Africa.

The Earth, photographed from Apollo 17. The symbol for the Earth appears at the top of this page.

Figure 14–2 The Martian Terra 5 presumably landed in the ocean, shown here off the coast of Massachusetts and New Hampshire.

Figure 14–3 At first, the Martians sighted the Great Wall of China. This view over Chicago and Lake Michigan shows more distinct signs of civilization.

Figure 14–4 A drive-in movie must look strange to a visiting alien.

Earth is accompanied by a remarkable moon, called Selene. Selene is not too much smaller than the planet Mercury, and ranks in size with the giant moons of Jupiter and Saturn. This is unusual because Earth is a martian planet rather than a giant planet, and planets ordinarily have moons that are only about one-thousandth their sizes. Selene, on the contrary, is approximately one-fourth the diameter of its primary, and as much as 1/81 of the primary's mass. The seasonal changes that we have detected on Earth have not been seen on Selene.

For the last few years, we have sent a series of rockets to Earth. The Terra 1 and Terra 2 did not succeed in traveling the long distance to Earth, but Terra 3 flew by at a distance of 8 billion centirams (remember that 1 ram is the length of the left antenna of Queen Schrip, who reigned from the year 15,363 to 16,437) and succeeded in getting a series of photographs from close up. They showed a planet that is mostly covered with maria, which correspond to the darker areas. Radar reflectivities indicate that they may be covered with dihydrogen oxide (H_2O). There are mountain areas and very few craters. The largest peaks (Fig. 14–1) are smaller than our own volcanoes, and the largest canyon, in a raised land mass that extends from slightly above the equator far into the southern hemisphere, is about the size of our own Great Canyon. Since Earth is a larger planet than ours, these canyons and mountains are smaller than ours with respect to the size of the planet.

Terra 4 went into orbit around Earth, and took a series of photographs over its ten phobon lifetime (one phobon, of course, is the length of time it takes for Phobos to orbit the Mars, and corresponds to about 1/100 of an Earth selenth). Four relatively smooth areas were chosen as prime landing sites.

Terra 5 attempted a crash landing on Earth last year, and succeeded in slowing its velocity as it passed through Earth's considerable atmosphere. But contact with it was lost a few seconds after landing; perhaps it was covered over by whatever material makes up the greenish areas (Fig. 14–2). Those of our scientists who say that this material is dihydrogen oxide claim that this mysterious disappearance supports their theory.

We have not been able to establish the presence of any intelligent life on the planet. Indeed, the presence of life would seem to depend on establishing that for Earth, as on the Mars, the seasonal blowing of dust that takes place provides shelter from solar radiation for individual Earthians, assuming that they, as we, cannot stand exposure to sunlight. We look for signs of Earthian work. Some of our intelligence analysts think that they have detected a long serpentine streak traversing one of the continents and some signs of a checkerboard pattern on a large scale, which could indicate the presence of agriculture. But these detections are marginal, and must be checked further. More recent observations seem to show traces of cities (Fig. 14–3).

A somewhat later look in the Martian archives might turn up the following ideas, as reported by Paul A. Weiss of the Rockefeller Institute.

A summary of our report is that we have discovered life on Earth! It is the discovery of the millennium for us Martians. From a height of 5 million rams, we could see streaks of light moving like waves across the landscape, often in two channels right next to each other and flowing in opposite directions. When we came closer, we saw that each luminous knot had an independent existence and had two white lights in front and two red lights

Color Plate 1 (left): The 4-m Mayall optical telescope on Kitt Peak, near Tucson, Arizona. (Kitt Peak National Observatory photo)

Color Plate 2 (top right): The 66-m radio telescope at the Australian National Radio Observatory at Parkes, N.S.W. (Photograph by the author)

Color Plate 3 (bottom): A. From top to bottom, the emission spectrum of ionized calcium, the Fraunhofer spectrum of the sun, a continuous spectrum, and the emission spectrum of a fluorescent lamp. Note the H and K lines of ionized calcium at the left, which are the strongest absorption lines in the visible part of the solar spectrum. (Bausch and Lomb photos)

B. Fraunhofer's original spectrum from 1814. Only the D and H lines retain their notation from that time. The C line is Hα.

Color Plate 4: 47 Tucanae, a prominent globular cluster visible from the southern hemisphere. (Cerro Tololo Inter-American Observatory photograph with the 4-m telescope)

Color Plate 5: The Pleiades, M45, is a galactic cluster in the constellation Taurus, the bull. Reflection nebulae are visible around the brightest stars. (Hale Observatories photograph with the 1.2-m Schmidt camera)

Color Plate 6: The Eta Carinae Nebula, NGC 3372, in the southern constellation Carina. Visually, this is the brightest part of the Milky Way. The central dark cloud superimposed on the brightest gas is the Keyhole Nebula. η Carinae is the brightest star left of the Keyhole. (Cerro Tololo Inter-American Observatory photo.)

Color Plate 7 (top): Two of the partial phases of the June 30, 1973 total solar eclipse. Note the clouds that covered part of the crescent shortly before totality.

Color Plate 8 (bottom): The diamond ring effect at the beginning of totality at the June 30, 1973 eclipse. These photographs were taken from Loiengalani, Kenya, by the author's expedition.

Color Plate 9 (top): The solar corona, photographed by the author's expedition to the 1973 solar eclipse. Note the equatorial streamers and polar plumes.

Color Plate 10 (bottom): The flash spectrum of the solar chromosphere, photographed at the 1970 total solar eclipse. (William C. Atkinson photo)

Color Plate 11 (top): An erupting prominence on the sun, photographed on August 9, 1973 in the ultraviolet radiation of ionized helium at 304 Å with the Naval Research Laboratory's slitless spectroheliograph aboard Skylab. Because of the technique used, radiation at nearby wavelengths appears as a background. A black dot visible on the solar surface slightly to the left of the base of the prominence may be the site of the flare that caused the eruption. Supergranulation is visible on the solar disk in the light of He II, and a macrospicule shows on the extreme right. (NRL/NASA photo)

Color Plate 12 (bottom): Contours of equal intensity computed for another eruptive prominence observed in the radiation of ionized helium with the NRL instrument aboard Skylab. Each color represents a different strength of emission. This view shows the prominence 90 minutes after its eruption, when the gas had moved 500,000 km from the solar surface. (NRL/NASA photo)

Color Graduation Scale

Minimum XUV Emission

Maximum XUV Emission

SIZE OF EARTH

Color Plate 13 (top): Isophotes showing the intensity of the white light solar corona, reduced from observations made by the High Altitude Observatory's coronagraph aboard Skylab. The solar photosphere has been artificially occulted. (HAO/NASA photo)

Color Plate 14 (left): Skylab did not pass into totality during the total solar eclipse of June 30, 1973, but this image shows the Moon as it crossed the solar corona as observed on that date from Skylab with the High Altitude Observatory's coronagraph. (HAO/NASA photo)

Color Plate 15 (right): On the few days when Comet Kohoutek was closest to its perihelion, it was invisible to observers on earth. The Skylab astronauts, fortunately, were able to sketch and photograph it. Here we see the image of Comet Kohoutek in the High Altitude Observatory's coronagraph on December 27, 1973. (HAO/NASA photo)

Color Plate 21 (top):   A planetary nebula, NGC 6781, in the constellation Aquila. (Hale Observatories photo with the 1.2-m Schmidt camera)

Color Plate 22 (bottom):   A computer display showing the distribution of x-radiation between 15 and 80 Å from the Cygnus loop, a supernova remnant. (Computer image by Richard J. Borken of the University of Wisconsin from rocket data he collected in collaboration with Saul A. Rappaport of M.I.T.)

Color Plate 23: Crab Nebula, M1, in Taurus is the remnant of the supernova of 1054 A.D. The red filaments radiate in the hydrogen lines; the white continuum is from synchrotron radiation. (Hale Observatories photo with the 5-m telescope)

Color Plate 24:   Veil Nebula, NGC 6992, part of the Cygnus loop, a supernova remnant. (Hale Observatories photo with the 1.2-m Schmidt camera)

Color Plate 25:   Orion Nebula, M42, is glowing gas excited by the Trapezium, four hot stars. The nebula contains stars in formation; it is 25 ly across and 400 parsecs away. (Hale Observatories photo with the 5-m telescope)

Color Plate 26: Apollo 17 view of the Taurus-Littrow Valley on the Moon. This huge, fragmented boulder had rolled a kilometer down the side of the North Massif to here. Scientist-astronaut Harrison Schmitt is at the left. The Lunar Rover is at the right. (NASA photo)

Color Plate 27: The lunar buggy carried by Apollo 17. (NASA photo)

Color Plate 28: Apollo 17 astronauts found a hint of orange in the lunar soil (picture's center), initially causing excitement since the possibility of volcanic origin was raised. (NASA photo)

Color Plate 29 (bottom, left): The blast-off of Apollo 17, a night launch. (Photo by the author)

Color Plate 30 (bottom, center): An Apollo 17 astronaut with an instrument package. (NASA photo)

Color Plate 31 (bottom, right): The crescent Earth seen from Apollo 11 in orbit around the Moon. (NASA photo)

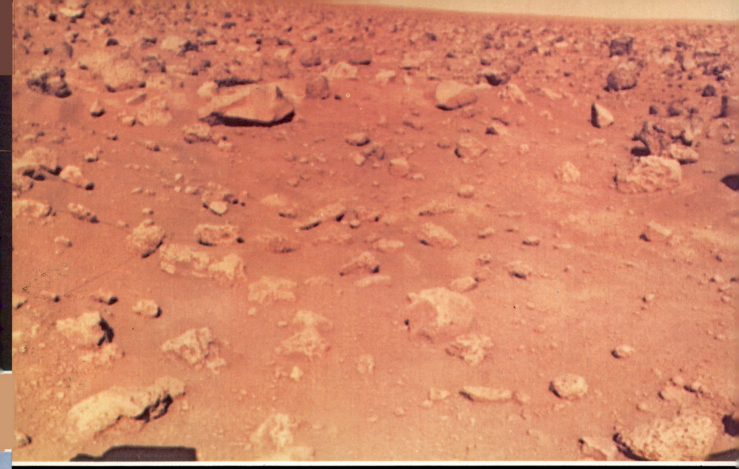

Color Plate 37:   Mars's Chryse Planitia as seen by Viking 1. The surface material is probably limonite (hydrated ferric oxide). Reddish dust suspended in the atmosphere makes the sky pink. (NASA photo)

Color Plate 38:   Mars's Utopia Planitia seen by Viking 2. Many of the rocks are porous and sponge-like, similar to some of Earth's volcanic rocks. (NASA photo)

Color Plate 39: Jupiter as observed by Pioneer 10. Bands, zones, and the Great Red Spot show clearly. (NASA photo)

Color Plate 40: Saturn, on March 11, 1974, showing bands on its disk and Cassini's division in its rings. (Lunar and Planetary Laboratory, University of Arizona photo)

Color Plate 41: Comet West, a bright comet visible in 1976. (Photo by Martin Grossmann)

Color Plate 42:   The Horsehead Nebula, NGC 2024, in Orion, is an absorbing region superimposed on one emitting red radiation characteristic of hydrogen. (Hale Observatories photo with the 1.2-m Schmidt camera)

Color Plate 43 (top):   M16, the Eagle Nebula in the constellation Serpens. Hydrogen radiation makes it appear red. The bright stars at the upper right are hot and young, and are part of a galactic cluster. The small, dark regions may be protostars. (Hale Observatories photo with the 5-m telescope)

Color Plate 44 (bottom):   M17, the Omega Nebula in Sagittarius. (Hale Observatories photo with the 1.2-m Schmidt camera)

Color Plate 45: The Trifid Nebula, M20, in Sagittarius is glowing gas divided into three visible parts by absorbing lanes. The blue nebula at the top is unconnected to the Trifid. (Hale Observatories photo with the 5-m telescope)

Color Plate 46: The Andromeda Galaxy, also known as M31 and NGC 224, the nearest spiral galaxy to the Milky Way. It is accompanied by two elliptical galaxies, NGC 205 (below) and M32 (above). (Hale Observatories photo with the 1.2-m Schmidt camera)

Color Plate 47:   M33, a spiral galaxy (class Sc) in the constellation Triangulum. Hot, blue stars outline the spiral arms while the central regions show cooler yellow and red stars. (Hale Observatories photo)

Color Plate 48:   Centaurus A, a powerful radio source whose optical image, NGC 5128, shows an elliptical galaxy around which a heavy zone of dust appears to be wrapped. The galaxy rotates perpendicularly to the dust lane. (Cerro Tololo Inter-American Observatory photo with the 4-m telescope)

Color Plate 49 (top):   NGC 7331, a spiral galaxy in Pegasus. This galaxy may be linked with the group of galaxies known as Stephan's Quintet. (Hale Observatories photo with the 5-m telescope)

Color Plate 50 (bottom):   Left is an image of 21-cm radiation from M81 in Ursa Major. Brightness represents intensity; color represents the Doppler shift: red = recession, violet = approach, green = no shift. (Westerbork data from A. H. Rots and W. W. Shane; imaged at NRAO.) Right is M81 observed optically at Palomar.

behind. They were phototactic, attracted by a flickering light source. But rather than rush into the source, they stopped just short of it and formed a crystalline array. Then they remained immobile, perhaps sleeping. Their luminous activity diminished (Fig. 14–4).

Inside these Earthians, we eventually discovered even smaller objects, probably parasites (Fig. 14–5). They are lodged, for the most part, in the interior of Earthians. They never stray far off from the Earthians, even when they are disgorged, so they must be dependent on the Earthians for sustenance. The parasites are more numerous in larger hosts (Fig. 14–6); since host size reflects host age, the accessory bodies obviously multiply inside the hosts as the latter grow.

We can show that the Earthians are alive, for they metabolize by taking in sustenance sucked in through tubes inserted in their rears (Fig. 14–7). They give off wastes mostly as gas and smoke. Some areas of Earth are positively fouled with the waste products. We also noted signs of grooming, with a wiping motion, but only in front (Fig. 14–8). Astonishingly, this was started and stopped in synchrony by all the members of the population, which may imply that the Earthians have brains.

OUTLINE AND SUMMARY OF CHAPTER 14

Aims: To understand the Earth on a planetary scale, and to realize the limitations of our studies of other planets from seeing how someone on another planet would interpret observations of the Earth

Terrain and atmosphere
Signs of life difficult to detect from afar

QUESTIONS

1. (a) Imagine that you are observing the Earth from an altitude of 100 kilometers. Estimate the angular size of a house, a football field, and an average size city. What features do you think would point to the existence of intelligent life? (b) What about from 1000 km up?

2. Consult an atlas and compare the size of the Grand Canyon in Arizona with the Rift Valley in Africa. How do they compare in size with the giant canyon on Mars, which is about the diameter of the United States?

3. Plan a set of experiments or observations that you, as a Martian scientist, would have an unmanned spacecraft carry out on Earth. What data would your spacecraft radio back if it landed in a corn field? In the Sahara? In the Antarctic? In Times Square?

Figure 14–5 If cars are considered to be the dominant life form on Earth, then the people would often be considered to live inside them.

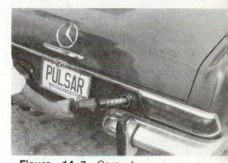

Figure 14–6 Larger vehicles have more people inside them.

Figure 14–7 Cars have many of the characteristics that we consider to imply that something is alive.

Figure 14–8 A car grooming itself, or at least grooming its windshield.

Mars

Mars has long been the planet of greatest interest to scientists and non-scientists alike. Its interesting appearance as a reddish object in the night sky and some of the past scientific studies that have been carried out have made Mars the prime object of speculation as to whether or not extraterrestrial life exists there.

In 1877, the Italian astronomer Giovanni Schiaparelli published the results of a long series of visual telescopic observations he had made of Mars. He reported that he had seen *canali* on the surface. When this Italian word for "channels" was improperly translated into "canals," which seemed to connote that they were dug by intelligent life, public interest in Mars increased.

We now know that the channels or canals Schiaparelli and other observers reported are not present on Mars—the positions of the *canali* do not even always overlap the spots and markings that are actually on the Martian surface (Fig. 15–1). But hope of finding life in the solar system springs eternal, and the latest studies have indicated the presence of considerable quantities of liquid water in Mars's past, a fact that leads many astronomers to hope that life could have formed during those periods.

15.1 CHARACTERISTICS OF MARS

Mars is a small planet, 6800 km across, which is only about half the diameter of Earth or Venus, although one-and-a-half times that of Mercury. Since Mars is much farther from the Sun than Mercury, it receives much less solar heat and so has been able to retain an atmosphere. The atmosphere is thin—at the surface its pressure is only 1 per cent of the surface pressure of Earth's atmosphere—but it might be sufficient for certain kinds of life.

Figure 15–1 A drawing of Mars and a photograph, both made at the close approach of 1926. We now realize that the "canals" seen in the past do not usually correspond to real surface features.

Taken by Viking Orbiter 2 as it approached the dawn side of Mars, in early August 1976. At the top, with water ice cloud plumes on its western flank, is Ascreaus Mons, one of the giant Martian volcanoes. In the middle is the great rift canyon called Valles Marineris, and near the bottom is the large, frosty crater basin called Argyre. The south pole is at the bottom. The symbol for Mars appears at the top of this page.

Figure 15–2 Photographs of Mars in violet (top) and in infrared (bottom) light. Mars has much less surface contrast in the violet.

From the Earth, we have relatively little trouble in seeing through the Martian atmosphere to inspect the planet's surface (Fig. 15–2), except when a Martian dust storm is raging. At other times, we are limited mainly by the turbulence in our own thicker atmosphere, which causes unsteadiness in the images and limits our resolution to 60 km at best (in other words, we cannot see detail smaller than 60 km even under the best observing conditions).

Unlike the orbits of Mercury or Venus, the orbit of Mars is outside the Earth's, so it is much easier to observe in the night sky (Fig. 15–3). We can simply observe it in the direction opposite from that of the Sun. We are then looking at the lighted face. When three celestial bodies are in a line, the alignment is called a *syzygy*, and the particular case when the planet is on the opposite side of the Earth from the Sun is called an *opposition*. These oppositions occur at intervals of 780 days on the average, equivalent to 26 Earth months.

Mars revolves around the Sun in 687 Earth days, equivalent to 23 Earth months. The axis of its rotation is tipped at a 25° angle from the plane of its orbit, nearly the same as the Earth's 23 1/2° tilt. Because the tilt causes the seasons, we know that Mars goes through a year with four seasons just as the Earth does, except that the Martian year is 23 Earth months long.

From the Earth, we have long watched the effect of the seasons on Mars. In the Martian winter, in a given hemisphere, there is a polar cap. As the Martian spring comes to the northern hemisphere, for example, the north polar cap shrinks and material at more temperate zones darkens. The surface of Mars is always mainly reddish, with darker gray areas that appear blue-green for physiological reasons of color contrast (see Color Plate 18). In the spring, the darker regions spread. Half a Martian year later, the same thing happens in the southern hemisphere.

One possible explanation that previously seemed obvious for these changes is biological; Martian vegetation could be blooming or spreading in the spring. But there are other explanations, too. The theory that presently seems most reasonable is that each year at the end of northern-hemisphere springtime, a global dust storm starts, covering the entire surface with light dust. Then the winds get as high as hundreds of kilometers per hour and blow fine, light-colored dust off the slopes. This exposes the dark areas underneath. Gradually over the next Martian year, dust is also stripped away in certain places, such as near craters and other obstacles.

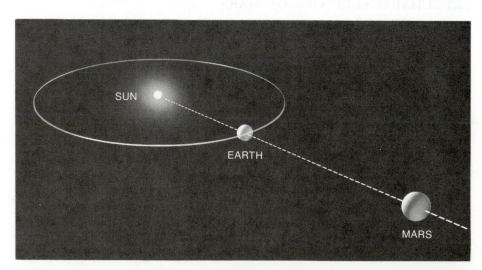

Figure 15–3 When Mars is at opposition, it is high in our nighttime sky.

Finally, a global dust storm starts up again to renew the cycle. If the dust were of certain kinds of materials, like limonite ($2Fe_2O \cdot 3H_2O$), the reddish color would be explained.

15.2 SPACE OBSERVATIONS OF MARS

Mars has been the target of a series of spacecraft launched by both the United States and the Soviet Union. The earliest of these spacecraft were flybys, which attempted to photograph and otherwise study the Martian surface and atmosphere in the few hours they were in good position as they passed by (Table 15–1).

In 1971, the United States sent out a spacecraft not just to fly by Mars for only a few hours, but actually to orbit the planet and send back data for a year or more. This spacecraft, Mariner 9, went into orbit around Mars in November 1971, after a 5-month voyage from Earth.

But when the spacecraft successfully reached Mars, our first reaction was as much disappointment as it was elation. While Mariner 9 was en route, a tremendous dust storm had come up, almost completely obscuring the entire surface of the planet. Only the south polar cap and 4 dark spots were visible.

In the next section, we will discuss the Vikings, the most recent spacecraft to visit Mars.

The storm began to settle after a few weeks, with the polar caps best visible through the thinning dust. Finally, three months after the spacecraft arrived, the surface of Mars was completely visible, and Mariner 9 could proceed with its mapping mission. It mapped the entire surface at a resolution of 1 km, and further photographed selected areas at a resolution 10 times better than this limit. The spacecraft was to last another year, and so more than completed its assigned tasks—in fact, we even got the scientific bonus of being able to see how the dust acted at the end of the storm. These observations gave us information both about the winds aloft and about the heights and shapes of the surface features. And studies of the temperature changes on Mars that resulted from the dust storm are helping us assess the possible results of some kinds of atmospheric pollution on Earth's temperatures.

Now that we have the data taken after the dust storm had ended, we can classify four major "geological" areas on Mars: volcanic regions, canyon areas, expanses of craters, and terraced areas near the poles.

"Geo-" is from the Greek for Earth. Strictly speaking, "geology" applies only to the Earth, but the term is now more widely used, especially for the application of geological principles to the Moon and planets.

TABLE 15–1 UNMANNED PROBES LAUNCHED TO MARS

Spacecraft	Launched by	Arrival Date	Comment
Mariner 4	USA	July 1965	Flyby
Mariner 6	USA	July 1969	Flyby
Mariner 7	USA	August 1969	Flyby
Mariner 9	USA	November 1971	Orbiter
Mars 2	USSR	November 1971	Orbiter; lander lost
Mars 3	USSR	December 1971	Orbiter and lander
Mars 4	USSR	February 1974	Flew by
Mars 5	USSR	February 1974	Orbiter
Mars 6	USSR	March 1974	Orbiter; lander failed
Mars 7	USSR	March 1974	Orbiter; lander failed
Viking 1	USA	June 1976	Orbiter and lander
Viking 2	USA	August 1976	Orbiter and lander

Figure 15–4 Olympus Mons. Its height profile, measured from a radar on Mariner 9, is also shown.

Mauna Kea is shown in Fig. 14–1.

Even through the storm, there had been an indication of four dark spots on the Martian surface. When the dust cleared, these proved to be volcanoes. A chief surprise of the Mariner 9 mission was the discovery of extensive areas of vulcanism on Mars. The largest of the volcanoes, which corresponds in position to the surface marking long known as Nix Olympica, "the snow of Olympus," is named Olympus Mons, "Mount Olympus." It is a huge volcano—600 kilometers across at its base and about 25 kilometers high (Fig. 15–4). The tallest volcano on Earth is Mauna Kea, in the Hawaiian islands, if we measure its height from its base deep below the ocean. Mauna Kea is only 9 kilometers high, taller than Everest. And remember, it is on the Earth, a much bigger planet than Mars. Olympus Mons is crowned with a crater 65 kilometers wide; Manhattan Island could be easily dropped inside the crater. The volcanoes on Mars would release oxygen and hydrogen into the atmosphere in the form of water and carbon dioxide.

Besides the volcanoes, another surprise on Mars was the discovery of systems of canyons. One tremendous canyon (Fig. 15–5)—about 5000 kilometers long—is as big as the United States and comparable in size to the Rift Valley in Africa, the longest geological fault on Earth. Here again the size of the canyon with respect to Mars is proportionately larger than any geologic formations on the Earth.

Perhaps the most amazing discovery on Mars was the presence of sinuous channels. These are on a smaller scale than the *canali* that Schiaparelli had seen, and are entirely different phenomena. Some of the channels show tributaries (Fig. 15–6), and the bottoms of some of the channels show the characteristic features that stream beds on Earth have. Even though water cannot exist on the surface of Mars under today's conditions, it is difficult to think of other ways to explain them satisfactorily than to say that the channels were cut by running water in the past.

This is particularly interesting because biologists feel that water is necessary for the formation and evolution of life. The presence of water on Mars, therefore, even in the past, may indicate that life could have formed and may even have survived. If we could discover life on another planet that arose independently of life on Earth, comparison of the life forms would tell us what is important to cells and to life, and what things may be of peripheral importance. For us on the Earth, the implications of these discoveries for advances in medicine, not to mention theology, are obvious.

If there had been water on Mars in the past, and in quantities great enough to cut the river beds, where has it all gone? One possible solution is

Figure 15–5 The tremendous canyon, now named Valles Marineris—Mariner Valleys—is about 5000 km long, nearly the diameter of the United States. The canyon is 120 km wide and 6 km deep.

Figure 15–6 Mariner 9 photographed stream beds that showed many signs of the past presence of running water.

Figure 15–7 This mosaic of Mariner 9 views showed the north polar cap, which is shrinking as a result of Martian springtime.

that it has been bound into the rocks by an interaction with the carbon dioxide that is known to make up the largest part of the Martian atmosphere. Or the water could have been absorbed by the surface soil.

Some of the water is bound in the polar caps (Fig. 15–7). At the time of Mariner 9, we thought that the polar caps were mostly frozen carbon dioxide—"dry ice"; the presence of water was controversial. In the winter these caps extend down to latitude 50°.

If the water trapped in the polar caps or under the surface of Mars was released at some time in the past, as liquid water on the surface or as water vapor in the atmosphere, it may have been sufficient to make Mars verdant. Perhaps the polar caps melted, releasing quantities of water because dust that settled on the ice caused more sunlight to be absorbed or a widespread episode of vulcanism caused a general heating. Climatic change could also have resulted from a change in the orientation of Mars's axis of rotation, from a change in the solar luminosity, or from a variation in atmospheric composition (which would affect the greenhouse effect).

The composition of the atmosphere is measured by looking at the spectrum. This can even be done from the Earth, but problems arise because the spectrum of the Earth's own atmosphere interferes, particularly at the wavelengths of the water vapor lines.

We found that the Martian atmosphere is composed of 90 per cent carbon dioxide (CO_2) with small amounts of carbon monoxide (CO), oxygen (O, O_2, and O_3), and water (H_2O). The surface pressure is 1 per cent of that which exists on the Earth's surface, but the pressure decreases with altitude more slowly than it does on Earth.

Scientists have been concerned for many years about the effects of our spacecraft landing on other planets. This discussion has particular immediacy for Mars. Perhaps a spacecraft would bring along micro-organisms that would contaminate the planet's atmosphere. When later spacecraft come to explore, they might find life forms that had come from the contamination introduced earlier rather than from life indigenous to the planet.

Different sides are taken in the controversy over how the possibility of contamination should affect our future plans. Some scientists feel that we should not send landers until we are ready for more detailed exploration. Others believe that the risk of contamination is low and say we are ready right now to land on other planets. As a compromise, certain sterilization procedures have been worked out for spacecraft to Mars and the other planets. The spacecraft are baked under high temperatures for some time.

15.3 VIKING!

In the summer of 1976, two U.S. spacecraft named Viking reached Mars after flights of about 10 months. Each spacecraft contained two parts: an orbiter and a lander. The orbiter served two roles; it not only used its cameras and other instruments to map and analyze the surface but also served as a relay station for the radio signals from the lander to Earth. The lander's role was surely no less significant: its major task was to sample the surface and decide whether there was life on Mars!

Viking 1 reached Mars in June of 1976 after a flight that led to spectacular large-scale views of the Martian surface (Color Plate 35). Viking 1 orbited Mars for a month while it radioed back pictures to scientists on Earth who were trying to determine a relatively safe spot to land.

The orbiter part of the mission alone was extremely successful. Observations were obtained at higher resolution than that provided by Mariner 9 of the giant volcanoes (Fig. 15–8) and of the huge 5000-km-long valley. Even the atmosphere could be observed (Fig. 15–9).

The first choices for a landing site, made long in advance based on Mariner 9 pictures, had to be rejected when Viking's pictures revealed dangerous boulders or craters that could overturn the spacecraft. But finally a site was selected, and on July 20, 1976, exactly seven years after the first manned landing on the Moon, Viking 1's lander descended safely onto a plain called Chryse.

The views showed rocks of several kinds (Color Plates 36–38), covered with reddish material that is probably limonite, an iron oxide compound. Sand dunes were also visible (Fig. 15–10). The sky on Mars turns out to be pink, from reddish dust suspended in the air as a result of one of Mars's frequent dust storms.

A series of experiments aboard the lander was designed to search for signs of life. A long arm was deployed (Color Plate 36) and a shovel at its end dug up a bit of the Martian surface (Fig. 15–11). The soil was dumped into three experiments that searched for such signs of life as respiration and metabolism. The results were astonishing. The biological experiments sent back signals that were similar to those that would be caused on Earth by biological rather than by mere chemical processes. Still, scientists searched hard for non-biological explanations. It is possible that some strange chemical process mimics life in these experiments.

Figure 15–8 Olympus Mons photographed by the Viking 1 orbiter. Clouds extend up most of its 25-km height. The volcanic crater, 80 km across, extends into the stratosphere and was cloud free. Mars's limb shows several haze layers.

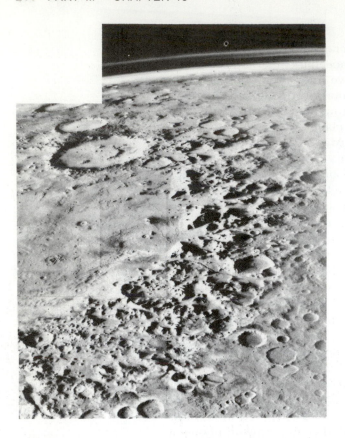

Figure 15–9 This oblique view from the Viking 1 orbiter shows Argyre, the smooth plain at left center. The Martian atmosphere was unusually clear when this photograph was taken, and craters can be seen nearly to the horizon. The brightness of the horizon results mainly from a thin haze. Detached layers of haze can be seen to extend from 25 to 40 km above the horizon, and may be crystals of carbon dioxide.

Figure 15–10 This 100° view of the Martian surface was taken from the Viking 1 lander, looking northeast at left and southeast at right. It shows a dune field with features similar to many seen in the deserts of Earth. From the shape of the peaks, it seems that the dunes move from upper left to lower right. The large boulder at the left is about 8 meters from the lander and is 1 × 3 meters in size. The boom that supports Viking's weather station cuts through the center of the picture. ("Chance of precipitation," the local newscaster would say, "0 per cent.")

One important experiment was much more negative in evaluating the chance that there is life on Mars. It analyzed the soil and looked for traces of organic compounds. On Earth, many organic compounds left over from dead forms of life remain in the soil; the life forms themselves are only a tiny fraction of the organic material. On Mars, who knows? Perhaps life forms evolved that efficiently used up their predecessors. Still, the absence of organic material from the Martian soil is a strong argument against the presence of life on Mars.

The Viking 2 lander descended on September 3, 1976, on a site that was thought to be much more favorable to possible life forms than was the Viking 1 site. Again, Viking 2 found a variety of types of rocks (Fig. 15–12 and Color Plate 38) and a pink sky. The atmosphere at this second site, Utopia Planitia, contains three or four times more water vapor than has been observed near Chryse. Viking 2's experiments sent back data similar to Viking 1's, with roughly positive results for the biological experiments and negative results for the chemical ones.

Since the Viking results were sent back, scientists on Earth have devised a chemical way that the Viking results can be explained. Thus we cannot say that the presence of biological activity has been shown.

Anyway, we mustn't always ask for a yes or no answer. Even if the life signs detected by Viking come from chemical rather than biological processes, we have still learned of some fascinating new chemistry going on. When life arose on Earth, it probably took up chemical processes that had been previously in existence. Similarly, if life arose on Mars in the past or would normally arise there in the future (assuming our visiting there doesn't contaminate Mars and ruin the chances for the beginning of indigenous life), we might expect the life forms to use the chemical processes that already existed. So even if we haven't detected life itself, we may well have learned important things about its origin.

Viking 2 has the additional scientific value that its seismograph can measure marsquakes. The seismograph on Viking 1 had been the only instrument on that craft that did not work on the Martian surface. The time-delay information that would have been afforded by two working seismographs has been lost, but still the seismic information from Viking 2 is exciting. Some marsquakes have been detected.

The presence of two orbiters circling Mars allowed a division of duties. One continued to relay data from the two Viking landers to Earth, while the orbit of the other was changed so that it passed over Mars's north pole. The main polar cap, made of dry ice, had retreated as a result of summertime and only a residual ice cap was left. From the amount of water vapor detected with its spectrograph, and from the fact that the temperature was too high for this residual ice cap to be frozen carbon dioxide, the conclusion has been reached that it is made of water ice. This important result indicates the probable presence of lots of water on the Martian surface in the past—in accordance with the existence of the stream beds—and seems to raise the probability that life has existed or does exist on Mars.

What's the next step? There are no definite plans, but the Viking missions have been so spectacularly successful that it seems reasonable that further exploration of Mars will be given priority. Carl Sagan, a Cornell astronomer who not only has worked with the Viking observations but also has become the foremost spokesman to the public about the project and its results, points out the frustration at not being able to see over the horizon. He asks for a roving vehicle that could go over many kilometers of the Martian surface. Enough spare parts exist from Viking that this could be done at a cost that is reasonable for such complex space experiments. So perhaps we can hope for a Martian rover in the 1980's. Manned space flights to Mars would be much more expensive, and at present seem much farther off.

The more we learn about it, the more interesting Mars becomes.

Figure 15–11 The first trench, 7 cm wide × 5 cm deep × 15 cm long, excavated by Viking 1 to collect Martian soil for the life-detection experiments. From the fact that the walls have not slumped, we can conclude that the soil is as cohesive as wet sand.

An unmanned mission to bring soil back to Earth could be launched in the late 1980's if we start planning now.

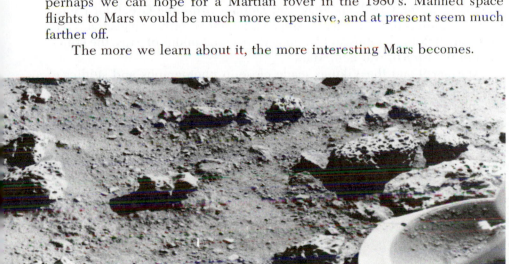

Figure 15–12 The first photograph of the surface of Mars taken from Viking 2 on September 3, 1976. A wide variety of rocks are seen to lie on a surface of fine-grained material. One of the lander's footpads can be seen at the lower right.

Figure 15–13 *(A)* A Viking 1 mosaic of photographs of Phobos. The largest two craters on Phobos have been named after Asaph Hall and after Angelina Stickney, his wife, who encouraged the search. *(B)* A close-up view of Phobos from Viking Orbiter 2. Phobos's north pole is near upper left. It is heavily cratered as was expected but, surprisingly, shows striations and chains of small craters. Similar chains of small craters on Earth's Moon, on Mars, and on Mercury were formed by secondary cratering from a larger impact.

Figure 15–14 This photograph of Deimos was obtained by Viking Orbiter 1 from a range of 3300 km. About half of the side facing the camera is illuminated, and that lighted portion measures about 12 × 8 km. While Mariner 9 pictures taken nearly five years ago from greater distances showed only a few large craters on Deimos, the improved resolution here reveals a heavily-cratered and presumably very old surface. At least a dozen craters are prominent—the two largest measuring 1.3 km and 1 km. Craters as small as 100 meters pockmark the surface. Still more recent photographs, taken when the Viking 2 orbiter went within 65 km of Deimos, show a surprisingly smooth surface. Thus it, and Phobos as well, may be covered with dust.

15.4 SATELLITES OF MARS

Mars has two moons. In a sense, these are the first moons we have met in our study of the solar system, for Mercury and Venus have no moons at all, and the Earth's Moon is so large relative to its parent that we may consider the Earth and Moon as a double planet system. The moons of Mars, Phobos (from the Greek for Fear) and Deimos (from the Greek for Panic), are mere chunks of rock, only 27 and 15 kilometers across, respectively. (In Greek mythology, Phobos and Deimos were companions of Ares, the equivalent of the Roman war god, Mars.)

Mariner 9 and the Vikings made closeup photographs and studies of Phobos and Deimos (Figs. 15–13 and 15–14). Each turns out to be not at all like our Moon, which is a spherical, planet-like body. Phobos and Deimos are just cratered chunks of rock. Chunks may have been broken off on impact with meteoroids.

SUMMARY AND OUTLINE OF CHAPTER 15

Aims: To study Mars, and to assess the chances of finding life there

Observations from the Earth (Section 15.1)
 Rotation period

Surface markings
Dust storms
Seasonal changes
Space observations
Many space probes sent from the U.S. and U.S.S.R.
Mariner 9 orbiter mapped all of Mars (Section 15.2)
Volcanic regions indicating geological activity
Canyons larger than Earth counterparts
Craters and terraced areas near the poles
Signs of water evident
Underlying layer of polar caps
Sinuous channels, some with tributaries
Channel bottoms show markings similar to terrestrial stream beds
Viking (Section 15.3)
Orbiters and landers
Closeup photographs of surface rocks
Analysis of soil
Both positive and negative results for tests for life
Seismograph to study marsquakes
Residual polar cap is made of water ice
Satellites of Mars (Section 15.4)
Phobos and Deimos: small, irregular chunks of rock
Close-up photographs from Mariner 9 and Viking

QUESTIONS

1. Outline the features of Mars that make scientists think that it is a good place to search for life.

2. If Mars were closer to the Sun, would you expect its atmosphere to be more or less dense than it is now? Explain.

3. From the relative masses and radii (see Appendix 5), verify that the density of Mars is about 85 per cent that of Earth.

4. Approximately how much more solar radiation strikes a square meter of the Earth than a square meter of Mars?

5. Compare the tallest volcanoes on Earth and Mars relative to the diameters of the planets.

6. What evidence is there that there is, or has been, water on Mars?

7. Compare the fractional abundances of various major gases in the atmospheres of Mars and the Earth.

8. How does the absolute abundance of carbon dioxide in the atmosphere of Mars, that is, the total amount of carbon dioxide, compare with the absolute abundance of carbon dioxide in the Earth's atmosphere.

9. List the various techniques for determining the composition of the atmosphere of Mars.

10. List the evidence from Viking for and against the existence of life on Mars.

11. (a) Aside from the biology experiments, list three types of observations made from the Viking landers. (b) What are two types of observations made from the Viking orbiters?

12. Why aren't Phobos and Deimos regular, round objects like the Earth's Moon?

TOPICS FOR DISCUSSION

1. Discuss the problems of sterilizing spacecraft, and whether we should risk contaminating Mars by landing spacecraft.

2. Discuss the additional problems of contamination that manned exploration would bring. Analyze whether, in this context, we should proceed with manned exploration in the next decade or two.

Jupiter

The planets beyond the asteroid belt—Jupiter, Saturn, Uranus, and Neptune—are very different from the four "terrestrial" planets. These *giant planets*, or *Jovian planets*, not only are much bigger and more massive, but are also much less dense. This suggests that the internal structure of these giant planets is entirely different from that of the four terrestrial planets. The ninth planet, Pluto (to be discussed in Section 17.3), represents still another different type.

Jovian is the adjectival form of Jupiter.

Jupiter, also called Jove, was the chief Roman deity.

16.1 FUNDAMENTAL PROPERTIES

The largest planet, Jupiter, dominates the Sun's planetary system. Jupiter is 5 A.U. from the Sun and revolves once every 12 years. It alone contains two-thirds of the mass in the solar system outside of the Sun, 318 times as much mass as the Earth. Jupiter has at least fourteen moons of its own and is a miniature planetary system in itself. It is often seen as a bright object in our night sky, and observations with even a small telescope reveal bands of clouds across its surface and show four of its moons.

Jupiter's density is only 1.3 grams/cm^3, not much greater than the 1 gram/cm^3 density of water. This tells us that any core of heavy elements (such as iron) that Jupiter may have does not make up as substantial a fraction of Jupiter's mass as the cores of the inner planets make up of their planets' masses. Jupiter, rather, is mainly composed of the lighter elements hydrogen and helium, sufficiently compressed to reach this density (Fig.

Figure 16–1 The composition of Jupiter.

The north pole of Jupiter, photographed from the Pioneer 11 spacecraft. The Great Red Spot is also visible. The symbol for Jupiter appears at the top of the page.

Figure 16–2 Jupiter, photographed from the Earth, shows belts and zones of different shades and colors. By convention (that is, the set of terms we arbitrarily choose to use) the bright horizontal bands are called *zones* and the dark horizontal bands are called *belts.*

Figure 16–3 Jupiter's differential rotation, which causes the points that were aligned at time T_1 to be separated at time T_2, spreads out the clouds into bands. The rate at which this happens is exaggerated in this drawing.

16–1). Its chemical composition is closer to that of the Sun and other stars than it is to that of the Earth.

The bands on the surface of Jupiter appear in subtle shades of orange, brown, gray, yellow, cream, and light blue, and are beautiful to see (Fig. 16–2). They are in constant turmoil; the shapes and distribution of bands continually change in a matter of days.

Jupiter rotates very rapidly, once every 10 hours. Undoubtedly, this rapid spin rate is a major reason for the colorful bands, which are clouds spread out parallel to the equator. Interestingly, the clouds do not rotate as would the surface of a solid body: the clouds at the equator rotate slightly faster than the clouds nearer the poles, completing their rotation about five minutes earlier each time around. Thus every fifty days, the clouds at the equator have made an additional rotation. Jupiter isn't solid; it probably has no crustal surface at all. Its gas just gets denser and denser, eventually liquefying, at deeper and deeper levels.

The most prominent feature of the visible cloud surface of Jupiter is a large reddish oval known as the *Great Red Spot* (see Fig. 16–2 and Color Plate 39). It is about 14,000 km × 30,000 km, many times larger than the Earth, and drifts about slowly with respect to the clouds as the planet rotates. The Great Red Spot is a relatively stable feature, for it has been visible for at least 150 years.

The fact that Jupiter emits radio waves indicates that Jupiter, even more so than the Earth, has a strong magnetic field and a strong *radiation belt* (actually, a belt filled with magnetic field in which particles are trapped, a large-scale version of the Van Allen belts of Earth), because we can account for the presence of the radio radiation only with mechanisms that involve such magnetic fields and belts. Some of the radio emission comes from interactions of high-energy particles passing through space with Jupiter's magnetic field.

A   **B**

Figure 16–4 *(A)* Jupiter with the four Galilean satellites. These satellites were named Io, Europa, Ganymede, and Callisto by Simon Marius, a German astronomer who independently discovered them. *(B)* Galileo's sketches of the changes in the positions of the satellites.

16.2   JUPITER'S MOONS

Jupiter has fourteen satellites. Four of the innermost satellites are very nicely named Io, Europa, Ganymede, and Callisto. At 5216 km in diameter, Ganymede is one of the biggest satellites in the solar system (Saturn's Titan is larger), and has a strong gravitational field.

These four moons are called the *Galilean satellites* (Fig. 16–4). They were discovered by Galileo in 1610 when he first looked at Jupiter with his telescope, and they have played a very important role in the history of astronomy. In particular, it was the fact that these particular satellites were noticed to be going around another planet, like a solar system in miniature, that supported Copernicus's heliocentric model of our solar system. Not everything revolved around the Earth!

Recently, in 1974, a study of the outermost Jovian moons was carried out with the Schmidt telescope at Palomar. An extra dot, the thirteenth Jovian satellite (Fig. 16–5), was found. In 1975, a fourteenth was detected.

Figure 16–5 The photograph on which the thirteenth moon of Jupiter, shown with an arrow, was discovered. The moon is named Leda. The telescope was set to track across the sky along with Jupiter; thus stars show as trails. The trails here correspond to anonymous—unnamed and unnumbered—stars between about 14th and 20th magnitudes.

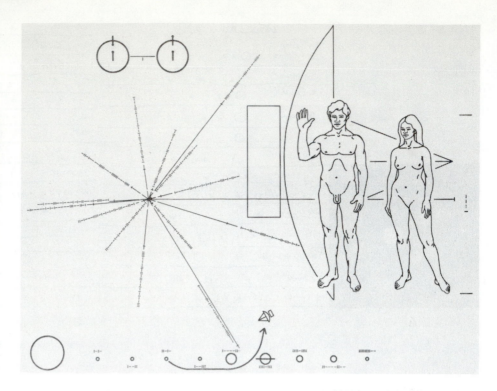

Figure 16–6 The plaques borne by Pioneers 10 and 11. A man and a woman are shown standing in front of an outline of the spacecraft, for scale. The spin-flip of hydrogen, described in Section 21.3, is shown at top left, also for scale, because even travelers from another solar system would know that the wavelength of the transition is 21 cm. The sun and planets, including distinctively ringed Saturn, and the spacecraft's trajectory from Earth, are shown at bottom. The directions and periods of several pulsars are shown at left. Numbers are given in the binary system.

Figure 16–7 An artist's conception of the Pioneer spacecraft.

16.3 SPACECRAFT OBSERVATIONS

We have seen that each of the planets we have discussed so far has recently been visited by spacecraft that have greatly advanced our knowledge about them. Jupiter is no exception. Two spacecraft, Pioneer 10 and Pioneer 11, gave us our first close-up views of the colossal planet in 1973 and 1974. The Pioneers did not go into orbit around Jupiter; in fact they are traveling fast enough to escape the solar system. By 1980, Pioneer 10 will be as far away from the Sun as Uranus is; beyond that, radio signals from the satellite will be too weak for us to detect. But until that time, the spacecraft will beam back valuable information on interplanetary conditions. In about 80,000 years the spacecraft will be about one parsec (about three light years) away from the solar system. A plaque that gives some information about the Pioneer's origin and about life on Earth was affixed to the spacecraft in case some interstellar traveler from another solar system might pick it up (Fig. 16–6).

The Great Red Spot shows very clearly in many of the images (Color Plate 39). It is a gaseous island about 35,000 km across, whose top is about 8 km above the rest of the atmosphere. It is now generally believed that the Spot is the vortex of a violent, long-lasting cyclonic storm.

There have been several suggestions as to the causes of the color of the Spot. Some of the possibilities are predicated on the idea that the Spot is a giant storm in which material from lower atmospheric layers may be rapidly brought up, changing color in the process because solar radiation acts on it or because it interacts with the water vapor in Jupiter's atmosphere.

Figure 16–8 The current model of the interior of Jupiter.

Data from the Pioneers increased our understanding not only of the atmosphere but also of the interior of Jupiter (Fig. 16–8). It appears that most of the interior is in liquid form. The central temperature may be between 13,000 and 35,000 K, with a pressure of 100 million times the pressure of the Earth's atmosphere measured at our sea level. Because of this high pressure, Jupiter might even have an interior composed of ultra-compressed hydrogen surrounding a rocky core consisting of 20 Earth masses of iron and silicates. The heavy elements should have sunk to the center, but we have no direct evidence that they are there. The liquid hydrogen is probably in a state called "metallic" because it would conduct heat and electricity; these properties are basic to our normal definition of "metal." (The unusual use of the word "metal" by stellar astronomers to mean all elements heavier than helium is not relevant here.) This metallic region is probably where Jupiter's magnetic field is generated. The metallic region makes up 75 per cent of Jupiter's mass.

Jupiter radiates twice as much heat as it receives from the Sun. No other planet comes even close to doing this. This fact is the key to understanding Jupiter's complex and beautiful cloud circulation pattern. There must be some internal energy source—perhaps the energy remaining from Jupiter's collapse from a primordial gas cloud 20 million km across to a protoplanet 700,000 km across, 5 times the present size of Jupiter. This catastrophic phase of collapse started, it has been suggested, when the temperature grew sufficiently high to break up hydrogen atoms. The rapid phase may have taken only 3 months to occur, following the 70,000 years it had previously taken to shrink from a more diffuse cloud. Jupiter is undoubtedly still contracting. Jupiter lacks the mass necessary, however, to have heated up enough to have become a star. If Jupiter had been about 75 times more massive, nuclear reactions would have been able to start. Then it would have become a star, and we would have been in a double star system (some of which are described in Section 4.1).

Pioneer 10 took spectacular pictures, but took them only in the plane of Jupiter's rotation, so that the photographs were taken from the same point of view that we have from Earth. But Pioneer 11 took a different approach to Jupiter. It came in over the polar region. It took this path for two reasons. First, when Pioneer 10 swept by Jupiter in late 1973, its instruments narrowly escaped crippling damage during its passage through the intense Jovian radiation belts. Pioneer 11 came three times closer to Jupiter, passing 40,000 km (0.6 Jupiter radii) above the cloud tops. Kepler's second law shows that Pioneer 11 was traveling much faster than Pioneer 10, and so spent less time in the radiation zone. It survived. Another reason it took this path was scientific. We can never see the polar regions from Earth, as Jupiter's axis is nearly perpendicular to our orbit. So scientists were anxious to get their first look at Jupiter's poles (page 248). The results were noteworthy; at high latitudes the circulation pattern of cloud bands that we see at the equator is destroyed and the bands break up into eddies. With information like this, we can now study "comparative atmospheres": those of the Earth, Venus, Mars, and Jupiter.

As a result of the Pioneer missions, we now know that Jupiter possesses a tremendous magnetic field, more intense than many scientists had expected (Fig. 16–9). (The existence of this magnetic field was known because of Jupiter's radio emission.) The magnetic field of Jupiter interacts with the solar wind as far as 7,000,000 km from the planet and forms a shock wave, as does the bow of a ship plowing through the ocean. At the height of Jupiter's clouds the magnetic field is 10 times that of the Earth, which itself has a strong field.

At the time that Pioneer 10 reached the distance of Saturn's orbit in 1976, it passed into a zone where the solar wind was absent. This is undoubtedly the effect of a long magnetic tail that Jupiter has. The tail may be a billion kilometers long and results from the stretching of some of Jupiter's

Figure 16–9 Jupiter's magnetosphere, the region of space occupied by the planet's magnetic field. It rotates at hundreds of thousands of miles per hour along with the planet (like a big wheel with Jupiter as the hub). The inner magnetosphere is shaped like a doughnut with the planet in the hole. The highly unstable outer magnetosphere is shaped as though the outer part of the doughnut had been squashed. The outer magnetosphere is "spongy." It pulses in the solar wind like a huge jellyfish, and often shrinks to one-third of its largest size.

ROTATION AXIS   MAGNETIC AXIS

RADIATION BELTS

SPACECRAFT TRAJECTORY

magnetic field lines by the solar wind. These field lines prevent the solar wind from entering the tail region.

16.4 RECENT OBSERVATIONS OF JUPITER'S MOONS

One of the side benefits of sending a probe to Jupiter is that merely tracking the rocket's trajectory tells us about the gravitational influences on it. Thus new improved values have been derived for the mass of the Galilean satellites, revising the old values by as much as 22 per cent.

All these Galilean satellites have been photographed and otherwise studied from the Pioneers. The four satellites all have surface temperatures of only 130 K to 155 K. From Pioneer 10, Ganymede showed a bright polar feature that may be a polar cap, plus other features that are probably craters and maria (Fig. 16–10).

Figure 16–10 Pioneer 10 observations of Ganymede. The three marked round spots resemble the appearance of maria.

16.5 FUTURE OBSERVATIONS OF JUPITER

The Pioneers are lighter spacecraft than the Mariners and spin around one axis, while the Mariners are completely stabilized. In 1977 NASA sent two Mariner-type craft, called Voyager, to Jupiter and Saturn. They are scheduled to arrive at Jupiter in 1979. Jupiter's large mass makes it a handy source of energy to use in order to send probes to more distant planets. Just as a sling, like David's, transfers energy to a stone, some of Jupiter's energy can be transferred to a spacecraft through a gravitational interaction. We will get more observations of Jupiter as a bonus in much of our exploration of the outer reaches of the solar system because of the use of this *gravity assist* method. NASA is planning a Jupiter Orbiter Probe mission for arrival at Jupiter in the mid-1980's. It will carry an orbiter that will take high-resolution pictures of Jupiter and four of its moons, and a probe that will descend into the Jovian atmosphere.

"And David put his hand in his bag, and took thence a stone, and slung it, and smote the Philistine [Goliath] in his forehead."
First Samuel 17:49

SUMMARY AND OUTLINE OF CHAPTER 16

Aims: To study Jupiter, the dominant planet of the solar system, and to study an example of a class of planets very different from the inner, terrestrial planets

Fundamental properties (Section 16.1)
    Highest mass and largest diameter of any planet
    Low density: 1.3 grams/cm^3

Composition primarily of hydrogen and helium
Colored bands on surface; Great Red Spot
Intense bursts of radio radiation
Strong magnetic field and radiation belt
Jupiter's moons (Section 16.2)
At least 14 satellites
Galilean satellites are largest
Moons still being discovered
Spacecraft observations (Section 16.3)
Pioneer 10 and 11 flybys in 1973 and 1974
Many photographs taken of cloud structure at higher resolution
Great Red Spot believed to be giant cyclonic storm
No solid surface detected: gaseous atmosphere and liquid interior
Jupiter radiates 2 to 3 times the heat it receives from the Sun (confirmed previous observations
Pioneer 11 photographed the Jovian polar regions, a task impossible from Earth
Tremendous magnetic field detected
Recent observations of Jupiter's moons (Section 16.4)
Future observations of Jupiter (Section 16.5)
Voyager spacecraft will arrive in 1979

QUESTIONS

1. Why does Jupiter appear brighter than Mars despite its greater distance from the Earth?

2. Assume that the Jovians put a dye in their atmosphere such that there is a green line running from the north pole to the south pole. Sketch the appearance of this line a few days later.

3. (a) How did we first know that Jupiter has a magnetic field? (b) What did the recent studies show?

4. It has been said that Jupiter is more like a star than a planet. What facts support this statement?

5. What advantages did Pioneers 10 and 11 have for photography over the 5-m telescope on Earth?

6. Aside from photography, what are two other types of observations made from the Pioneers to Jupiter?

7. How does the interior of Jupiter differ from the interior of the Earth?

8. (a) Describe the "gravity assist" method. (b) What does it help us do?

TOPICS FOR DISCUSSION

1. Although many of the most recent results about Jupiter came from Pioneers 10 and 11, prior ground-based studies had told us many things. Discuss the status of our pre-1973 knowledge of

Jupiter, and specify both some things about which space research did not add appreciably to our knowledge and some things about which space research led to a major revision of our knowledge.

2. If we could somehow set up a space station suspended in Jupiter's clouds, the astronauts would find a huge gravitational force on them. What are some of the effects this would have? (The novel *Slapstick* by Kurt Vonnegut (Delacorte Press, 1976) deals with some of the problems that would arise if gravity were different in strength from what we are used to.)

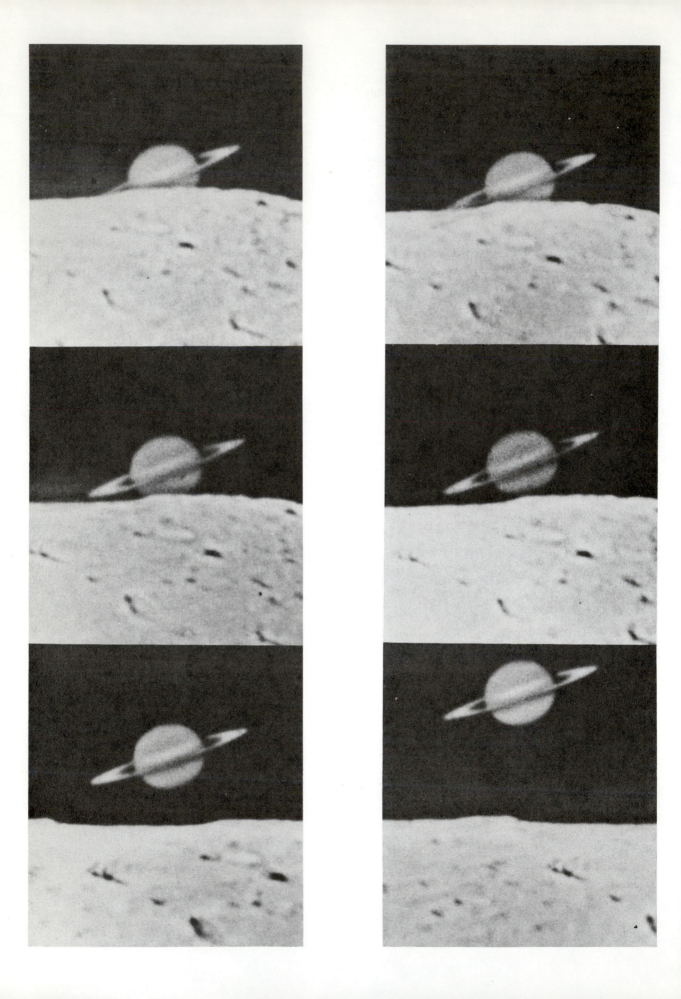

♄ ♅ ♆ ♇ 17

The Outer Planets —Saturn, Uranus, Neptune, and Pluto

Saturn, Uranus, and Neptune are all giant planets, in many ways similar to Jupiter. Both Saturn and Uranus are now known to have rings, although we can see only Saturn's rings directly. The outermost planet, Pluto, is quite different from the four giant planets. We shall discuss these planets in turn.

17.1 SATURN

Saturn is shown in Color Plate 40.

Saturn is the most beautiful object in our solar system, and possibly even the most beautiful object we can see in the sky. The glory of its system of rings makes it stand out even in small telescopes.

Saturn, like Jupiter, Uranus, and Neptune, is a giant planet. Saturn is 9.5 A.U. from the Sun, and so, as we can deduce from Kepler's third law or by simply observing, it has a lengthy year, equivalent to 30 Earth years.

The giant planets are characterized by low densities. Saturn has the lowest density of any planet in our solar system: only 0.7 gm/cm³, which is 70 per cent the density of water. Thus, if we could find a big enough bathtub, Saturn, like Ivory Soap, would float (Fig. 17–1). It could have a core of heavy elements making up 20 per cent of its interior.

The rings of Saturn are either the bits of a moon that were torn apart by Saturn's gravity or chunks that failed to accrete into a moon at the time when the planet and the other moons were forming. These bits or chunks

Figure 17–1 Saturn's density is lower than that of water.

Saturn, emerging from an occultation by the Moon. The symbols for Saturn, Uranus, Neptune, and Pluto appear from left to right, respectively, at the top of this page.

Figure 17–2 The rings of Saturn, photographed on January 19, 1973. (New Mexico State University Observatory photograph)

Each rock in the ring independently follows Kepler's laws, so the ones closer to Saturn orbit more quickly than the ones farther out.

Artificial satellites that we send up to orbit around the Earth are constructed of sufficiently rigid materials that they do not break up even though they are within the Earth's Roche limit.

formed individual rocks and spread out in concentric rings around Saturn. There is a sphere for each planet, called *Roche's limit*, or *the Roche limit*, inside of which blobs of gas are separated from each other; an agglomeration of blobs cannot form and be held together by the blobs' own gravity. The forces that tend to tear the blobs apart from each other are called *tidal forces*; they arise because some blobs are closer to the planet than others and are thus subject to higher gravity. (The differing gravitational forces tend to pull the blobs apart; similarly, differential gravitational forces from the Moon and the Sun cause the ocean tides on Earth.) The radius of the Roche limit varies with the amount of mass in the parent body. The Sun also has a Roche limit, but all the planets lie outside it. All the moons of the various planets lie outside the respective Roche limits, but Saturn's rings lie inside Saturn's Roche limit, so it is not surprising that the material in the rings is not collected into a single orbiting body.

There are several concentric rings around Saturn. The brightest ring is separated from a fainter outer ring by what is called *Cassini's division* (Fig. 17–2). Two fainter rings are inside the brightest ring. The gaps in the

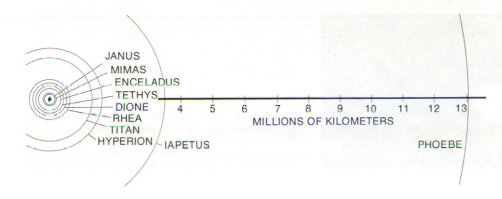

Figure 17-4 Saturn's moons.

ring structure probably result from gravitational effects from Saturn's satellites, and finer structure surely exists in the rings than the few divisions that are visible from the Earth.

We can see that the rings are not solid objects, for stars occulted by the rings can be seen shining dimly through. Even though the rings are 275,000 km across, they are very thin from top to bottom: not more than 10 km and perhaps less. Some astronomers think they may be only 10 to 100 meters thick from top to bottom, a good trick for something hundreds of thousands of km across! Radar waves were first bounced off the rings in 1973. The results of the radar experiments show that the particles in the ring are probably rough chunks of ice or rock at least a few centimeters and possibly a meter across.

Like Jupiter, Saturn rotates very quickly on its axis. It also has delicately colored bands of clouds.

Saturn has ten moons (Fig. 17-4). The tenth moon, the innermost, was discovered in 1966 when Saturn's rings were seen edge-on, and thus did not interfere with the view of the space nearest the planet. The new moon was named Janus, after the Roman deity who presided over beginnings and endings. (Janus is the figure commonly represented with two faces oriented in opposite directions.)

All the moons, except Titan, range from about 130 km to 1600 km across. Planet-size Titan, however, is a different kind of body. An atmosphere has been detected on Titan, and methane has been found in it. Because of this atmosphere, a greenhouse effect may have warmed the surface of Titan, and so Titan has become one of the more interesting places in the solar system on which we can search for life—some astronomers think that there may even be more chance of finding life there than on Mars.

An occultation occurs when one celestial object passes in front of, occults, another.

Figure 17-3 The rings of Saturn photographed at various times. The rings extend far out in the equatorial plane of Saturn, and range from about 70,000 to over 135,000 km from the planet's center. They are inclined to Saturn's orbit by 27°. Over a 30-year period, we sometimes see them from a vantage point of 27° above their northern side, sometimes from 27° below their southern side, and at intermediate angles at intermediate times. When seen edge on, they are all but invisible. (Lowell Observatory photographs.)

Figure 17–5 Saturn and seven of its ten moons. We can see (1) Mimas, (2) Enceladus, (3) Tethys, (4) Dione, (5) Rhea, (6) Titan, and (8) Iapetus. Only (7) Hyperion, (9) Phoebe, and the innermost moon, (0) Janus, cannot be seen in this photograph. A faint star is visible at top.

Pioneer 11, the second of the two craft that passed by Jupiter, has been targeted to fly to Saturn, and will arrive there in September 1979 (Fig. 17–6). Jupiter's gravity sped up the spacecraft in a manner opposite to the way that Venus's gravity slowed down Mariner 10 en route to Mercury. Because of this gravity assist, the spacecraft is able to go all the way to Saturn even though we only expended enough power at launch to send it to Jupiter.

Current U.S. plans for further study of Saturn are for Voyagers 1 and 2 to reach Jupiter in 1979, and Saturn in 1980 and 1981.

17.2 URANUS AND NEPTUNE

Figure 17–6 The trajectories of Pioneers 10 and 11 are shown with dotted lines and the orbits of the Earth, Saturn, and Jupiter are shown with solid lines. The positions of the planets and the two spacecraft are shown for each January 1st.

There are two other giant planets beyond Saturn: Uranus (U'ranus) and Neptune. Both Uranus and Neptune are large planets, about 50,000 km across and about 15 times more massive than the Earth. The densities of Uranus and Neptune are low (1.2 and 1.7 g/cm³, respectively). Their albedos are between 60 per cent and 70 per cent in the visible part of the spectrum, which indicates that they are covered with clouds.

Figure 17-7 Uranus, photographed from the ground in blue light, shows no surface markings.

Uranus was discovered to be a planet by William Herschel in England in 1781, but it had been plotted on sky maps for about a hundred years prior to that with the thought that it was just another star. It revolves around the Sun in 84 years, at an average distance of more than 19 A.U. from the Sun. Even at the most favorable possible place in its orbit, which has an eccentricity of almost 5 per cent, it is never closer than 17 A.U. (2.5 billion km) to the Earth. Thus Uranus never gets any larger in the sky than 3.6 seconds of arc, and studying its surface structure from the surface of the Earth is very difficult (Fig. 17-7). Our most accurate measurements of Uranus's diameter were obtained from a telescope sent aloft in a balloon, but even with this resolution no detail on the surface can be seen.

The other planets rotate such that their axes of rotation are roughly parallel to their axes of revolution around the Sun; Uranus is different, for its axis of rotation is roughly perpendicular to the other planetary axes. Uranus's axis of rotation lies in the plane of its orbit (Fig. 17-8). Sometimes, one of Uranus's poles faces the Earth, 21 years later its equator crosses our field of view, and then another 21 years later we face the other pole. Thus there are strange seasonal effects on Uranus, and the differences in the heating pattern from year to year may be the reason that there are no cloud bands. When we understand just how the heating affects the clouds, we will be closer to understanding our own Earth's weather systems. Of course, even though there is an effect because the angle at which sunlight hits different parts of the surface of Uranus varies with the

Uranus, in Greek mythology, was the personification of Heaven and ruler of the world, the son and husband of Gaea, the Earth. Neptune, in Roman mythology, was the god of the sea, and the planet Neptune's astronomical symbol reflects that origin.

Figure 17-8 Its axis of rotation lies in the plane of Uranus's orbit.

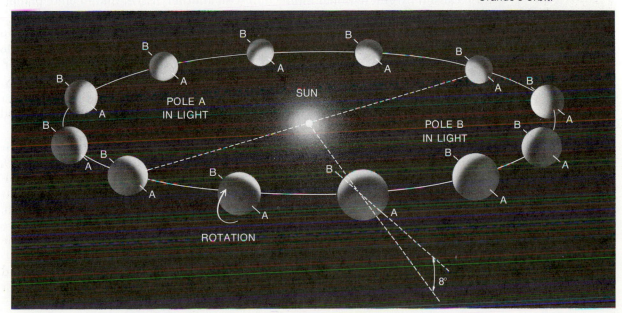

POLE A
IN LIGHT

SUN

POLE B
IN LIGHT

ROTATION

8°

Figure 17–9 At a 1977 occultation of a star by Uranus, observers were surprised to find that the star winked out briefly 5 times before the main occultation and a symmetric 5 times afterward. This means that Uranus is surrounded by rings, as is Saturn. Only 3 of the rings are shown in this drawing.

Figure 17–10 The moons of Uranus.

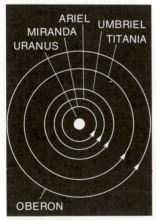

Figure 17–11 Uranus and its moons.

season, we must recall that Uranus is very distant from the Sun, so the intensity of sunlight is never very great. Uranus is always very cold, perhaps 90 K (−185° C).

In 1977, Uranus occulted a faint star. The occultation was predicted to be visible only from the Indian Ocean southwest of Australia, so a team of Cornell scientists flew in NASA's Kuiper Airborne Laboratory, an instrumented airplane, to observe it. Surprisingly, about half an hour before the predicted time of occultation, a few slight dips were detected in the brightness of the star. (The equipment had been turned on early because the exact time of the occultation was uncertain.) Similar dips were recorded, in the reverse order, about half an hour after the occultation. Furthermore, ground-based observers detected similar dips, even though the star was never completely occulted for them. These dips indicate that Uranus is surrounded by at least five rings (Fig. 17–9). They are called, from the innermost outward, by the first five Greek letters. Note that we have detected the rings only by this occultation; they are presumably too faint to be observed directly.

Uranus has five moons (Fig. 17–10), each with a beautiful name. From the innermost to the outermost, they are Miranda, Ariel, Umbriel, Titania, and Oberon. The innermost, Miranda, was discovered fairly recently, in 1948, and is the smallest. The other four moons range from 400 to 1000 km across. Very little is known about them.

At present it is planned for the second of the Voyagers, launched in 1977, to pass not only Jupiter and Saturn, but also Uranus, arriving there in 1985 or 1986. The definite decision need not be made until we see in 1980 how well the first of the Voyagers has succeeded in studying Saturn. The option will remain until then to target the second Voyager for another close-up of Titan instead of sending it on toward Uranus.

Neptune is even farther away than Uranus, 30 A.U. compared to about 20 A.U. (actually 19 A.U.). Its orbital period, by Kepler's third law, is thus 165 years in length. Its discovery was a triumph of the modern era of Newtonian astronomy. Neptune had not been known until mathematicians analyzed the deviations from an elliptical orbit shown by Uranus. These deviations are small, but were detectable, and could have been caused by gravitational interaction with another, as yet undetected, planet.

The story is one of astronomy's most famous anecdotes. John C. Adams in England (Fig. 17–12) predicted positions for the new planet in 1845, but the astronomy professor at Cambridge did not bother to try to observe this prediction of a recent college graduate. Through a misunderstanding, the Astronomer Royal did not take Adams's calculations seriously. The story then continues in France, where a year later Urbain Leverrier was independently working on predicting the position of the undetected planet.

When the Astronomer Royal saw in the scientific journals that Leverrier's work was progressing well (Adams's work had not been made public), for nationalistic reasons he began to be more responsive to Adams's calculations. But the search for the new planet, though begun in Cambridge, was carried out half-heartedly. Neither did French observers take up the search. By this time it was 1846. Leverrier sent his predictions to an acquaintance at Berlin, who enthusiastically began observing and discovered Neptune within hours.

Neptune's orbit is so large that it takes a very long time to travel around the Sun, and it has not yet made a full orbit since it was discovered. It never

Figure 17–12 "1841. July 3. Formed a design in the beginning of this week, of investigating, as soon as possible after taking my degree, the irregularities in the motion of Uranus..."

Memoranda.

1841. July 3. Formed a design, in the beginning of this week, of investigating as soon as possible after taking my degree, the irregularities in the motion of Uranus, wh. are

appears larger than 2 seconds of arc across in our sky, so is always very difficult to study. Even measuring its diameter accurately is hard. In 1968, Neptune occulted (passed in front of) an eighth magnitude star, a star about as bright as Neptune itself. From observations of the rate at which the star dimmed, astronomers could deduce information about Neptune's upper atmosphere and eventually the atmosphere's temperature and pressure structure. Further, from the length of time that the star was hidden by the disk of Neptune, they deduced Neptune's diameter.

Neptune and Uranus appear greenish in a telescope. Only hydrogen and methane have thus far been detected with spectrographs. Neptune has two moons (Fig. 17–13). Triton (a sea god, son of Poseidon) is large, perhaps 6000 km across. Its orbit is 350,000 km in diameter. A second moon, Nereid (a sea nymph), is small.

Our hope for getting better information about Neptune and its moons is that a spacecraft that might be launched to Uranus in the mid-1980's could continue on to Neptune, but such a mission is not currently scheduled.

Figure 17–13 Neptune and its satellites. Triton is close to Neptune, and Nereid is at upper right.

NEPTUNE   NEREID

TRITON

10,000,000 KM

Figure 17-14 Pluto's orbit is so eccentric that it sometimes goes inside the orbit of Neptune. Pluto will enter the mean orbit of Neptune on November 28, 1978, and exit from within the mean orbit of Neptune on May 12, 2000.

Figure 17-15 Small sections of the plates from which Tombaugh discovered Pluto. On February 18, 1930, Tombaugh noticed that one dot among many had moved between January 23, 1930 *(top)* and January 29, 1930 *(bottom)*.

17.3   PLUTO

Pluto, the outermost known planet, is a deviant. It has the most eccentric orbit—an orbit that is most different from a circle. Its orbit also has the greatest inclination with respect to the ecliptic plane, in which the other planets revolve. Pluto's orbit is inclined by 17°, while Mercury's is inclined by only 7° and the other planets are all inclined by 4° or less. In a drawing of planetary orbits (see Fig. 10–25) Pluto sticks out.

Pluto will reach perihelion, its closest possible distance from the Sun, in 1989. Its orbit is so eccentric that for many years at the end of this century, up to 1998, the planet will be on the part of its orbit that is inside the orbit of Neptune (Fig. 17–14). Thus in a sense, Pluto will be the eighth planet for a while (though, of course, the semi-major axis of Pluto's orbit is greater than that of Neptune).

The discovery of Pluto was a result of a long search for an additional planet which, together with Neptune, was causing perturbations in the orbit of Uranus. Finally, in 1930, Clyde Tombaugh, after a year of diligent study of photographic plates at the Lowell Observatory, found the dot of light that is Pluto. From its slow motion with respect to the stars from night to night, it was identified as a new planet (Figs. 17–15 and 17–16).

Even such basics as the mass and diameter of Pluto are very difficult to determine. It is hard to deduce the mass of Pluto because the procedure requires measuring Pluto's effect on Uranus, a more massive body. Moreover, Pluto has made less than one revolution around the Sun since its discovery. As recently as 1968, Pluto was thought to have a mass 91 per cent that of the Earth. The latest studies of the orbit of Uranus indicate that Pluto may have a mass only 11 per cent that of the Earth, but these observations are very uncertain. On the basis of current data, we are not able to reliably determine Pluto's mass; the value given may be off by a factor of two or more.

The best method for determining the radius of Pluto, as it is for Neptune, is to observe a stellar occultation. It was predicted that Pluto would pass near a 15th magnitude star in 1965, and so the passage was observed very closely by several observatories to see if the star would be occulted. But the star was never hidden from view, and from this fact astronomers knew that Pluto appeared smaller in the sky than the minimum angular separation (Fig. 17–17). Since we know the distance to Pluto, simple trigonometry gives a limit to the radius of Pluto. This observation showed that Pluto had to be smaller than 6800 kilometers across, which confirmed that Pluto was closer in size to the terrestrial than to the Jovian planets.

Pluto could be substantially smaller than this. In 1976, infrared spectral studies carried out at Kitt Peak showed the presence of methane ice on Pluto's surface. Since ice has a high albedo, the planet would not have to be very large to reflect the amount of light that we measure.

We can calculate the effect that Pluto would have on the orbits of Uranus and Neptune by using the latest value of mass instead of the value that had been deduced in the original discovery predictions. The calculations show that Pluto's mass is far too small to cause the perturbations in Uranus's orbit that originally led to Pluto's discovery. These earlier meas-

Figure 17–16 Pluto's motion can be seen in these photographs taken on successive nights.

urements were apparently inaccurate. It is now thought by many astronomers, therefore, that the prediscovery prediction was wrong and that the discovery of Pluto was purely the reward of hard work in conducting a thorough search in a zone of the sky near the ecliptic.

Pluto is so far away that there are no current plans to send spacecraft there. So it will be a long time before we know very much more than the very little that we now know.

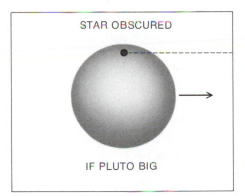

STAR OBSCURED

IF PLUTO BIG

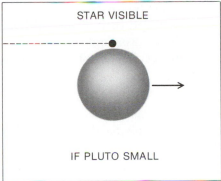

STAR VISIBLE

IF PLUTO SMALL

Figure 17–17 The fact that a 15th magnitude star was not occulted by Pluto gives us a limit for how large the diameter of Pluto could be.

SUMMARY AND OUTLINE OF CHAPTER 17

Aims: To study the three other giant planets: Saturn, Uranus, and Neptune; to understand Saturn's rings; and to see how little we know about Pluto

Saturn (Section 17.1)
  Giant planet with system of rings
  Lowest density of planets: 0.7 g/cm³
  Ring system
        27° inclination to orbit
        Orbiting chunks of rock and ice
        Inside Roche's limit for Saturn
        Very thin from top to bottom
        Radar waves bounced off rings indicate that they are made of rocks
            of certain sizes and roughness
  Rapid rotation period
  No evidence of solid surface
  Magnetic field present
  Internal heating
  10 moons, including Titan, largest moon in solar system

Atmosphere detected on Titan giving rise to speculation as to the chance of finding life there

Pioneer 11 en route to 1979 flyby

Uranus (Section 17.2)

Large planet with low density and high albedo

Discovered by Herschel in 1781

Atmosphere of methane and molecular hydrogen

No surface structure can be seen

Axis of rotation is in the plane of its orbit and retrograde

5 known moons

Rings discovered at an occultation

Neptune (Section 17.2)

Large planet with low density and high albedo

Discovery by Adams and Leverrier part science and part luck

Atmosphere of methane and molecular hydrogen

Diameter measured more accurately as the result of a stellar occultation

2 known moons

Pluto (Section 17.3)

Most eccentric orbit in solar system

Greatest inclination to ecliptic

Discovered in 1930 after diligent search along the ecliptic

Radius, mass (and thus density) very difficult to determine

Near occultation of star indicates that diameter is less than 6800 km

QUESTIONS

1. What are the similarities between Jupiter and Saturn? What are the differences?

2. What is the angular size of the Sun as viewed from Saturn? How many times smaller is this than the angular size of the Sun we see from the Earth?

3. When Jupiter and Saturn are closest to each other, what is the angular size of Jupiter as viewed from Saturn? How does this compare with the angular size of Jupiter as viewed from the Earth?

4. What conditions in the early moments of the solar system would have led to other planets in addition to Saturn and Uranus having rings?

5. Why are the moons of the giant planets more appealing for exploration than the planets themselves?

6. What is strange about the direction of rotation of Uranus, and how might that affect Uranus's weather?

7. Why do we know so little about Uranus and Neptune compared with Jupiter and Saturn?

8. a) Explain how the occultation of a star can help us learn the diameter of a planet. b) How did an occultation tell us about the rings of Uranus?

9. The inaccuracies in measuring the diameter of Neptune mean that we know only that Neptune's diameter is probably between 46,900 km and 51,500 km. That is, it is known to ±5 per cent (±2300/49,200). To what percentage of accuracy is the density known?

10. What fraction of its orbit has Neptune traversed since it was discovered?

11. What fraction of its orbit has Pluto traversed since it was discovered?

12. What evidence suggests that Pluto is not a "normal" planet?

13. The position of Pluto in the sky was accurately known, so why were astronomers unsure whether it would occult a particular star in 1965?

14. Summarize the evidence that suggests that Pluto is not a giant planet.

15. What are NASA's current plans for exploring the outer planets?

REVIEW QUESTIONS ON THE PLANETS

1. List the planets in order of
   (a) increasing size,
   (b) increasing density,
   (c) increasing distance from the Sun.

2. Which planets probably have significant internal energy sources?

3. Mars and Venus are the planets potentially "most like" the Earth, but there are substantial differences. To what degree do you think the various differences can be ascribed to their different distances from the Sun, and to what extent must other explanations be invoked?

4. Which two planets come closest together?

5. If you were an astronomer and could set up an observatory on any planet, which would you choose? To what degree would your decision be influenced by the part of the spectrum in which you are interested?

6. Of all the moons in the solar system, the one whose existence is hardest to understand is the Earth's. Why?

7. Which planets have radiation belts like the Earth's? What property of the planet determines whether or not it will have radiation belts?

8. Classify the planets according to those whose atmosphere is thicker or thinner than that of the Earth.

9. Which of the three photographs in Set A shows (a) the Moon, (b) Mars, and (c) Mercury? Which of the three close-up views in Set B shows (a) the Moon, (b) Mars, and (c) Mercury? How did you know in each case?

1   2   3   A

1   2   3   B

Comets, Meteoroids, and Asteroids

Besides the planets and their moons, there are many other objects in the family of the Sun. The most spectacular, as seen from Earth, are some of the comets. Bright comets have been noted throughout recorded history, instilling in observers great awe of the heavens.

It has been realized since the time of Tycho Brahe (Fig. 10–17), who studied the comet of 1577, that comets are phenomena of the solar system rather than of the Earth's atmosphere. From the fact that the comet did not show a parallax when observed from different locations on Earth, Tycho deduced that the comet was at least three times farther away from the Earth than the Moon.

Asteroids and meteoroids are other residents of our solar system. We shall see how, along with the comets, they may prove to be storehouses of information about the solar system's origin.

18.1 COMETS

Every few years, a bright comet fills our sky with its tremendous tail (Fig. 18–1). From a small, bright area called the *head*, a *tail* may extend gracefully over one-sixth (30°) or more of the sky. Although the tail may give an impression of motion, because it extends out to only one side, the comet does not move visibly across the sky as we watch. With binoculars or a telescope, however, an observer can accurately note the position of the head with respect to nearby stars and detect that the comet is moving at a

Comets have long been seen as omens.
"When beggars die, there are no comets seen;
The heavens themselves blaze forth the death of princes."

Shakespeare,
Julius Caesar

The "long hair" that is the tail led to the name comet, *which comes from the Latin for "long-haired star,"* aster kometes.

Comet West, a bright comet that was visible to pre-dawn observers in the northern hemisphere in 1976.

271

Figure 18–1 Comet Ikeya-Seki over Los Angeles in 1965, observed from one of the solar towers at the Mount Wilson Observatory. Photographs like this are taken with ordinary 35-mm cameras; this one was a 32-sec exposure on Tri-X film at f/1.6.

slightly different rate from the stars as comet and stars rise and set together. A photograph of even a few seconds duration will show the relative motion. By the next night, the comet may have moved 2°, four times the diameter of the Moon. (Both its right ascension and its declination change.)

Within days or weeks this bright comet will have faded below naked-eye brightness, though it can be followed for additional weeks with binoculars and then for additional months with telescopes.

The tail of a comet is generally directed away from the Sun. Thus, if we see the comet setting in the western sky after sunset, its tail would extend easterly, directed upward toward the zenith and away from the Sun. Conversely, if we see the comet rising in the eastern sky before dawn, its tail would extend westerly toward the zenith.

Most comets are much fainter than the one described above. Half a dozen or so new comets are discovered every year, and most become known only to astronomers. An additional few are "rediscovered"—that is, from the orbits derived from past occurrences it can be predicted when and approximately where in the sky a comet will again become visible. Up to the present time, over 600 comets have been discovered, 5 in 1976 alone. The returns of four already known comets were also detected in that year.

Box 18.1 Discovering a Comet

A comet is generally named after its discoverer, the first person to see it (or the first two or three people if they independently find it without too much time separating the first observation). Comets are also assigned letters and numbers. First the letters are assigned in order of discovery in a given year. Then, a year or two later when all the comets that passed near the Sun are likely to be known, Roman numerals are assigned in order of their perihelion passage. The two comets, for example, discovered by Lubos Kohoutek in March 1973, were both named Comet Kohoutek. The one that eventually went close to the Sun was first known as 1973f. It was assigned the number 1973XII at the end of 1974, a year after it passed perihelion.

Many discoverers of comets are amateur astronomers, including some amateurs who examine the sky each night in hope of finding a comet. To do this, one must know the sky very well, so that one can tell if a faint, fuzzy object is a new comet or a well-known nebula. It is for that reason that Messier made his famous eighteenth century list of nebulae (described in Chapter 23).

If you find a comet, telegraph the International Astronomical Union Central Bureau for Astronomical Telegrams, at the Smithsonian Astrophysical Observatory in Cambridge, Massachusetts (telegraph address—RAPID SATELLITE CAMBMASS). Alternatively, telephone (617) 864-5758, though this is less desirable. In any case, you should identify the direction of the comet's motion, its position, and its brightness. Don't forget to identify yourself with your name, address, and telephone number. If you are the first (or may be even the second or third) to find the comet, it will be named after you.

The record for discovering comets belongs to the former caretaker of the Marseilles Observatory, Jean Louis Pons, who discovered 37 of them between 1801 and 1827. Several Japanese amateur astronomers have found many comets: Minoru Honda has found a dozen. After work, he spends many hours each night scanning the sky. But the comet may become sufficiently bright for discovery only while it is daylight or cloudy in Japan. So there is hope for less dedicated observers.

Many comets, particularly the ones discovered by professional astronomers, are discovered not by eye but only by examination of photographs taken with telescopes. The photographs are usually those taken for other purposes, and so the discovery of a comet is serendipity—a fortuitous extra discovery—at work.

18.1a The Composition and Origin of Comets

At the center of the head of a comet is the *nucleus*, which is at most a few kilometers or so across. It is composed of chunks of matter. The most widely accepted theory of the composition of comets, advanced in 1950 by Fred L. Whipple of the Harvard and Smithsonian Observatories, is that the

Figure 18–2 The head of
Halley's Comet in 1910.

Figure 18–3 Part of the spectrum of Comet Tago-Sato-Kosaka, photographed in 1969 at the European Southern Observatory in Chile. Molecules give not just single lines but rather bands (groups) of lines; one of the bands of CN is visible at the left. Bands of CH and C_3 are seen in the rest of the photograph.

nucleus is like a *dirty snowball.* The nucleus may be ices of such molecules as water (H_2O), carbon dioxide (CO_2), ammonia (NH_3), and methane (CH_4), with dust mixed in.

This model explains many observed features of comets, including the fact that the orbits of comets do not appear to accurately follow the laws of gravity. When sunlight evaporates the ices, the molecules are expelled from the nucleus. This action generates an equal and opposite reaction, the same force that runs jet planes. Since the comet nucleus is rotating, the force is not always directly away from the Sun, even though the evaporation is triggered on the sunny side. In this way, comets show the effects of non-gravitational forces in addition to the effect of the force of the solar gravity.

The nucleus itself is so small that it is impossible to observe directly from Earth. It is surrounded by the *coma* (pronounced cō′ ma), which may grow to be as large as 100,000 km or so across. The coma shines partly because its gas and dust are reflecting sunlight toward us and partly because gases liberated from the nucleus are excited enough that they radiate. The spectrum of a comet head shows sets of lines from simple molecules (Fig. 18–3).

The nucleus and coma together are the head of a comet (Fig. 18–2).

AUGUST 22, 1957   AUGUST 24, 1957

Figure 18–4 In Comet Mrkos, the straight *ion tail,* extending toward the top, and the *dust tail,* gently curving toward the right, were clearly distinguished.

The tail can extend as much as 1 A.U. (150,000,000 km), and so comets can become the largest objects in the solar system. But the amount of matter in the tail is very small—the tail is a much better vacuum than we can make in laboratories on Earth.

Many comets actually have two tails (Fig. 18–4). Both extend generally in the direction opposite to that of the Sun, but are different in appearance. The *dust tail* is caused by dust particles that had been impurities in the ices of the nucleus, released when the ice was vaporized. The dust particles are left behind in the comet's orbit; they are blown slightly away from the Sun by the pressure caused by photons of solar light hitting the particles (this is known as "radiation pressure"). As a result of the comet's continued orbital motion, the dust tail usually curves smoothly behind the comet. The *gas tail* (also called the *ion tail*) is composed of ions (atoms that have lost an electron) blown out more or less straight behind the comet by the solar wind (which was described in Section 5.7). As puffs of ionized gas are blown out and as the solar wind varies, the ion tail takes on a structured appearance. Each puff of matter can be seen. The magnetic fields carry only the ionized matter along with them; the neutral atoms are left behind in the coma.

With the second Orbiting Astronomical Observatory, we became able to observe in comets the Lyman alpha line of hydrogen, which is in the ultraviolet. In this way, it was discovered in 1970 that a huge hydrogen cloud (Fig. 18–5) surrounds the head. This hydrogen cloud may be a million miles in diameter! It probably results from the break up of water molecules by ultraviolet light from the Sun.

A comet—head and tail together—contains less than a billionth of the mass of the Earth. It has been said that comets are as close as something can come to being nothing.

It is now generally accepted that there are hundreds of millions of incipient comets surrounding the solar system in a sphere perhaps 50,000 A.U. (almost 1 light year) in radius. This sphere is known as the *Oort comet*

Figure 18–5 *(A)* These contours of intensity result from a photograph of Comet Kohoutek taken by the Skylab astronauts outside the Earth's atmosphere. The photograph was taken in the Lyman alpha line of hydrogen at 1216 Å in the ultraviolet, and shows the hydrogen halo. This halo was about 1° across, about 2,500,000 km in diameter, at the time of this observation. *(B)* These contours result from a photograph that was taken immediately after Fig. 18–5 *A*, but through a filter that did not pass the Lyman alpha line. Thus the hydrogen halo does not show, though the tail does. The tail was about 2° across, about 5,000,000 km in diameter. Hot stars from the background constellation, Sagittarius, also show.

A

B

cloud after Jan H. Oort, the Dutch astronomer who advanced the theory in 1950. The total mass of matter in the cloud is only 10 to 100 times the mass of the Earth. Occasionally one of the incipient comets leaves the Oort cloud, perhaps because gravity of a nearby star has tugged it out of place, and the comet approaches the Sun. Its orbit is a long ellipse (that is, an ellipse of high eccentricity). If the comet passes near the Jovian planets, its orbit is altered by the gravity of these massive objects. Because the Oort cloud is spherical, comets are not limited to the plane of the ecliptic and come in randomly from all angles.

As the comet gets closer to the Sun, the solar radiation that reaches it begins to vaporize the molecules in the nucleus. We have not generally been able to detect the molecules in the nucleus directly—the *parent molecules*—though we would clearly love to do so, but we have mostly detected in the head the simpler molecules into which the parent molecules break down—the *daughter molecules.*

The tail begins to form, and as more and more of the nucleus is vaporized, the tail grows longer and longer. Still, even though the tail can be millions of kilometers long, it is so tenuous that only 1/500 of the mass of the nucleus may be lost. Thus a comet may last for many passages around the Sun.

The comet is brightest and its tail is generally longest at about the time it passes perihelion (the closest point in its orbit to the Sun). However, because of the angle at which we view the tail from the Earth, it may not appear the longest at this stage.

Following perihelion, as the comet recedes from the Sun, its tail fades; the head and nucleus receive less solar energy and fade as well. The comet may be lost until its next return, which could be as short an interval as 3.3 years (as it is for Encke's Comet) or as long as 80,000 years (as it is for Comet Kohoutek) or more. Periodic comets lose their mass reappearance after reappearance, and the comet eventually disappears. We shall see in Section 18.2 that some of the meteoroids are left in its orbit.

Because new comets come from the places in the solar system that are farthest from the Sun and thus coldest, they probably contain matter that is unchanged since the formation of the solar system 4.6 billion years ago. So the study of the constituents of comets is important for understanding the early stages of solar system formation.

18.1b Halley's Comet

In 1705, the English astronomer Edmund Halley (Fig. 18–6) applied a new method developed by his friend Isaac Newton to determine the orbits of comets from observations of the positions of the comets in the sky. He reported that the orbits of the bright comets that had appeared in 1531, 1607, and 1682 were about the same. Because of this, and because the intervals between appearances were approximately equal, Halley suggested that we were observing a single comet orbiting the Sun, and predicted that it would again return in 1758. The reappearance of this bright comet on Christmas Night of that year, 16 years after Halley's death, was the proof of Halley's hypothesis (and Newton's method); it has since been known as Halley's Comet. It seems probable that the bright comets reported every 74 to 79 years since 240 B.C. (and possibly even before then) were earlier appearances. The fact that Halley's Comet has been observed at least 29

Figure 18–6 Edmund Halley.

APRIL 26   APRIL 27   APRIL 30   MAY 2   MAY 3   MAY 4   MAY 6   MAY 15

Mark Twain was born during the appearance of Halley's Comet in 1835, and often said that he came in with the comet and would go out with it. And so he did, dying during the 1910 return of Halley's Comet.

times endorses the calculations that have been made that show that less than one per cent of a cometary nucleus's mass is lost at each perihelion passage.

Halley's Comet went especially close to the Earth during its 1910 return (Fig. 18–7), and the Earth actually passed through its tail. Many people had been frightened that the tail would somehow damage the Earth or its atmosphere, but the tail had no noticeable effect. It was known to most scientists even then that the gas and dust in the tail were too tenuous to harm our environment.

MAY 23   MAY 28   JUNE 3   JUNE 6   JUNE 9   JUNE 11

Figure 18–7 Halley's comet in 1910.

Halley's Comet reappears every 74 to 79 years; Jupiter and Saturn perturb its orbit enough to cause the variation. Since we can count on its reappearance in 1985 in time for its 1986 perihelion (Fig. 18–8), we can make plans long in advance to observe it. The possibility of launching a spacecraft to fly near it or even through its tail is being discussed, though no definite plans yet exist.

The 1986 reappearance will not be as spectacular as its 1910 passage, however, for the Earth will be in a part of its orbit such that the comet is not seen broadside. Thus the tail will appear foreshortened, and is not expected

Figure 18–8 *(A)* The orbit of a comet.

Illustration continued on following page.

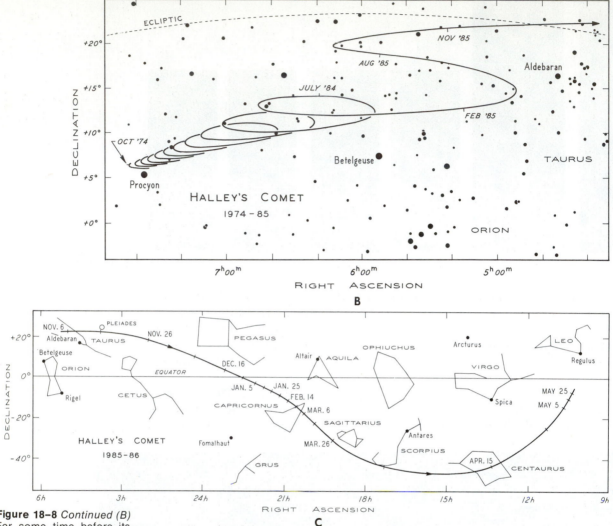

B

C

Figure 18–8 *Continued (B)* For some time before its perihelion, Halley's Comet will follow an orbit that seems to circle in the sky as a result of the Earth's revolution around the Sun. *(C)* When Halley's Comet draws close to us, its apparent motion across the sky will grow more rapid. Observers can use these charts to search for Halley's Comet in 1985, when we expect to be able to see it coming.

to appear as spectacular as, for example, 1975n, Comet West (which is shown in the photograph opening this chapter and is discussed in the following section).

18.1c Comet Kohoutek and Comet West

In March of 1973, Lubos Kohoutek was studying faint asteroids on photos he had taken at the Hamburg Observatory in Germany where he works. On two of these plates he discovered faint comets—about 16th magnitude—which were named for him when he sent word of the discoveries to the IAU Central Bureau for Astronomical Telegrams.

Further analysis of his plates of the second comet, including a further set of plates he took later in the month and a two-month-old plate that he found also showed the comet, revealed that he had made a most unusual discovery: the comet would not reach perihelion for about nine months and would pass very close to the Sun—within only 21 million km (.14 A.U.)—thus becoming very bright. Because of these predictions the second comet, 1973f, became known the world over as *Comet Kohoutek.*

DEC 29, 1973 PERIHELION +1 DAYS | DEC 30, 1973 PERIHELION +2 DAYS | DEC 31, 1973 PERIHELION +3 DAYS | JAN 2, 1974 PERIHELION +5 DAYS

Figure 18–9 Sketches of Comet Kohoutek and its sunward spike by the astronauts aboard the third Skylab mission.

Never before had a potentially bright comet been discovered this far from the Sun—at the distance of Jupiter's orbit—or this long before perihelion. Thus it was impossible to make an accurate prediction of the amount of brightening, but it seemed that it might brighten from 16th to −5th magnitude, a factor of 250,000,000. On the basis of this prediction, scientists readied all kinds of experiments that required months to prepare, and reserved time on large optical and radio telescopes for the period of anticipated brightness. The long advance notice even allowed the schedule of the third crew of astronauts due to visit Skylab to be delayed so that they would be aloft during Comet Kohoutek's perihelion passage. Anyone who has ever worked with a space experiment will especially realize how major an event it is to change such a schedule.

After a few months, the Earth in its orbit swung around to the side of the Sun opposite to Kohoutek, and we could temporarily no longer follow the comet's development. When the comet eventually re-emerged from the Sun's glare it had not brightened quite as much as had been expected, and so the astronomers revised toward fainter levels their predictions of the comet's eventual maximum brightness. Somehow these revised and updated predictions never became as widely circulated as the original hopes that the comet would be brighter than Venus.

Comet Kohoutek was expected to be at its brightest within a few days of its perihelion passage. During that time it was too close to the Sun to be seen by anyone except the astronauts in Skylab, who photographed (see Color Plate 15) and sketched the comet. The shape and size of the comet and the tail appeared to change rapidly, and these Skylab data on this phase of the development of the comet are unique. The comet was actually almost as bright as Venus at this time, but it was too close to the Sun for anybody except the astronauts to see it.

A day after perihelion, the astronauts discovered a brilliant spike—an anti-tail—pointing toward the Sun, the opposite direction from that in which comet tails point (Fig. 18–9). This rare phenomenon had previously been seen to such an extent in only one other modern comet, and to any extent in only a dozen comets. It had been thought that sunward spikes were caused only by effects of perspective, and that such spikes were actually directed away from the Sun. But analysis of the Kohoutek observations showed that the spike contained meteoroid particles as large as 1 mm across, far larger than the dust in a comet tail. These particles are controlled by different mechanisms from those that control the particles in the tail, and the spike was probably actually pointing in the general direction of the Sun.

A week after perihelion the comet drew far enough away in the sky

Figure 18–10 A photograph of Comet Kohoutek with the 1.2-meter Schmidt telescope on Palomar Mountain. Although the comet was too faint to be seen well with the naked eye, it was a fine photographic object and was bright enough to enable unique scientific observations to be made.

from the Sun that its tail could be seen by observers on Earth. Unfortunately, it was just barely bright enough to be seen with the naked eye, and was not the spectacular object to which many had been looking forward. In particular, it could not be seen by observers in cities, where light pollution made the sky brighter than the comet's tail. Nevertheless, the tail stretched gracefully over many degrees of sky (Fig. 18–10). Observers with optical telescopes and radio telescopes on the ground were able to get good spectra and discover molecules that had never before been seen in comets. Radio astronomers reported the detection of two molecules—methyl cyanide (CH_3CN) and hydrogen cyanide (HCN)—that had previously been detected only in interstellar space (as will be discussed in Section 21.5). If these reports could be fully accepted, they would indicate that comets were formed far from the Sun, because otherwise these molecules would have been torn apart. In particular, the observations would tend to favor the model that the comets formed at interstellar distances, even though it seems that they have always been bound to the solar system.

Why did Kohoutek appear so bright so early on? It has been hypothesized that Kohoutek may have been a "virgin comet," a comet making its first passage around the Sun. Thus, an especially great amount of material was vaporizing when the comet was first discovered, making it appear brighter than average for that distance from the Sun and from the Earth. Once that layer of material escaped, lower layers were more ordinary. We can think of the Sun vaporizing layer after layer of the material of a comet's nucleus, similar to the way in which we can peel off layer after layer from an onion. The idea that the comet was shedding dust was borne out by infrared observations; the fact that a subsequent comet's infrared brightness dropped by a factor of a hundred within a few weeks lends support to the layered-onion theory.

Comet Kohoutek is now receding into space, and will go out as far as 4000 A.U. While it is far from the Sun, we can see from Kepler's second law that it will move very slowly. It won't be back in our vicinity for another 80,000 years.

Astronomers often say that comets are unpredictable. Comet Kohoutek, which was a popular bust and an astronomer's delight, bore out that idea.

Bright comets ordinarily appear with much less notice than Kohoutek had. In late 1975, Richard West, of the European Southern Observatory, discovered on a plate taken in Chile a comet that within a couple of months became an object very easy to see with the naked eye. Before perihelion, Comet West was visible only to observers in the southern hemisphere, but after perihelion it was visible in the predawn sky to northern hemisphere observers (see Color Plate 41). For a few days its tail extended over 30°. Bits of dust in the nucleus were ejected at an irregular rate, which led to the delicate structure in the tail that is visible in the photograph opening this chapter. Indeed, the nucleus itself broke into at least four pieces. On the basis of the orbit, it might not be back for a million years; it might even have been set on a course of ejection from the solar system, in which case it will never return.

18.2 METEOROIDS

There are many small chunks of matter in interplanetary space, ranging up to tens of meters across. When these chunks are in space, they are called *meteoroids*. When one hits the Earth's atmosphere, friction slows it down and heats it up—usually at a height of about 100 km—until all or most of it is vaporized. Such events result in streaks of light in the sky, which we call *meteors*. (Meteors are popularly known as *shooting stars*.) The brightest meteors can reach magnitude −15 or even −20, which is brighter than the full moon. We call such bright objects *fireballs* (Fig. 18–11). We can sometimes even hear the sounds of their passage and of their breaking up into smaller bits. When a fragment of a meteoroid survives its passage through the Earth's atmosphere, the remnant that we find on Earth is called a *meteorite*. We now also refer to objects that hit the surfaces of the Moon or of other planets as meteorites.

There are several major kinds of meteorites. Most of the meteorites that are found have a very heavy content—about 90 per cent—of iron; the rest is nickel. These *iron meteorites* (or, for short, *irons*) are thus very dense—that is, they weigh quite a lot for a given volume.

Most meteorites that hit the Earth are stony in nature and are often referred to simply as *stones*. Most of these stony meteorites are of a type called *chondrites*, because they contain rounded particles called "chondrules." Because stony meteorites resemble ordinary rocks, they are not usually discovered unless their fall is observed. That explains why we more usually discover irons. But when a fall is observed, most meteorites recovered are stones. The stony meteorites have a high content of silicates; only about 10 per cent of their mass is iron and nickel.

The largest meteorite crater on Earth may be a depression over 400 km across deep under the Antarctic ice pack. This is comparable with the size of lunar craters. Another very large crater, in Hudson's Bay in Canada, is filled with water. Most meteorite craters on Earth are either disguised in such ways or have eroded away. A large crater that is obviously meteoritic in origin is the Barringer Crater in Arizona (Fig. 18–12). It is the result of what was perhaps the most recent large meteor to hit the Earth, for it was formed only 25,000 years ago.

Every few years a meteorite is discovered on Earth immediately after its fall (Fig. 18–13). The chance of a meteorite's landing on someone's house is very small, but it has happened! Often the positions of fireballs in the sky are tracked in the hope of finding fresh meteorite falls. The newly discov-

Figure 18–11 A fireball observed in 1970 by the "Prairie Network" of the Smithsonian Astrophysical Observatory, a network of wide-field cameras spaced around the mid-west of the United States in hope of pinpointing a meteor's path and thus permitting the meteorite to be found. A shutter in the camera rotates so that breaks occur in the image of the meteor trail at intervals of 1/20 sec. This permits timing of the motion. The fireball pictured here was photographed by two of the cameras of the Network, which led to the discovery of a meteorite at Lost City, Oklahoma. The trail persisted for about 8 seconds, though the short breaks in the trail from which this could normally be measured do not show because of film overexposure except at the top.

TABLE 18–1 METEORITES

Composition	Seen Falling	Finds
Irons	6%	66%
Stony-irons	2%	8%
Stones	92%	26%

Figure 18–12 The Baringer meteor crater in Arizona.

Figure 18–13 (*A*) A bright fireball was observed and photographed on February 5, 1977. Analysis indicated that it probably landed near Innisfree, Alberta, Canada. Scientists flew out to search the snow-covered wheat fields and found this 2 kg meteorite. It is only the third meteorite recovered whose previous orbit around the Sun is known. (*B*) A meteor crater 2 meters wide and 5.5 meters deep, one of a hundred from the fall in Kirin Province of China on March 8, 1976. A fireball was seen before the meteorites landed. The largest meteorite is the largest stony meteorite ever recovered; it weighs 1770 kg (3894 lbs).

ered meteorites are rushed to laboratories for chemical analysis of their constituents before the Earth's atmosphere or human handlers can contaminate them.

As we can measure the ages of the meteorites through study of the ratios of radioactive and non-radioactive isotopes contained in them, we know that the meteoroids were formed at times up to 4.6 billion years ago, the time of the beginning of the solar system. The analysis of the abundances of the elements in meteorites thus tells us about the solar nebula from which the solar system formed. In fact, up to the time of the first land-

A
B

TABLE 18-2   METEOR SHOWERS*

Name	Date of Maximum	Duration Above 25% of Maximum	Approximate Limits	Number per Hour at Maximum
Quadrantids	Jan 4	1 day	Jan 1–6	110
Lyrids	April 22	2 days	April 19–24	12
Eta Aquarids	May 5	3 days	May 1–8	20
Delta Aquarids	July 27–28	—	July 15–Aug 15	35
Perseids	Aug 12	5 days	July 25–Aug 18	68
Orionids	Oct 21	2 days	Oct 16–26	30
Taurids	Nov 8	—	Oct 20–Nov 30	12
Leonids	Nov 17	—	Nov 15–19	10
Geminids	Dec 14	3 days	Dec 7–15	58

*The number of sporadic meteors per hour is 7 under perfect conditions. The visibility of showers depends mostly on how bright the Moon is on the date of the shower, which depends on its phase. Meteors are best seen with the naked eye; using a telescope or binoculars merely restricts your field of view.

ing on the Moon, meteorites were the only extraterrestrial material we could get our hands on.

Meteors often occur in *showers*, that is, times when the rate at which meteors are seen is far above average. On any clear night a naked-eye observer may see a few *sporadic* meteors an hour, that is, meteors that are not part of a shower. (Just try going out to a field in the country and watching the sky for an hour.) During a shower many meteors may be visible to the naked eye each minute. Meteor showers occur at the same time each year (Table 18.2), and probably represent the passage of the Earth through the orbits of defunct comets. The meteoroids in the orbit of a former comet may be the products of the decay of the comet. Meteorites usually do not result from showers, so presumably the meteoroids that cause a shower are very small.

The rate at which meteors are seen usually increases after midnight on the night of a shower, because that side of the Earth is then facing and plowing through the oncoming interplanetary debris. If the Moon is gibbous or full, then the sky is too bright to see the shower well. The meteors in a shower are seen in all parts of the sky, but their trajectories all seem to emanate from a single point in the sky called the *radiant*. A shower is usually named after the constellation that contains its radiant. The existence of a radiant is just an optical illusion because perspective makes a set of parallel paths approaching you appear to emanate from a point.

18.3   ASTEROIDS

The nine known planets were not the only bodies to result from the agglomeration of planetesimals 4.6 billion years ago. Thousands of *minor planets*, called *asteroids*, also resulted. They are detected by their small motions in the sky relative to the stars (Fig. 18–14).

Space is full of meteoroids of all sizes, with the smallest particles being the most abundant. Most of the small particles, less than about 1 mm across, come from comets. Most of the large particles, more than about 1 cm across, come from collisions of asteroids in the asteroid belt. It is generally these larger meteoroids that become meteorites.

Figure 18–14 Asteroids leave a trail on a photographic plate when the telescope is tracking at a sidereal rate, that is, with the stars.

Most of the asteroids are found in elliptical orbits whose average size is between the sizes of the orbits of Mars and Jupiter. The zone where most asteroids are found is called the *asteroid belt*. Indeed, it was not a surprise when the first asteroid was discovered, on January 1, 1801, the first day of the nineteenth century, because Bode's law (described in Section 10.5) predicted the existence of a new planet in approximately that orbit.

This first asteroid was discovered by a Sicilian monk, Giuseppe Piazzi, and was named Ceres after the patron saint of Sicily. Though the discovery of Ceres was a welcome surprise, even more surprising was the subsequent discovery in the next few years of three more asteroids. Bode's law hadn't called for **them!** The next asteroids weren't discovered until 1845, and over 2000 are now named and numbered. They are assigned numbers when their orbits become well determined.

Though most asteroids are located between the orbits of Mars and Jupiter, from 2.2 to 3.2 A.U. from the Sun (Fig. 18–15), there is plenty of

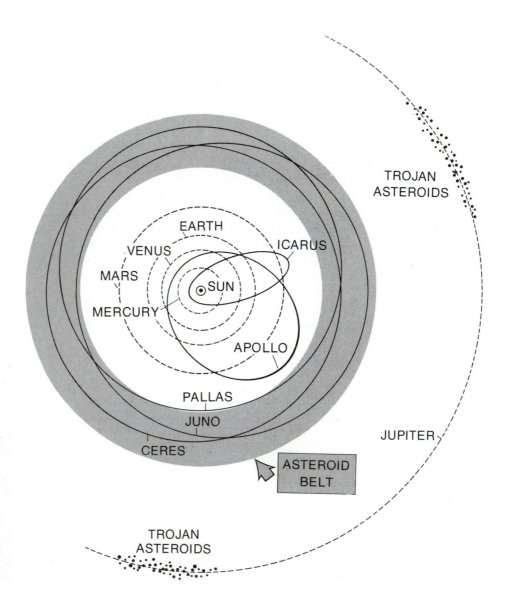

Figure 18–15 Some asteroid orbits. The asteroid belt, in which most asteroids lie, is shaded.

empty space among them. Asteroids rarely come within a million kilometers of each other. Still, asteroids may collide with each other occasionally, forming meteoroids.

Pioneers 10 and 11, en route to Jupiter and beyond, traveled through the asteroid belt for many months and showed that the amount of dust among the asteroids is not increased over the amount of interplanetary dust in the vicinity of the Earth. Particles the size of dust grains are only about three times more plentiful in the asteroid belt, and smaller particles are less plentiful. So the asteroid belt will not be a hazard for space travel to the outer parts of the solar system.

Some of the asteroids are not part of the asteroid belt, including even some that have orbits whose perihelia (the plural of perihelion) are inside the orbit of the Earth.

When the solar system was forming, proto-Jupiter probably became so massive that it disrupted a large region of space around it. The planetesimals in that region may have been sent into eccentric orbits, and may have collided at high velocity. In this way, they broke each other apart and sent the fragments off in orbits of high eccentricity and inclined to the ecliptic plane. Small asteroids may only be 1 km or less across, but the largest asteroids known (Fig. 18–16) are of the size of some of the moons of the planets. All the asteroids together contain less mass than does the Moon.

Many of the meteorites that we find on Earth may be chips of asteroids. If this is the case, then we already have some kinds of asteroid material on hand to analyze.

Considerations of albedos and spectra lead us to the conclusion that asteroids are made of different materials from each other, and represent the chemical compositions of different regions of space. The asteroids at the inner edge of the asteroid belt are mostly stony in nature, while the ones at the outer edge are darker (as a result of containing more carbon). Most of the small asteroids that pass near the Earth—Icarus, Eros, Toros, Geographos, and Alinda—are among the stony group. Three of the largest asteroids—Ceres, Hygiea, and Davida—are among a dark group containing lots of carbon. A third group may be mostly composed of iron and nickel. The differences may be a direct result of the variation of the solar nebula with distance from the protosun. Differences in chemical composition also show that we must discard the old theory that the asteroids represent the breakup of a planet that once existed between Mars and Jupiter.

The asteroids and their meteoritic offspring may provide the data to show us the details of how the solar system was formed.

SUMMARY AND OUTLINE OF CHAPTER 18

Aims: To study the non-planetary members of our solar system, and to see how the conditions they have been under may provide us with our best information about the origin of the solar system

Comets (Section 18.1)
    The head (the nucleus + the coma together); the tail

Figure 18–16 The sizes of the larger asteroids and their relative albedos.

CERES

PALLAS

200 KM

VESTA

HYGIEA   INTERAMNIA

DAVIDA   PSYCHE

BAMBERGA   FORTUNA   JUNO

Dust tail is sunlight reflecting off particles blown back by radiation pressure; ion tail (gas tail) is sunlight re-emitted by ions blown back by the solar wind

Dirty snowball theory: nucleus composed of rock covered with ices

Coma composed of gases vaporized from nucleus; spectrum shows molecules

Ultraviolet space observations revealed huge hydrogen cloud around coma

Origin of comets: Oort comet cloud; comets detached from cloud by gravity of passing star

Comet orbits are ellipses

Halley's Comet followed for last 29 returns, but 1986 return will not be as spectacular as some past passages

Comet Kohoutek did not become the popular display that had been hoped for, but nonetheless gave unique data because of its relatively high brightness and because it was discovered long before it passed perihelion, giving time for planning

Comet West became very bright with more typical notice of 2 months

Meteoroids (Section 18.2)

Meteoroids: in space; *meteors*: in the air; *meteorites*: on the ground

Most meteorites that hit the Earth are chondrites, which resemble stones and are therefore often undetected. Iron meteorites are more often found.

Meteorites may be chips of asteroids and thus give us material with which to investigate the origin of the solar system.

Meteorites provide us with much of our information about abundances in the solar system. We could use them to map out inhomogeneities in the solar nebula if we only knew what parts of the solar nebula the different meteorites came from.

Some meteors arrive in showers, others are sporadic

Asteroids, also called minor planets (Section 18.3)

Most in asteroid belt, 2.2 to 3.2 solar radii, between orbits of Mars and Jupiter

QUESTIONS

1. In what part of its orbit does a comet travel head first?

2. Why is the Messier catalogue important for comet hunters?

3. Would you expect comets to follow the ecliptic? Explain.

4. How far is the Oort comet cloud from the Sun, relative to the distance from the Sun to Pluto?

5. The energy that we see as light from a comet comes from what source or sources?

6. Many comets are brighter after they pass close to the Sun than they are on their approach. Why should that be?

7. Which part of a comet has the most mass?

8. Explain why Comet West, seen in the photograph opening this chapter, showed delicate structure in its tail. Are we observing mainly the dust tail or the ion tail?

9. Why was the large cloud of hydrogen that surrounds the head of a comet not detected until 1970?

10. What is the dirty snowball theory of comets? What is the relation of this to the layered onion model of Comet Kohoutek?

11. Why do most meteorites not reach the surface of the Earth?

12. Why do some meteor showers last only a day while others can last several weeks?

13. Why are meteorites important in our study of the solar system?

14. Does the asteroid belt fit Bode's law? How might this be interpreted in terms of the history of the solar system?

Life in the Universe

We have discussed the nine planets and the dozens of moons in the solar system, and have found most of them to be places that seem very hostile to terrestrial life forms. Yet some locations besides the Earth— Mars with its signs of ancient running water and perhaps Titan—have characteristics that allow us to convince ourselves that life may have existed there in the past or might even be present now.

Our first real attempt to search for life on another planet has led to fascinating results. The Viking landers, described in Section 15.3, have carried out a series of biological and chemical experiments with Martian soil, and some of these experiments have provided data that cannot be readily interpreted in terms of chemical processes that are common on the Earth. Whether the proper interpretation is that there is biological activity —life—on the surface of Mars, or that chemical processes can provide the complete explanation must await further experimentation on Earth and probably further exploration and experimentation on Mars. Later in this chapter, we shall discuss the criteria that we use to decide whether or not to accept such radical new conclusions as the existence of extraterrestrial life.

Since it seems reasonable that life, as we know it, anywhere in the Universe would be on planetary bodies, let us first discuss the chances of life arising elsewhere in our solar system. Then we will consider the chances that life has arisen in some more distant parts of the galaxy in which we live or elsewhere in the Universe.

Recently the study of life in other locations in the Universe, a study now called *exobiology*, has become a respectable science. The billion dollars spent to send the Viking probes to Mars equipped to search for signs of life is a sure sign that the possibilities of finding life are taken very seriously indeed.

Since life may well preferentially arise on planets, this chapter is included in Part III of this book, which discusses the solar system.

19.1 THE ORIGIN OF LIFE

It would be very helpful if we could state a clear, concise definition of life, but unfortunately that is not possible. Biologists state several criteria that are ordinarily satisfied by life forms, reproduction, for example. Still, there exist forms on the fringes of life—viruses, for example, which need a host organism in order to reproduce—and scientists cannot always agree whether some of these things are "alive" or not.

In writing science fiction, authors sometimes conceive of beings that show such signs of life as the capability for intelligent thought, even though the beings may share few of the criteria that we ordinarily recognize. In Fred Hoyle's novel *The Black Cloud*, for example, an interstellar cloud of gas and dust is as alive as—and smarter than—you or I. But we can make no concrete deductions if we allow such wild possibilities, and exobiologists prefer to limit the definition of life to forms that are more like "life as we know it."

This rationale implies, for example, that extraterrestrial life is based on complicated chains of molecules that involve carbon atoms. Life on Earth is governed by deoxyribonucleic acid (DNA) and ribonucleic acid (RNA), two carbon-containing molecules that control the mechanisms of heredity. Chemically, carbon is able to form "bonds" with several other atoms simultaneously, which makes these long carbon-bearing chains possible. In fact, we speak of compounds that contain carbon atoms as *organic*.

Carbon is one of the more abundant elements on a cosmic scale (see Table 5–1), although its abundance is far below those of hydrogen or helium. Another reasonably abundant element that can make several chemical bonds simultaneously is silicon, and some scientists have theorized that silicon-based life may exist somewhere. Serious consideration of life forms seems limited to life based on chains of carbon or silicon atoms.

If we accept this limitation, then we ask how hard it is to build up the long organic chains. To the surprise of many, an experiment was performed twenty-five years ago that showed that the construction of organic molecules was much easier than had been supposed.

At the University of Chicago, at the suggestion of Harold Urey, Stanley Miller put several simple molecules in a glass jar. Water vapor (H_2O), methane (CH_4), and ammonia (NH_3) were included along with hydrogen gas. Miller exposed the mixture to electric sparks, simulating the lightning that may exist in the early stages of the formation of a planetary atmosphere. After a few days, he found that long chains of atoms had formed in the jar, and that these organic molecules were even complex enough to include simple amino acids, the building blocks of life.

It was later found that exposing a similar mixture to ultraviolet light

Figure 19–1 It is all in the point of view. © 1974 United Features Syndicate, Inc.

had the same effect of causing the formation of organic molecules. It seems reasonable that the ultraviolet radiation reaching the ground in the earlier stages of the Earth's atmosphere was much stronger that that striking us today, because the ozone layer may not yet have formed. (In fact, strong ultraviolet radiation has reached the surface of Mars, and continues to reach it even now.) Thus it seems reasonable that ultraviolet radiation and lightning may have led to the formation of amino acids from simpler molecules in the primitive atmosphere of the Earth.

It seems clear, however, that mere amino acids or even DNA molecules are not life itself. A jar containing a mixture of all the atoms that are in a human being is not the same as the human being in person. This is the vital gap in the chain; astronomers certainly are not qualified to say what supplies the "spark" of life.

Still, it seems reasonable to many astronomers that since it is not very difficult to form complex molecules, life may well have arisen not only on the Earth but also in other locations in our solar system. Even if life is not found in our solar system, there are so many other stars in space that it would seem that some of them could have planets around them, and it would seem that life could have arisen on some of the planets independently of the origin of life on the Earth. The interesting nature of the results of the biological experiments on Viking add support to this view.

19.2 OTHER SOLAR SYSTEMS?

We have seen how difficult it was to detect even the outermost planets in our own solar system. At the four light year distance of Proxima Centauri, the nearest star to the Sun, any planets would be too faint for us to observe directly.

But even though we cannot hope to see the light reflected by even a giant planet, there is a faint hope of detecting such a planet by observing its gravitational effect on the star itself. One such method is the same that has been used to discover white dwarfs (see Section 7.4). We study the proper motion of a star—its apparent motion across the sky with respect to the other stars (as discussed in Section 2.15)—and see if the star follows a straight line or appears to wobble slightly. The part of astronomy that involves the measurement of the positions of objects in the sky is called **astrometry.**

The most likely stars to observe in a search for planets are the stars with the largest proper motion, for these are likely to be the closest to us and thus any wobble would cover the greatest angular extent we could hope for. The star with the largest proper motion, Barnard's star, is only 1.8 parsecs (6 light years) away from the Sun. It is the fourth nearest star to the Sun; only the members of the Alpha Centauri triple system are nearer.

At the Sproul Observatory, a major astrometric center, Peter van de Kamp has studied observations of Barnard's star taken from 1937 on, and in 1962 reported that he had discovered a wobble in its proper motion (Fig. 19–2). He interpreted this wobble to indicate the presence of a giant planet, larger even than Jupiter, in a very elliptical orbit. After further observations and analysis, he reported that the data were even better explained if there were two giant planets present; then the orbits could be more nearly circular. Van de Kamp also found evidence, though less confidently, for a planet around another of the nearest stars, ε Eridani.

Within the last few years, however, other observers have reported that

SEPTEMBER 1, 1948

MARCH 20, 1949

0 1 2 3 4 5'

S

W — E

N

SEPTEMBER 8, 1949

Figure 19–2 The photograph is a composite of three photos of Barnard's star, taken at the Sproul Observatory at intervals of approximately 6 months. The stars at the top and right have small proper motions and small parallaxes, and appear in the same position on each of the three negatives. Barnard's star, on the other hand, shows its proper motion of over 10 arc sec per year. It also shows a lateral displacement caused by parallax, the result of the Earth's orbit around the Sun. The displacement caused by the possible planets does not show at this scale.

Astrometry has been undergoing a revolutionary change in methods recently, with the advent of new machines that automatically measure positions on photograph plates. Photocells can be used to "center" an automatic measuring engine on the center of a star image, and the position can be recorded directly by a computer, or at least on punched cards. These new procedures greatly increase not only the accuracy but also the speed of studying the astrometric plates that are taken.

their study of other long-term observations of Barnard's star did not show the wobble in its proper motion. A restudy of the Sproul astrometric plates by van de Kamp shows that for a period of a few years the measured values were different from the trend of preceding and following years; the difference may have started when the telescope's lens was remounted in 1949. A restudy of the data with this in mind significantly reduced the amount of wobble that may be present.

At present, the matter is unresolved. One of the other observers, George Gatewood, now finds that his own data contain the suggestion of a deviation from straight-line proper motion for Barnard's star, but not conclusively so. He admits the possibility that the current data could imply the presence of planets, though this current indication of a wobble does not seem sufficiently convincing that he would have reported it in the absence of the historical context. Van de Kamp's new analysis of his own data still indicates a wobble that can best be interpreted in terms of planets orbiting Barnard's star, but this is a different wobble from the one he had reported previously.

For the moment, the data seem to indicate the presence of two planets around Barnard's star, one with the approximate mass of Jupiter and the other with the approximate mass of Saturn. Any Earth-sized planets would not have enough gravity for us to be able to detect them. At present at least some astronomers feel less certain about these results than they did a few years ago. The extrapolation based on proper motion studies that almost all stars have solar systems is less certain than it had been.

A new report has been made by Sarah Lee Lippincott, also of the Sproul Observatory, that another star—called Cin. 18, 2354 (from its number in Volume 18 of the Cincinnati Observatory Publications)— shows

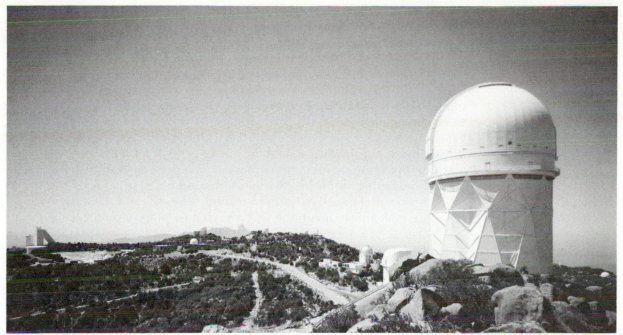

Figure 19–3 A spectroscopic study of the radial velocities of solar-type stars, carried out at the Kitt Peak National Observatory, shown here, has indicated that many have companions. One of the 0.9-m telescopes, barely visible at the center, was used in the study.

an astrometric wobble that would indicate the presence of a planet. So Barnard's star is not the only good candidate.

A new spectrographic survey carried out at Kitt Peak seems to endorse the idea that many stars have planets. One third of the 123 solar-type stars under study showed variations over time in their radial velocities, even though the invisible companions have masses that are too small to allow them to be stars. Some of these may be planets; just how many, it is difficult to say.

If we accept the nebular theory for the formation of planets, then it comes as no surprise that many stars have planets. Still, we like to see direct observational backing for our theories. Since the nearest three systems are multiple—our Sun has planets, the next nearest star is a triple system and the next star out seems to have planets—we might reasonably extrapolate that planets are a common rather than an uncommon phenomenon. If this is true, there certainly would be billions of planets in our galaxy alone, and billions more in other galaxies.

19.3 THE STATISTICAL CHANCES FOR EXTRATERRESTRIAL INTELLIGENT LIFE

Instead of phrasing one all-or-nothing question about life in the Universe, we can break down the problem into a chain of questions. This procedure was developed by Frank Drake, a Cornell astronomer, and extended, among others, by Carl Sagan of Cornell and Joseph Shklovskii, a Soviet astronomer.

First we ask what the probability is that stars at the centers of solar systems are suitable to allow intelligent life to evolve. For example, an O star would probably stay on the main sequence for too short a time to allow the evolution of intelligent life.

Second, we ask what are the chances of a suitable star having planets. The import of the previous section is that the chances are probably pretty high, but that we are less certain of this conclusion than we had thought.

Third, we need planets with suitable conditions for the origin of life. A planet like Jupiter might be ruled out, for example, because of its lack of a solid surface and because of its high surface gravity, though alternatively one could consider a liquid region, if it were at a suitable temperature, to be as advantageous as were the oceans on Earth to the development of life here.

Fourth, one has to consider the fraction of systems on which life begins. This is the biggest uncertainty, for if this fraction is zero (with the Earth being a unique exception), then we can get nowhere with this entire line of reasoning. Still, the discovery that amino acids can be formed in laboratory simulations of primitive atmospheres, and the discovery of complex molecules containing half a dozen or more atoms even in interstellar space (see Section 21.3) indicate to many astronomers that it is not as difficult as had been thought to form complicated molecules.

At this point we have already reached an interesting result, for if one tries to assess (or, really, guess) fractions for each of the above steps in the chain of reasoning, and multiplies the fractions together, one finds that there may be billions of planets in this galaxy on which life may have evolved, and that the nearest one may be within dozens of light years of the Sun.

Box 19.1 The Probability of Life in the Universe

The discussion we have been carrying out follows the lines of an equation written out to estimate the number of civilizations in our galaxy that would be able to contact each other. In 1961, Frank Drake of Cornell wrote the equation:

$$N = R_* f_p n_e f_l f_i f_c L,$$

where R_* is the rate at which stars form in our galaxy, f_p is the fraction of these stars that have planets, n_e is the number of planets per solar system that is suitable for life to survive (for example, an Earth-like atmosphere), f_l is the fraction of these planets on which life actually arises, f_i is the fraction of these life forms that develop intelligence, f_c is the fraction of the intelligent species that choose to communicate with other civilizations and develop adequate technology, and L is the lifetime of such a civilization. The largest uncertainty is f_l, which could be essentially zero or could be close to 1. The result is also very sensitive to the value one chooses for L—is it 1 century or a billion years? Do civilizations destroy themselves? Depending on the alternatives one chooses, one can find that there are dozens of communicating civilizations within 100 light years or that the Earth is unique in having one. Most people choose their values such that the result comes out closer to the former than to the latter situation.

If we want to have meaningful conversations with the aliens, however, we must have a situation where not just life but intelligent life has evolved. We cannot converse with algae or paramecia. Furthermore, the life must have developed a technological civilization capable of interstellar communication. These considerations reduce the probabilities somewhat, but it has still been calculated that there are likely to be technologically advanced civilizations of intelligent life within a few hundred light years of the Sun.

What are the chances of our visiting or being visited by representatives of these civilizations? As we have seen (Fig. 16–6), Pioneers 10 and 11 are even now carrying plaques out of the solar system in case an alien interstellar traveler should happen to encounter these spacecraft. Still, it seems unlikely that humans can travel the great distances to the stars, unless some day we develop spaceships to carry whole families and cities on indefinitely long voyages into space. We must recall, however, that the Sun has another 4 or 5 billion years to go on the main sequence, and recorded history on Earth has existed for only about 5000 years, one millionth of the age of the solar system. There is plenty of time in the future for interstellar space travel to develop. Gerard O'Neill of Princeton has recently led a series of investigations that seem to show that we could start to think of setting up cities in space. (A major purpose of these cities would be to collect solar energy and beam it to Earth.)

We can hope even now to carry out interstellar communication by means of radio signals. Much thought has been given to what signals to send out, and to what is the best frequency to use, as will be discussed in the following section. What is the chance of hearing another civilization or having our own messages heard? Note that we have known the basic principles of radio for only a hundred years, and that powerful radio and television transmitters have existed for less than 50 years. Even now we are sending out signals into space on the normal broadcast channels. A wave bearing the voice of Caruso is expanding into space, and at present is 50 light years from Earth. And once a week a new episode of *Charlie's Angels* is carried into the depths of the Universe.

Now one comes to the important question of the lifetime of the technological civilization itself. We now have the capability of destroying our civilization either dramatically in a flurry of hydrogen bombs or more slowly by, for example, altering our climate or increasing the level of atmospheric pollution. It is a sobering question to ask whether the lifetime of a technological civilization is measured in decades, or whether all the problems that we have, political and otherwise, can be overcome, leaving our civilization to last for millions or billions of years.

19.4 THE SEARCH FOR LIFE

For planets in our solar system, we can search for life by direct exploration, but how would you go about trying to detect radio signals from creatures located outside our solar system? It would be too overwhelming a task to listen for signals at all frequencies in all directions at all times. One must make some reasonable guesses.

There are a few frequencies in the radio spectrum that seem especially fundamental. Neutral hydrogen in space, for example, has a basic spectral line at 1420 MHz, a frequency over ten times higher than stations at the high end of the normal FM band. (This hydrogen radiation is of great importance in studying our galaxy, and will be discussed in Chapter 21.) We might conclude that creatures on a far-off planet would decide that we would be most likely to listen near this frequency because it is so fundamental. On this theory, they would broadcast at about 1420 MHz.

On the other hand, perhaps the great abundance of radiation from hydrogen itself would clog this frequency, and one or more of the other frequencies that correspond to strong natural radiation should be preferred. If we ever detect radiation at any frequency, I am sure that we will immediately start to wonder why we did not realize that this frequency was the obvious choice. If only we now had hindsight on this matter.

In 1960, Frank Drake used a telescope at the National Radio Astronomy Observatory to listen for signals from two of the nearest stars. He was searching for any abnormal kind of signal, a sharp burst of energy, for example, such as pulsars emit. He was able to devote to this investigation, known as Project Ozma (after the princess of the land of Oz in L. Frank Baum's stories) only 200 hours of observing time, distributed over a few months.

The observations were later extended in a more methodical search called Ozma II. Starting in 1973, Ben Zuckerman of the University of Maryland and Patrick Palmer of the University of Chicago systematically monitored the 500 nearest stars of supposedly suitable spectral type. They used computers to search the data they recorded for any sign of special signals. Needless to say, nothing significant turned up; the world will know it rapidly if anything ever does.

A grandiose proposal has been made to build a huge array of over 1000 radio telescopes, each 100 meters (100 yards) in diameter, covering an area about 10 kilometers on a side. The total area collecting radiation would be almost that of a single telescope with a 10-kilometer diameter! The array would be fantastically sensitive. This plan, called Project Cyclops, would cost 5 billion dollars. Its construction would depend on a decision that the search for interstellar life in this way was an important national priority, a decision that seems unlikely to be made in the foreseeable future. Of course, valuable astronomical research could be carried out with the Cyclops array when it was not being used for interstellar monitoring, but

Figure 19–4 The Arecibo telescope in Puerto Rico.

nobody is about to spend 5 billion dollars on straight astronomy either. Still, the amount is not out of line with the cost of space research.

In 1974 the giant radio telescope at Arecibo, Puerto Rico (Fig. 19–4), was upgraded, in that the precision and quality of the mesh of the bowl-like surface of the dish was improved. The telescope is 305 meters across, the largest telescope on Earth. At the rededication ceremony, a powerful signal was sent out into space at the hydrogen frequency, bearing a message from the people on Earth (Fig. 19–5). This is the only major signal that we have purposely sent out as our contribution to the possible interstellar dialogue.

The signal from Arecibo was directed at the globular cluster M13 in the constellation Hercules (see page 74) on the theory that the presence of 300,000 closely packed stars in that location would increase the chances of our signal being received by a civilization on one of them. But the travel time of the message (at the speed of light) is 24,000 years to M13, so we certainly could not expect to have an answer before twice 24,000, or 48,000 years, have passed. If anybody (any**thing**) is observing our Sun when the signal arrives, the radio brightness of the Sun will increase by 10 million times for a 3-minute period.

Not only astronomers in the United States but also astronomers in the Soviet Union are interested in "communication with extraterrestrial intelligence," which is becoming known by the acronym CETI, and in the simpler and less expensive "search for extraterrestrial intelligence," SETI. The first phase of the Soviet program will last until 1985, and includes scanning a wide range of frequencies in all directions in the sky for suitable strange signals. In the second phase, which is scheduled to last from 1980 to 1990, radio telescopes in space would continue the search, perhaps from a position behind the Moon where the Moon's bulk prevents interference by radio signals from the Earth. In late 1976, NASA announced a SETI program, a major part of which will be the construction of a special computer to analyze the incoming radio waves for signals from far-off civilizations.

19.5 UFO'S AND OCCAM'S RAZOR

But why, you may ask, if most astronomers accept the probability that life exists elsewhere in the Universe, do they not accept the idea that unidentified flying objects (UFO's) represent visitations from these other civilizations (Fig. 19–6)? The answer to this question leads us not only to explore the nature of UFO's but also to consider the nature of knowledge and truth. The discussion that follows is a personal view, but one that is shared by many scientists.

First of all, most of the sightings of unidentified flying objects that are reported can be explained in terms of natural phenomena. Astronomers are experts on strange effects that the Earth's atmosphere can display, and many UFO's can be explained by such effects. For example, I know that every time Venus shines brightly on the horizon, in my capacity as local astronomer I will get a lot of telephone calls asking me about the UFO. A planet or star low on the horizon can seem to flash red and green because of atmospheric refraction, a phenomenon not generally realized by nonscientists. Atmospheric effects can affect radar waves as well as visible light.

Sometimes other natural phenomena — flocks of birds, for example — are reported as UFO's. One should not accept explanations that UFO's are

A

B

Figure 19–5 (A) The message sent to M13, with a translation into English added. (B) The message was sent as a string of 1679 consecutive characters, in 73 groups of 23 characters each. There were two kinds of characters, each represented by a frequency; the two kinds of characters are reproduced here as 0's and 1's.

Figure 19–6 Gort, from *The Day the Earth Stood Still.* (Museum of Modern Art/Twentieth-Century Fox, Inc.)

Figure 19–7 (*A*) A UFO fabrication: an aluminum plate topped by a cottage cheese container on which black dots have been placed to simulate portholes. The landing gear is constructed from hemispheres of ping-pong balls. (*B*) A UFO fabrication.

A

B

flying saucers from other planets before more mundane explanations—including exaggeration and fraud—are exhausted.

The topic is of tremendous public interest. As the result of public pressure, a committee was appointed under governmental auspices a few years ago to investigate UFO's. A distinguished physicist, the late E. U. Condon, headed the group, which worked from 1966 to 1969. They asked the leading flying saucer groups to send them the information about the cases that seemed to prove most convincingly that UFO's really have an extraterrestrial identification. After exhaustive analysis, the Condon group was able to account for these cases on the basis of more commonplace effects. Only when too little information is available—as when one has only a report that "something bright was seen in the northern sky for ten minutes and it moved around"—do the UFO reports become impossible to explain without invoking flying saucers.

Further, it has been shown that for many of the effects that have been reported, the UFO's would have been defying well-established laws of physics if their reported motions actually took place. Where are the sonic booms, for example, from rapidly moving UFO's? Scientists treat challenges to well-established laws of physics very seriously, as much of our science and technology is based on these laws. We can ordinarily explain new discoveries in terms of these known laws. If the laws of physics are seriously challenged by UFO's, then we would have to start questioning other applications we make of these laws, such as the safety of skyscrapers, or the lift of airplanes.

The results of a survey of members of the American Astronomical Society, released in 1977, indicated that only 5 per cent of Ph.D. respondents who had spent 51 to 365 hours studying the UFO problem chose "a technological device not of terrestrial origin" as the explanation of UFO's. Since only about half of the members answered the questionnaire, and a study of their reasons for not answering showed that the non-responding group was on the whole even more negative to the study of UFO's, the actual percentage of astronomers even considering that UFO's are flying saucers containing extraterrestrials is certainly tiny. Further, less than half of the Ph.D. respondents indicated that "the UFO problem" certainly or probably "deserves scientific study." When this is weighted by the non-responding half, and interpreted in the knowledge that no question was asked about what priority should be given to such studies in the face of other demands for our time and money, it seems clear that the number of those who even believe that other than cursory further effort should be put into UFO studies is small.

I conclude, mainly from my own extensive conversations on the subject, that almost every professional astronomer feels that UFO's can be so obviously and completely explained by natural phenomena that they are not worthy of more of our time. Astronomers just know too much about the properties of the atmosphere to give credence to most of the UFO reports or to think that further investigation is a worthwhile investment of time or money that could be put to better pursuits.

It is mainly non-scientists who feel that UFO's are worth investigating. And it should be pointed out that there is a lot of money involved in reporting sightings of UFO's—since the topic is so popular there are books and articles to be written, and interviews to be given, all of which have a financial return. One must be skeptical of the motives of those who push reports of UFO sightings or want their organizations to continue UFO investigations.

Some individuals may ask why we reject the identification of UFO's with flying saucers, when that explanation is "just as good an explanation as any other." Let us go on to discover what we mean by "truth" and how that applies to the above question.

We regard as *true* the simplest explanation that satisfies all the data we have about any given thing. This principle is known as *Occam's Razor;* it is named after a fourteenth-century British philosopher. Without it, we would always be subject to such complicated doubts that we would accept nothing as known.

Sometimes we call something true that might be more accurately described as a *theory.* The scientific method is based on hypotheses and theories. A *hypothesis* is an explanation that is advanced to explain certain facts. When it can be shown that the hypothesis actually explains most or all of the facts known, then we call it a *theory.* We usually test a theory by seeing whether it can predict things that were not previously observed, and then by trying to confirm whether the predictions are valid. We might call the theory, if it passes these tests, a *well-established theory.* Others might prefer to reserve the word *theory* for this stage and call the previous stages *hypotheses* or *conjectures.* The dividing line between hypotheses and theories is not always clear.

How does this view of truth tie in with our discussion of UFO's? Every moment we must make decisions about what to do, based on what we think will happen next. We wouldn't even take a step if we thought that a yawning chasm would open up before us in the midst of our living rooms. Scientists have assessed the probability of UFO's being flying saucers from other

Figure 19–8 The two UFO's are really lights shining through the hole in a phonograph record viewed obliquely. The background was added later.

Occam's Razor, sometimes called the Principle of Simplicity, is a razor in the sense that it is a cutting edge that allows a distinction to be made among theories.

Figure 19–9 Cartoon by Charles Schulz. © 1975 by United Features Syndicate, Inc.

worlds, and most have decided that the probability is so low that the possibility is not even worth considering. We have better things to do with our time and with our national resources. We have so many other, simpler explanations of the phenomena that are reported as UFO's that we apply Occam's Razor and decide to call the identification of UFO's with extra-terrestrial visitation *false*. UFO's may be unidentified, but they are probably not flying, nor for the most part are they objects.

19.6 OF ASTRONOMY AND ASTROLOGY *(Optional)*

Neither astrology nor UFO's are at all connected with astronomy, except in a historical context, so neither really deserves a place in a text on contemporary astronomy. But since so many people associate astrology with astronomy, and since astrologers claim to be using astronomical objects to make their predictions, let us use our astronomical knowledge to assess astrology's validity.

Most professional astronomers would privately agree to sentiments something like those expressed in this section, or at least reach the same conclusion. Many astronomers don't even think that astrology is worth discussing, or at best, that discussions have no effect on those who "believe" and so are a waste of time. We will not take this latter position here.

Astrology is an attempt to predict or explain our actions and personalities on the basis of the positions of the stars and planets now and at the instants of our births. Astrology has been around for a long time, but it has never been shown to work. Believers may cite incidents that reinforce their faith in astrology, but no successful scientific tests have ever been carried out. If something happens to you that you had expected because of an astrological prediction, you would more certainly notice that this event occurred than you would notice the thousands of other unpredicted things that happened to you that day. And likewise with hindsight. We do enough things, have sufficiently varied thoughts, and interact with enough people that if we make many predictions in the morning, some of them are likely to be at least partially fulfilled during the day. We simply forget that the rest ever existed. Besides, we always have the easy out of being able to say, "well, we just didn't have a good astrologer."

Studies have shown that superstition actively constricts the progress of science and technology in various countries around the world and is therefore not merely an innocent force. It is not just that some people harmlessly believe in astrology. Their lack of understanding of scientific structure may actually impede the proper scientific training of people needed to work on the problems of our age, including the problems of pollution and the energy crisis. A recent paper even shows how widespread superstitious beliefs can impede smallpox-prevention programs. Thus many scientists are not content to ignore astrology, but actively oppose its dissemination. Further, if large numbers of citizens do not understand the scientific method and the difference between science and pseudoscience, how can they intelligently vote on or respond to scientific questions that have societal implications?

A major reason why scientists in general and astronomers in particular don't believe in astrology is that they cannot conceive of a way in which it would work. The human brain is so complex that it seems most improbable that any celestial alignment can affect people, including newborns, in an

Gloria: Oh Ma, you don't believe in astrology, do you?
Edith: No, but it's fun to know what's going to happen to you, even if it don't.
from All in the Family

overall way. The celestial forces that are known cannot be sufficient to set personalities nor influence day to day events. And even if people reply that they do not think astrology is true but merely find it interesting, many scientists feel that so many strange and exciting explainable things are going on in the Universe that we wonder why anybody should waste time with far-fetched astrological concerns.

Moreover, astrology just doesn't work. Bernie I. Silverman, a psychologist at Michigan State University, tested specific values: Do Libras and Aquarians rank "Equality" highly? Do Sagittarians especially value "Honesty"? Do Virgos, Geminis, and Capricorns treasure the value "Intellectual"? Several astrology books agreed that these and other similar examples are values typical of those signs. Although believers often criticize objections on the ground that these group horoscopes are not as valuable or accurate as individualized charts, surely some general assumptions and rules hold in common.

The subjects, 1600 psychology graduate students, did not know in advance just what was being tested. They gave their birthdates, and the questioners determined their astrological signs. The results: no special correlation was apparent for any of the signs with the values they were supposed to hold. This was confirmed by statistical tests. As many subjects, regardless of their astrological signs, when asked to what extent they shared the qualities of each given sign, ranked themselves above average as below.

Furthermore, Silverman tested whether individuals from particular pairs of signs were especially compatible or incompatible. To do this, he sampled marriage and divorce rates in Michigan in 1967 and 1968. He compared these rates with the predictions of two astrologers as to compatible or incompatible combinations. His sample was large—2978 marriages and 478 divorces—so even slight effects should show up in his statistical treatment. And none did—those born under "compatible" signs neither married more frequently nor divorced less frequently than would be expected by chance, nor did those born under "incompatible" signs marry less frequently or divorce more frequently. Silverman concluded that "the position of the Sun in the zodiac at birth does not affect later personality," and in general that "astrology is invalid." Thus astrological predictions were shown by his studies to be no better than chance.

But if astrology is so meaningless, why does it still have so many adherents? Well, it could be the bandwagon effect, and Silverman had an ingenious test that endorsed this idea. He took twelve personality descriptions from astrology books, one for each astrological sign, and displayed them to two groups of individuals. The first group, composed of 51 subjects, was told to which astrological signs the descriptions pertained, and was asked to write their own signs on the covers of the questionnaires. More than half the members of this group thought that the descriptions listed under their own signs were, for each individual, among the four best descriptions of themselves out of the twelve choices. It would seem, if one considered only this phase of the experiment, that astrology was working.

Yet when the second group was given the twelve descriptions without mention of astrology, being told that they came from a book entitled "Twelve Ways of Life," their choices were random. Only 30 per cent chose their own sign's description as being in the group most closely describing them. So the idea that astrology can predict personality types seems to be the result of self-delusion. When people know what they are expected to be

Figure 19–10 The ecliptic is divided into the signs of the zodiac in this illustration from the *Epitoma In Almagestum Ptolemaei (Venice, 1496)* by Regiomontanus. This was the first translation into Latin of Ptolemy's *Almagest,* which contained the standard description of Ptolemy's earth-centered universe. The *Almagest* had been formerly available only in Arabic. "Almagest" means "the greatest."

like, they tend to identify themselves with the description. But that doesn't mean that they actually satisfy the description that astrology predicts for them better than any other description.

From an astronomer's view, astrology is meaningless, unnecessary, and impossible to explain if we accept the broad set of physical laws we have conceived over the years to explain what happens on the Earth and in the sky. Astrology snipes at the roots of all pure science. Moreover, astrology patently doesn't work. If people want to believe in it on an *a priori* basis, as a religion, or have a personal astrologer act as a psychologist, let them not try to cloak their beliefs in scientific astronomical gloss. The only reason people may believe that they have seen astrology work is that it is a self-fulfilling means of prophecy, conceived of long ago in times when we knew less about the exciting things that are going on in the Universe. Let's all learn from the stars, but let's learn the truth.

SUMMARY AND OUTLINE OF CHAPTER 19

Aims: To discuss the possibility of intelligent life existing elsewhere in the Universe besides the Earth, and to assess our chances of communicating with such life; to give an astronomer's view of UFO's and of astrology; to give scientific criteria for "truth," and definitions of "hypotheses" and "theories"

The origin of life (Section 19.1)
    Earth life based on complex chains of carbon-bearing (organic) molecules
    Carbon, as well as silicon, forms such chains easily
    Organic molecules are easily formed under laboratory conditions that simulate primitive atmospheres
Evidence for other solar systems (Section 19.2)
    Search for wobbles in proper motions of stars
    Evidence for a planet around Barnard's star once was positive, but is currently not as certain as had been thought previously
Statistical chances for extraterrestrial intelligent life (Section 19.3)
    Problem broken down into stages
    Major uncertainties include what the chance is that life will form given the component parts, and what the lifetime of a technological civilization is likely to be
The search for life (Section 19.4)
    Experiments on Viking spacecraft to Mars
    Projects Ozma and Ozma II
    Signal beamed toward globular cluster M13 from Arecibo radio telescope
    Soviet and American CETI and SETI plans
UFO's and Occam's Razor (Section 19.5)
    UFO's explainable as natural phenomena
    Application of Occam's Razor—the Principle of Simplicity—rules out the existence of UFO's
    Definitions of truth, hypothesis, and theory
Astronomy and astrology (Section 19.6)
    No scientific basis known for astrology

Belief in such pseudoscience impedes the advance of science and technology

Statistical arguments show that astrology doesn't work

QUESTIONS

1. What is the significance of the Miller-Urey experiments?

2. Assume that a star has a planet with a mass one tenth that of the star. Sketch the path of the star and the planet in the sky, omitting the parallax effect of the Earth's motion.

3. Does a star with a large proper motion necessarily have planets? Explain.

4. Discuss the evidence that Barnard's star has planets around it.

5. If one tenth of all stars are of suitable type for life to develop in the case that planets exist around them, and 1 per cent of all stars have planets, and 10 per cent of all planetary systems have a planet at a suitable distance from the star for a comfortable environment, what fraction of stars have a planet with conditions suitable for life? How many such stars would there be in our galaxy?

6. List the means by which we might detect extraterrestrial civilizations.

7. Describe how Occam's razor might have been applied during the debate over the geocentric and heliocentric pictures of the solar system.

8. Give a non-scientific example of a hypothesis, and suggest how it might be tested. At what stage would the hypothesis become a theory?

TOPICS FOR DISCUSSION

1. Comment on the possibility of scientists in another solar system observing the Sun, suspecting it to be a likely star to have populated planets, and beaming a signal in our direction.

2. Discuss the extent of the astrological beliefs of the members of your class. If any students believe in astrology, do their beliefs affect their daily lives?

Figure 19-11 From *Star Wars.* © Copyright 1977 Twentieth-Century Fox, Inc. All rights reserved.

Part IV

The Milky Way Galaxy

The Milky Way from Sagittarius to Cassiopeia.

(left) The Lagoon Nebula, M8, in Sagittarius.

The Milky Way Galaxy . . .

On the clearest nights, when we are far from city lights, we can see a hazy band of light stretched across the sky. This is the *Milky Way*—the aggregation of dust, gas, and stars that make up the galaxy in which the sun is located.

Don't be confused by the terminology: the Milky Way itself is the band of light that we can see from the earth, and the Milky Way Galaxy is the whole galaxy in which we live. Like other galaxies, our Milky Way Galaxy is composed of a hundred billion stars plus many different types of gas, dust, planets, etc. The Milky Way is that part of the Milky Way Galaxy that we can see with the naked eye in our nighttime sky.

The Milky Way appears very irregular in form when we see it stretched across the sky—there are spurs of luminous material that stick out in one direction or another, and there are dark lanes or patches in which nothing can be seen. This is just the manifestation of the splotchy distribution of the dust, stars, and gas.

Here on earth, we are inside our galaxy together with all of the matter we see as the Milky Way. Because of our position, we see a lot of matter when we look in the plane of our galaxy. On the other hand, when we look "upwards" or "downwards" out of this plane, our view is not obscured by matter, and we can see past the confines of our galaxy. We are able to see distant galaxies only by looking in parts of our sky that are away from the Milky Way, that is, by looking out of the plane of the galaxy.

As seen from the earth, the center of our galaxy is located in the direction of the constellation Sagittarius. This constellation never rises high in the sky seen from the United States, but in the southern hemisphere it can appear directly overhead. Since the denser clouds of stars, gas, and dust are in the direction of the galactic center, the Milky Way appears more spectacular in the sky when observed from points south of the equator.

The gas in our galaxy is more or less transparent to visible light, but the small solid particles that we call "dust" are opaque. So the distance we can see through our galaxy depends mainly on the amount of dust that is present. This is not surprising: we can see great distances through our gaseous air on earth, but let a small amount of particulate material be introduced in the form of smoke or dust thrown up from a road and we

The Origin of the Milky Way by Tinto circa 1578. (The National Gallery, Lor

find that we can no longer see very far. Similarly, the dust between the stars in our galaxy dims the starlight by scattering it in different directions besides absorbing some of it. This property of the dust is called *extinction*. The extinction of visible light in the plane of our galaxy is about 1 magnitude per thousand parsecs (1 magnitude per kiloparsec)—that is, a star a thousand parsecs away would appear about one magnitude fainter than it would if there were no interstellar dust, a star two thousand parsecs away would appear two magnitudes dimmer, and so on. Of course, the amount of extinction is really irregular, in that it varies from this average value as one looks in different directions.

The presence of so much dust in the plane of the Milky Way Galaxy actually prevents us from seeing very far toward its center—we can see only 1/10 of the way toward the galactic center itself. This explains the dark lanes across the Milky Way, which are just areas of dust, obscuring any emitting gas or stars. The net effect is that we can see just about the same distance in any direction we look in the plane of the Milky Way. It follows that these direct optical observations by themselves might fool us into thinking that we are in the center of the galaxy, or indeed in the center of the universe. Using this reasoning, scientists at the turn of the century thought that the earth was near the center of the galaxy.

It was not until the 1920's that the American astronomer Harlow Shapley realized that we were not in the center of the galaxy. He was studying the distribution of globular clusters and noticed that they were all in the same general area of the sky as seen from the earth. They mostly appear above or below the galactic plane and thus are not obscured by the dust. When he plotted their distances and directions (using the methods described in Part II), he saw that they formed a spherical halo around a point many thousands of light years away from us (see Fig. 20–1). Shapley's touch of genius was to realize that this point must be the center of the galaxy.

In recent years astronomers have also been able to use wavelengths other than optical ones, especially radio wavelengths, to study the Milky Way Galaxy. We shall first discuss the types of objects that we find in the Milky Way Galaxy, and then go on to describe how radio astronomy and other methods of observation have given us new information about our galaxy.

Components of the Milky Way

In Part II, we described stars, which are important constituents of any galaxy. In this chapter, we describe other material in our galaxy. We also describe the overall structure of the Milky Way Galaxy and how, from our location inside it, we detect this structure.

The current picture that we have of our own galaxy (Fig. 20–1) is that around the center is a *nuclear bulge* about 2000 parsecs in radius, with the galactic *nucleus* at its midst. The nucleus itself is only about 5 parsecs across and is best studied in the x-ray, radio, and infrared parts of the spectrum (see Chapter 22). The nuclear bulge has the shape of a flattened sphere and does not show spiral structure. Outside the bulge, the next 10,000 parsecs (10 kiloparsecs) or so to each side contain the spiral arms; the sun is about in the middle of this region, which is called the *disk*. The disk is only about 600 parsecs thick on the average, 2 per cent of its width. It is slightly bent at the ends, like a warped phonograph record, perhaps by interaction with the Magellanic Clouds. Older stars, including the globular clusters, form a *halo* around the disk. This halo is at least as large in diameter as is the disk, but extends far above and below the plane of our galaxy.

20.1 THE INTERSTELLAR MEDIUM

The gas and dust between the stars is known as the *interstellar medium.* The gas and dust are spread out throughout our galaxy, but in some regions are assembled into beautiful objects called nebulae.

"Nebula" is Latin for fog or mist.

A *nebula* (pl: *nebulae,* not "nebulas") is a cloud of gas and dust that we see in visible light. Sometimes we see the gas actually glowing, sometimes we see the dust appear as a dark silhouette against other glowing material, and sometimes we see the dust reflecting light from other objects toward us. These are called *emission nebulae, absorption* (or *dark*) *nebulae,* and *reflection nebulae,* respectively.

The North America Nebula (Fig. 20–2), gas and dust that has a shape in the sky similar to the shape of North America on the earth's surface, is a

The Milky Way in Sagittarius.

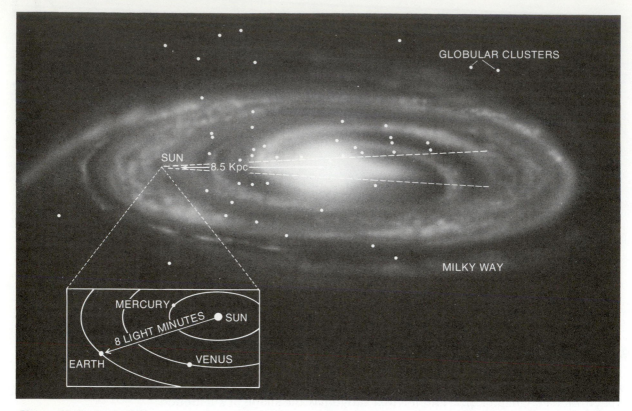

Figure 20–1 An artist's conception of the Milky Way Galaxy, with the actual positions of the globular clusters for which distances are known. Absorption by interstellar matter blocks our view of most objects between the dotted lines. From the fact that most of the clusters appear in less than half of our sky, Shapley deduced that the galactic center is in the direction indicated. The drawing shows the *nuclear bulge* surrounded by the *disk,* which contains the spiral arms. The globular clusters are part of this *halo.*

good example of both an emission and an absorption nebula. The red emission comes from glowing gas spread across the sky. This gas is not completely opaque to visible light; of the stars that are visible in the nebula, some are between the sun and nebula but others are behind the nebula.

The boundary of the nebula as we see it, however, is not just the point where the glowing gas stops. There is actually dark absorbing dust that forms some of the boundaries. For the North America Nebula, there is dust absorbing in the area that corresponds to the Gulf of Mexico. We can tell this because the number of stars that we see in that direction is very small. We think that the average number of stars we could see if there were no dust is fairly constant in a given area of the sky. Thus when we see fewer stars in a part of that area, we conclude that there must be dust absorbing the radiation from stars behind it.

The Horsehead Nebula in Orion (Fig. 3–23 and Color Plate 42) is another example of emission and absorption nebulae. A bit of absorbing dust intrudes onto emitting gas, outlining the shape of a horse's head for us. We can see that the horsehead is a continuation of a dark area in which very few stars are visible. A reflection nebula is also visible nearby.

The Great Nebula in Orion (Color Plate 12) is an emission nebula. It can readily be observed even with small telescopes in the winter sky, but only with long photographic exposures or large telescopes can we study the structure in the nebula in detail.

The clouds of dust surrounding some of the stars in the Pleiades (Color Plate 5) are examples of reflection nebulae—they merely reflect the starlight toward us without emitting much radiation of their own. Many other reflection nebulae are known (Fig. 20–3).

Figure 20-2 The North America Nebula, NGC 7000, in Cygnus.

The nebulae are particularly beautiful objects because of their interesting configurations, and because the different processes of emission and absorption cause color effects that, although they are too faint to be visible to the naked eye, can be captured on photographs. Other nebulae are shown in Color Plates 6, 43, 44, and 45.

The emission nebulae glow because the gas is heated by one of several mechanisms. Bright spectral lines are emitted at certain wavelengths. The

Figure 20-3 A reflection nebula, NGC 7129, in Cepheus.

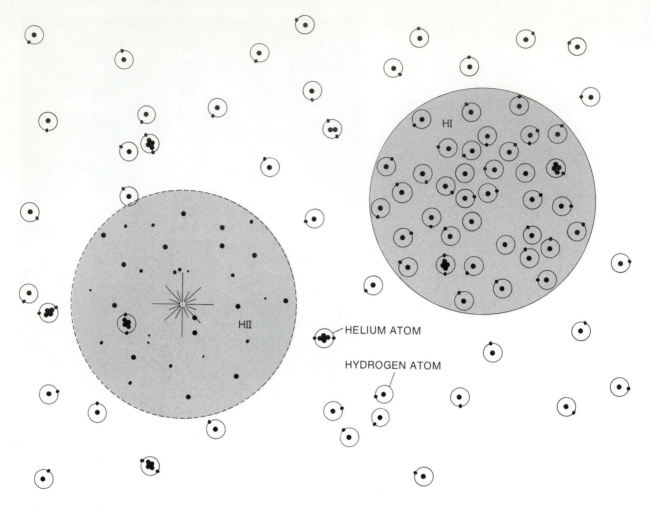

HI

HII

HELIUM ATOM

HYDROGEN ATOM

Figure 20–4 *H I regions* are regions of neutral hydrogen and *H II* regions are regions of ionized hydrogen. The protons and electrons that result from the ionization of hydrogen by a hot star, and the neutral hydrogen atoms, are shown schematically. The outlines of the regions are shown for illustrative purposes only; in space, of course, the regions are not outlined.

hydrogen that makes up the overwhelming proportion of the gas is ionized —that is, the hydrogen is separated into protons and electrons. Neutral hydrogen is known as H I (the "first state," with all its electrons), and ionized hydrogen as H II (the "second state," with one electron missing), so these ionized nebular regions are known as *H II regions* (Fig. 20–4).

We may consider interstellar space as filled with hydrogen at an average density of about 1 atom per cubic centimeter, although individual regions may have densities departing greatly from this average. Regions in which the atoms of hydrogen are predominantly neutral are called *H I regions*. Where the density of an H I region is high enough, pairs of hydrogen atoms combine to form molecules (H_2). The densest part of the molecular cloud associated with the Orion Nebula might have 10^6 or more hydrogen molecules per cubic centimeter. On the other hand, wherever a hot star provides enough energy to ionize hydrogen, an H II region results.

The interstellar gas is "invisible" in the visible part of the spectrum (except at the wavelengths of certain weak spectral lines); special techniques are needed to observe the gas in addition to observing the dust. Radio astronomy is the most widely used technique for studying the gas, and we shall discuss it at length in the next chapter.

20.2 THE SPIRAL STRUCTURE OF THE GALAXY

20.2a *Bright Tracers of the Spiral Structure*

When we look out past the boundaries of the Milky Way Galaxy, usually by observing in directions above or below the plane of the Milky Way, we can see a number of galaxies external to our own system. Some of these appear as though bright arms spiral off a central core. We shall discuss these *spiral galaxies* (Color Plates 46, 47, 49, and 50) in Chapter 23; in this section we shall discuss some of the evidence that our own Milky Way Galaxy also has spiral structure.

It is always difficult to tell the shape of a system from a position inside it. Think, for example, of being somewhere inside a maze made of tall hedges. We might be able to see through some of the foliage and be reasonably certain that there were layers and layers of hedges surrounding us, but we would find it difficult to trace out the pattern. If we could fly overhead in a helicopter, though, the pattern would become very easy to see.

Similarly, we have difficulty tracing out the spiral pattern in our own galaxy, even though the pattern would presumably be apparent to someone located outside our own galaxy. Still, by meticulously noting the distances and directions to objects of various types, we can tell something about the Milky Way's spiral structure.

One type of object that can be studied for this purpose is a galactic cluster, which is known to always be in the spiral arms. We can find the distance to a galactic cluster from its Hertzsprung-Russell diagram (Section 4.5a). We have now determined the distances to over 200 of these clusters.

In addition, it is known from studies of other galaxies that H II regions are also preferentially located in spiral arms. In studying the locations of the H II regions, we are really studying the locations of the O stars and the hotter B stars, since it is ultraviolet radiation from these hot stars that provides the energy for the H II regions to glow. These O and B stars, which we know must be relatively young because their lifetimes are short, must tend to be formed in the spiral arms; in the next section we shall discuss a theory that explains why this should be the case.

When the galactic clusters and H II regions of known distance are plotted, they appear to trace out three spiral arms (Fig. 20–5). The spacings between the arms and the widths of the arms appear consistent with spacings and arm widths that we can observe in other galaxies.

These observations are carried out in the visible part of the spectrum. Note that even when we are observing the O and B stars, which are very bright and therefore can be seen at distances that are relatively great compared to other stars, the optical map shows only regions of our galaxy close to the sun. The interstellar extinction prevents us from studying parts of the spiral arms farther away from the sun. Another very valuable method of studying the spiral structure in our own galaxy involves a spectral line of hydrogen in the radio part of the spectrum, and will be discussed in Section 21.3. Radio waves penetrate the interstellar dust, and we are no longer limited to studying the local spiral arms.

20.2b *Differential Rotation (Optional)*

The latest calculations indicate that the sun is approximately 8½ kiloparsecs from the center of our galaxy. (An older estimate of 10 kiloparsecs is often still used for calculations. It is much more convenient to calculate with a round number like 10 instead of 8½, and one can easily make the

Figure 20–5 The positions of young galactic clusters and H II regions in our own galaxy are projected on a photograph of the spiral galaxy NGC 1232, which has the same linear diameter as our own galaxy.

correction necessary in order to adopt other values for the distance of the sun from the galactic center.)

From spectroscopic observations of the Doppler shifts of globular clusters, which do not participate in the galactic rotation, or of distant galaxies, we can tell that the sun is revolving around the center of our galaxy at a speed of approximately 250 kilometers per second. At this velocity, it would take the sun about 250 million years to travel once around the center; this period is called the *galactic year*. But not all stars revolve around the galactic center in the same period of time. The central part of the galaxy rotates like a solid body. Beyond the central part, the stars that are farther out have longer galactic years than stars closer in, following Kepler's third law (which is described in Section 10.3).

If this system of *differential rotation*, with differing rotation speeds at different distances from the center, has persisted since the origin of the galaxy, we may wonder why there are still just a few spiral arms in our galaxy and in the other galaxies we observe. The sun could have made fifty revolutions during the lifetime of the galaxy, but points closer to the center would have made many more revolutions. Thus the question arises: why haven't the arms wound up very tightly?

20.2c Density-wave Theory (Optional)

The leading current solution to this conundrum is a theory first suggested by the Swedish astronomer B. Lindblad and elaborated mathematically by the American astronomers C. C. Lin and Frank Shu. They say, in effect, that the spiral arms we now see are not the same spiral arms that were previously visible. In their model, the spiral arm pattern is caused by a spiral *density wave*, a wave that moves through the stars and gas in the galaxy. This density wave is a wave of compression, not of matter being transported. It rotates more slowly than the actual material, and causes the density of material to build up as it passes by.

We can think of the analogy of a crew of workers painting the white line down the center of a busy highway. A bottleneck occurs at the location of the painters, and an observer in an airplane would see an increase in the number of cars at that place. As the sign painters continued slowly down the road at, say, 8 km/hr (5 mph), the airborne observers would seem to see the place of increased density move down the road at that slow speed. But they would not be seeing the individual cars, which could still speed down the highway at 80 km/hr (50 mph), slow down briefly as they cross the region of the bottleneck, and then resume their high speed. Similarly, we might only be viewing some galactic bottleneck at the spiral arms. The gas is compressed there, both from the increased density itself and, to a greater extent, from the shock wave formed by other gas piling into this slowed-up gas from behind. The compression causes heating, and also leads to the formation of protostars that collapse to become stars. Thus the spiral density wave leads to star formation, though the stars themselves in their revolution around a galactic center would move across the arms.

It may help to think of the analogy of a sound wave, which is also a compression wave. The sound wave travels through the air at a rapid rate, but the air itself doesn't move at that rate in the direction of the sound wave. The mass of air has no net velocity at all.

Similarly, the density wave can pass through the galactic matter even though the galactic matter does not share in its motion. As the density wave

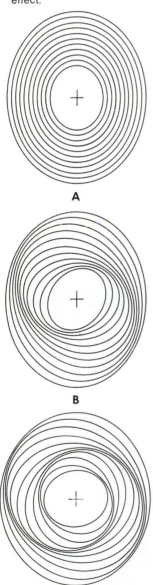

Figure 20–6 Each part of the figure includes the same set of ellipses; the only difference is the relative alignment of their axes. Consider that the axes are rotating slowly and at different rates. The compression of their orbits takes a spiral form, even though no actual spiral exists. The spiral structure of a galaxy may arise from an analogous effect.

A

B

C

passes a given point, stars begin to form there because of the increased density. The distribution of new stars can take a spiral form (Fig. 20–6). The new stars heat the interstellar gas so that it becomes visible. In fact, we do see young, hot stars and glowing gas outlining the spiral arms, which is one check of this prediction of the density-wave theory.

SUMMARY AND OUTLINE OF CHAPTER 20

Aims: To understand the Milky Way as the visible part of our galaxy and to discuss the various aspects of the matter that exists between the stars in our galaxy; to understand spiral structure; our galaxy has a nucleus, disk, and halo

Interstellar medium (Section 20.1)
  Nebulae
    Emission
    Absorption (dark)
    Reflection
    "Spiral"
    H I and H II regions
Spiral structure of the galaxy (Section 20.2)
  Bright tracers of the spiral structure (Section 20.2a)
    Difficult to determine shape of our galaxy because of obscuring dust and gas
    H II regions used to determine spiral structure
  Differential rotation (Section 20.2b)
    Sun approximately 8½ kpc from center of galaxy
    1 galactic year equals about 250 million years for the sun
    Central part of galaxy rotates as a solid body
    Outer part, including the spiral arms, is in differential rotation
Density-wave theory (Section 20.2c)
  Leading current solution to why spiral arms have not wound up very tightly
  We see the effect of a wave of compression; the distribution of mass itself is not spiral in structure

QUESTIONS

1. Why do we think that our galaxy is a spiral?

2. How would the Milky Way appear if the sun were close to the edge of the galaxy?

3. Sketch (a) a side view and (b) a top view of the Milky Way galaxy, showing the shapes and relative sizes of the nuclear bulge, the disk, and the halo.

4. Compare: (a) absorption (dark) nebulae, (b) reflection nebulae, and (c) emission nebulae.

5. How can something be both an emission and an absorption nebula? Explain and give an example.

6. If you see a blue star surrounded by a blue nebula, would you expect this nebula to be emission or reflection? Explain.

7. If you see a blue star surrounded by a red nebula, is this nebula emission or reflection? Explain.

8. Which spectral type of stars would you expect to best excite H II regions?

9. (a) What types of objects trace out spiral arms? (b) Why?

10. What is another system, besides the Milky Way Galaxy, that exhibits differential rotation?

The Interstellar Medium

From antiquity until 1930, observations in a tiny fraction of the electromagnetic spectrum, the visible part, were the only way that observational astronomers could study the universe. Most of the images that we have in our minds of objects in our galaxy (e.g., nebulae) are based on optical studies, since most of us depend on our eyes to discover what is around us.

If they had to make a choice, however, between observing only optical radiation or only everything else, many astronomers might choose "everything else." The rest of the electromagnetic spectrum carries more information in it than do the few thousand angstroms that we call visible light. In this chapter we will first discuss the basic techniques of radio astronomy, and then go on to see how radio astronomy joins with other observing methods to investigate interstellar space.

21.1 RADIO ASTRONOMY

One cannot always pinpoint the discovery of a whole field of research as decisively as that of radio astronomy. In 1931, Karl Jansky of the Bell Telephone Laboratories in New Jersey was experimenting with a radio antenna to track down all the sources of noise that might limit the performance of short-wave radiotelephone systems (Figs. 21–1 and 21–2). After a time, he noticed that a certain static appeared at approximately the same time every day. Then he made the key observation: the static was actually appearing four minutes earlier each day. This was the link between the static that he was observing and the rest of the universe. The stars and galaxies rise four minutes earlier every day, with respect to solar time. The

Figure 21–1 Karl Jansky lecturing about his discovery of a source of radio signals moving across the sky according to sidereal time.

The 100-meter radio telescope at Effelsburg, near Bonn, Germany. Note the tiny figures at the right-hand side of the base.

Figure 21–2 Karl Jansky with the rotating antenna with which he discovered radio astronomy.

Figure 21–3 Even though the holes in this radio telescope may look large to the eye, they are much smaller than the wavelength of radio radiation being observed. Thus the telescope is smooth and shiny to the incoming radio radiation.

four-minute difference arises from the fact that sidereal time and solar time differ, as described in Section 3.11. Jansky's static kept sidereal time, and thus was coming from outside our solar system! Jansky was actually receiving radiation from the center of our galaxy.

The principles of radio astronomy are exactly the same as those of optical astronomy: radio waves and light waves are exactly alike—only the wavelengths differ. Both are simply forms of electromagnetic radiation. But different technologies are necessary to detect the signals: light waves can blacken silver grains on a photographic plate or cause a response in the human eye, while radio waves have such long wavelengths that they pass undetected right through cameras or humans. Radio waves do cause electrical changes in antennas, and these faint electrical signals can be detected with instruments that we call radio receivers.

Radio waves are "received" by antennas not unlike television aerials. If we want to collect and focus radio waves just as we collect and focus light waves, we must find a means to concentrate the radio waves at a point at which we can place an aerial. Lenses to focus radio waves are impractically heavy, so refracting radio telescopes are not used. However, radio waves will bounce off metal surfaces. Thus we can make reflecting radio telescopes that work on the same principle as the 5-meter optical telescope on Palomar Mountain. The mirror, called a "dish" in common radio astronomy parlance, is usually made of metal rather than the glass used for optical telescopes. The dish needn't look shiny to our eyes, as long as it looks shiny to incoming radio waves. To appear shiny, the surface of a radio telescope has to be smooth to within a small fraction of the wavelength of radiation, one-twentieth or less (Fig. 21–3).

We want dishes as big as possible for two reasons: first, a larger surface area means that the telescope will be that much more sensitive. The second reason is also simple but takes a few more sentences to explain. The basic

Figure 21–4 A radio telescope system is similar to a home hi-fi system in that radio waves are electronically amplified. We tend to use the output of a hi-fi, though, to drive a loudspeaker, while astronomers usually record the strength of the signal received by a radio telescope in other ways. In radio astronomy, a reflector is usually used to concentrate the weak radio signals and focus them onto the actual antenna that converts the radio waves into an electric signal.

point is that measurements of the size of a telescope are most meaningful when they are in units of the wavelength of the radiation the telescope is being used to observe (Fig. 21–5). Thus an optical telescope one meter across used to observe optical light, which is about one two-millionth of a meter in wavelength, is two million wavelengths across. A radio telescope, even one 100 meters across (the size of the currently largest fully steerable dish, at the Max Planck Institute for Radio Astronomy in Bonn, Germany), is only 1000 wavelengths across when used to observe radio waves 10 centimeters in wavelength. The larger the telescope, measured with respect to the wavelength of radiation observed, the narrower is the width of the beam of radiation that the telescope receives. (The reflecting radio telescope dish receives radiation only from within a narrow cone. This cone is called the *beam;* see Fig. 21–6.) Radio astronomers normally work with broad beams and cannot resolve any spatial details smaller than the size of their beams. The 100-meter telescope at 10 cm has a beam that is four minutes of arc across; at longer wavelengths or with smaller telescopes one cannot even resolve an area the size of the moon. Obviously, this is looking at the sky with less than the optimum resolution, given that large optical telescopes ordinarily resolve 1 arc sec. However, under certain circumstances, one can now use the technique known as interferometry to improve the resolution. This technique will be discussed later on, in Section 23.4b.

100 meters/10 centimeters per wave = 1000 wavelengths.

Resolution is discussed in Section 3.1, where it is stated that resolution is proportional to wavelength and inversely proportional to telescope diameter.

Figure 21–5 It is more meaningful to measure the diameter of telescope mirrors in terms of the wavelength of radiation that is being observed than it is to measure it in terms of units like centimeters that have no particularly relevant significance. The radio dish at top is only 1¾ wavelengths across, while the mirror at bottom is over 10 wavelengths across, making it effectively much bigger.

Figure 21–6 The beams of radio telescopes are ordinarily relatively large compared with radio sources. Thus single-dish radio telescopes cannot resolve—detect detail on—radio sources to the same extent that optical telescopes resolve optical objects.

21.2 CONTINUUM RADIO ASTRONOMY

All radio astronomy research in the early days was of the continuum. (In radio astronomy, as in optical astronomy, that means that we consider the average intensity of radiation at a given wavelength without regard for variations in intensity over small wavelength ranges; that is we ignore any spectral lines.) When radio "continuum spectra" were measured, the continuum levels were measured at widely separated wavelengths: for example, at 5, 10, 20, and 50 cm. These continuum values, and the spectra derived from them, provided information about the mechanism that causes the continuum emission.

As radio astronomical observations got under way, it was immediately apparent that the brightest objects in the radio sky, that is, the objects that give off the most intense radio waves, are not identical with the brightest objects in the optical sky. The radio objects were named with letters and with the names of the constellations in which they are located. Thus Taurus A is the brightest radio object in the constellation Taurus; we now know it to be the Crab Nebula. Sagittarius A is the center of our galaxy; Sagittarius B is another radio source nearby, whose emission is caused by clouds of gas near the galactic center.

Many of the radio objects have been discovered to lie outside our galaxy, and we shall discuss them in subsequent chapters. Quasars (Chapter 24), for example, are an important group of extragalactic objects that radiate strongly in the radio region of the spectrum.

Continuum radio radiation can be generated by any of several processes. One of the most important is *synchrotron emission*, the process that produces the radiation from Taurus A (Fig. 21–7). Taurus A is a supernova remnant, and lines of magnetic field extend throughout the visible Crab

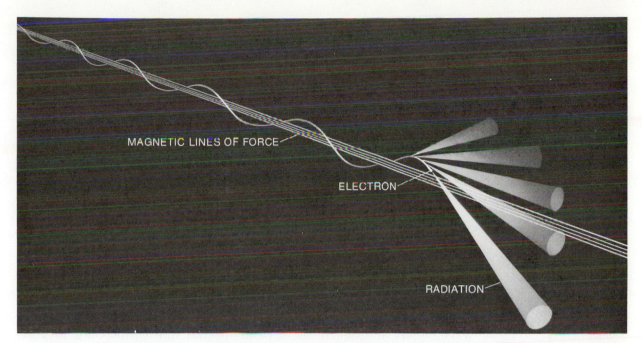

MAGNETIC LINES OF FORCE

ELECTRON

RADIATION

Figure 21-7 Electrons spiralling around magnetic lines of force at velocities near the speed of light (we say "at relativistic velocities") emit radiation in a narrow cone. This radiation, which is continuous and highly polarized, is called synchrotron radiation. Synchrotron radiation has been observed in both optical and radio regions of the spectrum.

When radiation is *polarized,* all the waves oscillate in the same direction (we might say all "up and down" rather than some "up and down" and some "side to side") as they travel forward together. We will not discuss polarization further, other than to say that we can observe whether a set of radio waves is polarized or not. If it is, that is a strong indication that the waves were emitted by the synchrotron process.

Nebula and beyond. Electrons, which are electrically charged, tend to spiral around magnetic lines of force. Electrons of high energy spiral very rapidly, at speeds close to the speed of light. We say that they move at "relativistic speeds," since the theory of relativity must be used for calculations when the electrons are going that fast. Under these conditions, the electrons radiate very efficiently. (This is the same process that generates the light seen in electron synchrotrons in laboratories on earth, hence the name synchrotron radiation.)

The intensity of synchrotron radiation is related not to the temperature of the astronomical body that is emitting the radiation, but rather to the strength of its magnetic field and to the number and energy distribution of the electrons captured in that field. Since the temperature of the object cannot be derived from knowledge of the intensity of the radiation, we call this radiation *non-thermal* radiation. Synchrotron radiation is but one example of such non-thermal processes. The synchrotron process can work so efficiently that a relatively cool astronomical body can give off a tremendous amount of such radiation, at a given wavelength, perhaps so much that it would have to be heated to a few million degrees before it would radiate as much *thermal* radiation at that wavelength (by thermal radiation we mean radiation that is consistent with Planck's law).

21.3   THE RADIO SPECTRAL LINE FROM INTERSTELLAR HYDROGEN

In about 1950, though radio astronomers were very busy with continuum work, there was still a hope that a radio spectral line might be discovered. This discovery would be important for many reasons, but especially because it would allow Doppler shift measurements to be made.

What is a radio spectral line? Just as an optical spectral line corresponds to a wavelength (or frequency) in the optical spectrum that is more (for an

Figure 21–8 Since all electromagnetic radiation travels at the same speed (the speed of light) in a vacuum, fewer waves of longer wavelength *(top)* pass an observer in a given time interval than do waves of a shorter wavelength *(bottom)*. If the wavelength is half as long, twice as many waves pass, i.e., the frequency is twice as high. The wavelength times the frequency is constant, with the constant being the speed of light: $\lambda\nu = c$.

FIXED POINT

emission line) or less (for an absorption line) intense than neighboring wavelengths or frequencies, a radio spectral line corresponds to a frequency (or wavelength) at which the radio noise is slightly more, or slightly less, intense. If a radio station broadcasted just a hum, it would appear as an emission line on our home radios; terrestrial transmissions are normally "modulated" (the voice or music is carried as a modulation on a steady "carrier" signal).

The most likely candidate for a radio spectral line that might be discovered was a line from the lowest energy levels of interstellar hydrogen atoms. This line was predicted to be at a wavelength of 21 cm. Since hydrogen is by far the most abundant element in the universe, it seems reasonable that it should produce a strong spectral line. Furthermore, since most of the interstellar hydrogen has not been heated by stars or by any other strong mechanism, it is most likely that this hydrogen is in its state of lowest possible energy.

This line at 21 cm comes not from a transition to the ground state from one of the higher states, not even from level 2 to level 1 as does Lyman alpha, but rather from a transition between the two sublevels into which the ground state of hydrogen is divided (Fig. 21–9).

An atom of hydrogen, you will recall, simply consists of an electron orbiting around a proton. (Actually, in quantum mechanics, there is an electron cloud with a certain set of probabilities that the electron is at given places at a given time, but for this astronomical purpose it is good enough to think in terms of the classical picture.) Both the electron and the proton have the property of spin; each one has angular momentum (Section 10.4) as if it were spinning on its axis. When we describe a spinning object, we describe the direction of the spin by what is called the *right-hand rule*. This rule can be explained as follows: if you wrap your right hand around the spinning object with your fingers pointing in the direction of the rotation, the direction that your thumb, held out from your hand, is pointing is defined as the direction of spin (Fig. 21–10). Thus if you spin a top on a table so that a dot on its side passes across your field of view from left to right, if you wrap your right hand about it you will see that your thumb is pointing up. We say that the "spin of the top is in the upward direction."

Box 21.1 Conversion from Frequency to Wavelength (Optional)

Although optical astronomers usually use wavelength units, radio astronomers usually use *frequency* units. If a wave travels at a constant speed—in this case at the speed of light—fewer peaks will cross a given point in a given time when the wavelength is longer. Thus the frequency at which peaks in the wave pass a given point is decreased. The converse is also true. The simple equation that links wavelength and frequency is wavelength × frequency = c. Wavelength is usually denoted with the Greek letter λ (lambda) and frequency is usually denoted with the Greek letter ν (nu), so the equation is written $\lambda\nu = c$ (Fig. 21–8). The energy E is simply expressed as $h\nu$, $E = h\nu$, where h is Planck's constant.

Students often find this dual terminology—sometimes in wavelengths and sometimes in frequencies—confusing. So do astronomers. After doing research in a given field, one actually thinks in the appropriate units; radio astronomers think in frequency units. So at meetings at which both optical and radio astronomers are present, there is considerable translating of units back and forth.

The one basic number that must be remembered is the speed of light: 3×10^{10} cm/sec. From that, frequency-wavelength conversions can be derived easily. For example, if a radio wave has a wavelength of 1 cm, its frequency is 3×10^{10} cm/sec ÷ 1 cm = 3×10^{10} hertz (hertz is the name for cycles per second, and is abbreviated Hz). Since M is the symbol for mega (10^{6}), and G is the symbol for giga (10^{9}), the frequency of this line would be 3×10^{10} Hz = 30×10^{9} Hz = 30 GHz (see Appendix 2). The 21-cm line has a frequency of 1420 MHz (which can also be written as 1.42 GHz).

Figure 21–9 21-cm radiation results from an energy difference between two sub-levels, called *hyperfine levels,* in the lowest principal energy state of hydrogen. The energy difference is much smaller than the energy difference that leads to Lyman α. So (because $E = h\nu = hc/\lambda$) the wavelength is much longer.

Figure 21–10 We characterize the *direction of spin,* according to a convention called the *right-hand rule,* as the direction in which the thumb is pointing when the fingers are wrapped around in the direction in which an object is spinning.

With this terminology, the spin of the electron can be either in the same direction as the spin of the proton or in the opposite direction. Intermediate orientations are prohibited according to the rules of quantum mechanics. If the spins are in opposite directions, the energy state of the atom is very slightly lower than the energy state occurring if the spins are in the same direction.

If an atom is sitting alone in space in the upper of these two energy states, with its electron and proton spins aligned in the same direction, it has a certain small probability of having the spinning electron spontaneously flip over to the lower energy state and emit a photon. We thus call this a *spin-flip* transition (Fig. 21–11 top). The photon corresponds to radiation at a wavelength of 21 cm (Fig. 21–12).

We have described how an *emission line* can arise at 21 cm. But what happens when continuous radiation passes through neutral hydrogen gas? In this case, some of the electrons in atoms in the lower state will absorb a 21-cm photon and flip over, putting the atom into the higher state. Then the radiation that emerges from the gas will have a deficiency of such photons and will show the 21-cm line in *absorption* (Fig. 21–11 bottom).

21-cm hydrogen radiation has proven to be a very important tool for studying our galaxy because it passes unimpeded through the dust that

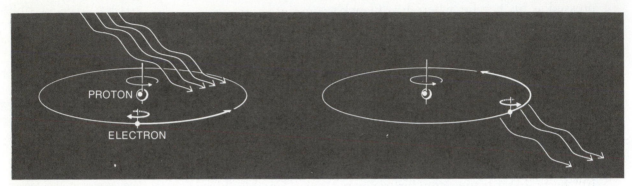

Figure 21–11 When the electron in a hydrogen atom flips over so that it is spinning in the opposite direction from the spin of the proton (*top*), an emission line at a wavelength of 21 cm results. When an electron takes energy from a passing beam of radiation to flip from spinning in the opposite direction from the proton to spinning in the same direction (*bottom*), then a 21-cm line in absorption results.

prevents optical observations very far into the plane of the galaxy. Using 21-cm observations, astronomers can study the distribution of gas in the spiral arms. We can detect this radiation from gas located anywhere in our galaxy, even on the far side, whereas light waves penetrate the dust clouds in the galactic plane only about 10 per cent of the way to the galactic center.

21.3a Mapping Our Galaxy (Optional)

But here again we come to the question that bedevils much of astronomy: how do we measure the distances? Given that we detect the 21-cm radiation from a gas cloud (since it is neutral hydrogen, it is technically called an H I region), how do we know how far away the cloud is from us?

Figure 21–12 The position of the hydrogen line, shown here in absorption, can be given in either frequency or wavelength units. These numbers are commonly rounded to 21 cm or 1420 MHz in conversation. Any Doppler shift, of course, shifts the absorption in wavelength and frequency.

The answer can be found by using a model of rotation for the galaxy. As we have already learned, the outer regions of galaxies rotate differentially, that is, the gas nearer the center rotates faster than the gas farther away from the center. It is as though the gas were revolving around a central gravitational mass in Keplerian orbits, with the period increasing as you go farther from the center. Note that because we can consider ourselves either as being in the galaxy and revolving around its center or as part of the galaxy and rotating around its center, there is a confusion between the words "rotation" and "revolution" here that cannot be avoided.

Figure 21–14 shows a simplified version of differential rotation. Because of the differential rotation, the distance between us and point A is decreasing. Therefore, from our vantage point at the sun, point A has a net velocity toward us. Thus its 21-cm line is Doppler shifted toward shorter wavelengths. If we were talking about light this shift would be in the blue direction; even though we are discussing radio waves we say "blueshifted" anyway. If we look from our vantage point at gas cloud C, we see a redshifted 21-cm line, because its higher speed of rotation is carrying C away from us. But if we look straight toward the center, clouds B_1 and B_2 are both passing across our line of sight in a path parallel to that of our own orbit. They have no net velocity toward or away from us. Thus this method of distance determination does not work when we look in the direction of the center, nor indeed in the direction of the anti-center.

Once we measure the redshift, we can deduce the net velocity. In the inner part of the galaxy, we can deduce the law of differential rotation with distance from the center since the highest velocity cloud along each line of sight must be the cloud along the line of sight that is closest to the center. We can get the distance of that cloud from the center by simple geometry. Once we work out the law of differential rotation, then we can tell how far away each cloud is from the center of the galaxy by figuring out where along our line of sight in a given direction the cloud would have the proper velocity to match the observations. Clouds farther from the galac-

Figure 21–13 Harold Ewen with the horn radio telescope he and Edward Purcell used to discover the 21-cm line in 1951.

Figure 21–14 Because of the differential rotation, the cloud of gas at point A appears to be approaching the sun, and the cloud of gas at point C appears to be receding. Objects at points B_1 or B_2 have no net velocity with respect to the sun, and therefore show no Doppler shifts in their spectra.

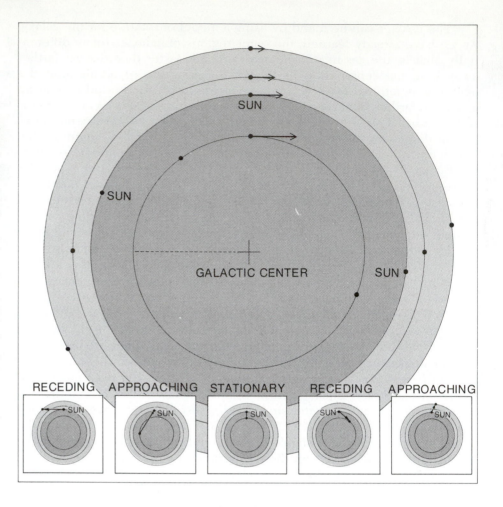

Figure 21–15 Because of differential rotation in our galaxy, stars or gas clouds that were lined up at one time (upper dots on top part of diagram) are spread out by the time they have gone about 1/4 of the way *(right)* or about 3/4 of the way *(left)* around the galaxy. The inserts show the net velocity that stars or gas clouds at different distances from the galactic center would have with respect to the sun. Objects interior to the sun's orbit around the galactic center (darker shaded area) are orbiting faster than does the sun, and objects farther out than the sun's orbit (lighter shaded area) are orbiting more slowly than does the sun. Thus measurements of the Doppler shift of 21-cm radiation can be used to construct a map of the hydrogen clouds, which in turn map out spiral structure.

tic center take longer to revolve than do clouds closer to the center, in a manner similar to Kepler's third law (Fig. 21–15). By observing in different directions, we can build up a picture of the spiral arms.

Of course, the whole sky cannot be seen from either the northern hemisphere or the southern hemisphere of the earth. Therefore maps must be made from observatories in both hemispheres and then correlated.

Figure 21–16 Two artists' impressions of the structure of our galaxy based on 21-cm data. *A* was compiled almost 20 years ago by Gart Westerhout of the University of Maryland, and *B* was recently compiled by Gerrit Verschuur of the University of Colorado. Differences between the two maps are probably not real and give an idea of the accuracy of the method. Because hydrogen clouds located in the directions either toward or away from the galactic center have no radial velocity with respect to us, we cannot find their distances.

A

B

The 21-cm maps show many narrow arms (Fig. 21–16) but no clear pattern of a few broad spiral arms like those we see in other galaxies. Is our galaxy really a spiral at all?

Even with all these questions and uncertainties waiting to be resolved, it is clear that the 21-cm radiation is at the base of our mapping efforts for our own galaxy, and has allowed us to make major advances in our understanding of the Milky Way Galaxy.

21.4 MOLECULAR HYDROGEN

While individual atoms of hydrogen, whether neutral or ionized, have been extensively observed in the interstellar gas, hydrogen molecules (H_2) have been observed only recently. Even though astronomers have long thought that molecular hydrogen could be a major constituent of the interstellar medium, it was simply not possible to observe hydrogen in molecular form. This is because at the low temperature of interstellar space, only the lowest energy levels of molecular hydrogen are excited, and no lines linking these levels fall in the optical or radio regions of the spectrum. We had to wait to observe from space in order to observe lines from H_2 because these lines occur in the far ultraviolet. Our atmosphere prevents these lines from reaching us on earth.

In 1972, a 90-cm (36-inch) telescope was carried into orbit aboard NASA's third Orbiting Astronomical Observatory, named "Copernicus" in honor of that astronomer's 500th birthday (which occurred a year later). The telescope, operated by a group of scientists from the Princeton University Observatory, was largely devoted to observing interstellar material. The observers could point the telescope at a star (directing it by remote control from the ground, of course) and look for absorption lines caused by gas in interstellar space as the light from the star passed through the gas en route to us.

Since the hydrogen molecule is easily torn apart by ultraviolet radiation, the observers did not find much molecular hydrogen in most directions. But whenever they looked in certain directions, in which it was known from other studies that there was a lot of interstellar matter, presumably including both gas and dust, they found a very high fraction of hydrogen in molecular form: more than 50 per cent. Presumably, in these regions the dust shields the hydrogen from being torn apart by ultraviolet radiation. There are also theoretical grounds for believing that the molecu-

Figure 21–17 From above the atmosphere of the earth, the Copernicus satellite looks toward a star that provides a more-or-less continuous spectrum. (A rapidly-rotating B star is usually chosen, because B stars have few lines and those lines are washed out by Doppler shifts resulting from the velocity of rotation.) When a cloud of gas is in the path, then absorption or emission from molecules or atoms in the cloud results.

LIMIT OF ATMOSPHERE

SATELLITE

EARTH

lar hydrogen is formed on dust grains (other molecules are probably formed on dust grains too, though this is less certain), so it seems reasonable that the high fraction of H_2 is found in the regions where there are more dust grains

21.5 RADIO SPECTRAL LINES FROM MOLECULES

A few interstellar optical lines from atoms and simple molecules have been studied; they can be observed with ordinary optical telescopes and spectrographs and have long been known.

Even though optical studies of the interstellar medium have been going on for decades, recently much more information has come from ultraviolet observations made from space and radio studies made from the earth. For several years after the 1951 discovery of 21-cm radiation, spectral-line radio astronomy continued with just the one spectral line. Astronomers tried to find others. One prime candidate was OH, the hydroxyl radical, which should be relatively abundant, as molecules go, because it is a combination of the most abundant element, hydrogen, with one of the most abundant of the remaining elements, oxygen. OH has four lines close together at about 18 cm in wavelength, and the relative intensities expected for the four lines have been calculated.

It wasn't until 1963 that other radio spectral lines were discovered, and four new lines were indeed found at 18 cm. But the intensity ratios were all wrong to be OH, according to the predicted values, and for a time we spoke of the "mysterium" lines. The idea of discovering a new element was not unprecedented—after all, unknown lines at the solar eclipse in 1868 were assigned to an unknown element, "helium," because they occurred only (as far as was known at that time) on the sun. Of course, the periodic table of the elements has been filled in during the last 100 years, so a new element wasn't really expected. Mysterium turned out to be OH after all, but with strange processes affecting the excitation of the energy levels of OH and thus amplifying certain of the lines at the expense of others, which are weakened. This in itself is very interesting. The process is that of masering. *Masers* (maser is an acronym for microwave amplification by stimulated emission of radiation) and lasers, their analogue using light instead of radio waves, are of great practical use on the earth. For example, masers are used as sensitive amplifiers. Masers were "invented" on earth not long before they were found in space. If astronomical research had been a little more advanced, a multi-billion dollar industry that now exists, and which may provide the key to solving our energy problems through laser-induced fusion, might have been set up earlier. This is another lesson about the value of basic research.

So the four "mysterium" lines turned out to be OH undergoing maser action. OH lines have now been detected not only in absorption against background emission, but also in emission. They are quite widespread and are used to tell us the physical conditions in the clouds from which the OH radiation comes.

The abundance of OH can be calculated from the measurements of OH radiation. It was very much less than that of isolated hydrogen or oxygen atoms, which seems reasonable since in the virtual vacuum of interstellar space there is little chance for oxygen and hydrogen atoms to interact. Still, radio astronomers were disappointed with the low OH densities: only one OH molecule for every billion H atoms. It seemed quite unlikely that the quantities of any molecules composed of three or more atoms would be great enough to be detected. The chance of three atoms getting together in the same place, it seemed, would be very small.

In 1968, however, scientists used Berkeley's Hat Creek Observatory to observe at the radio frequencies in the centimeter range that were predicted to be the frequencies of water (H_2O) and ammonia,(NH_3). The lines of these molecules proved surprisingly strong, and were easily detected. In fact, we had had the capability for some years to make equipment sensitive enough to detect them, but nobody had tried to make the observations.

At about this time, a group of radio astronomers used a telescope at the National Radio Astronomy Observatory in Green Bank, West Virginia, to search for interstellar formaldehyde (H_2CO) at a wavelength of 6 cm. They succeeded in detecting the formaldehyde molecule, the first molecule that contained two "heavy" atoms, that is, two atoms that were of an element other than hydrogen.

By this time it was apparent that the earlier notion that it would be difficult to form molecules in space was wrong. There has been much research on this topic, but the mechanism by which molecules are formed has not been satisfactorily determined. For some molecules, including molecular hydrogen, it seems that the presence of dust grains is necessary. In this scenario, one atom would hit a dust grain and stick to it (Fig. 21–18). It may be thousands of years before a second atom hits the same dust grain, and even longer before still more atoms hit. But these atoms may stick to the dust grain rather than bouncing off, which gives them time to join together. Complex reactions may take place on the surface of the dust grain. Then, somehow, the molecule must get off the dust grain back into space as a gas. Perhaps incident ultraviolet radiation or the energy released in the formation of the molecule allows the molecule to escape from the grain surface.

The dust serves a second purpose: it prevents high-energy (short-wavelength) radiation from breaking up the molecules. Thus we can best study conditions in dust clouds, which we think are the birthplaces of stars, by analyzing the emission from molecules.

After the discoveries of water, ammonia, and formaldehyde in interstellar space, further discoveries came one after another. Many different radio astronomers looked up the wavelengths of likely spectral lines from molecules containing abundant elements and were able to observe the radiation. Sometimes the lines were in absorption and sometimes in emission. The list of molecules discovered gradually expanded from three-atom molecules like ammonia and water, and four-atom molecules like formaldehyde, to even more complex molecules, like formic acid (HCOOH) with five atoms, methyl cyanide (CH_3CN) with six, and cyanodiacetylene (HC_5N) with seven. The discovery of methyl alcohol, CH_3OH, in 1970, was greeted jocularly, and cases of liquor were bet on the discovery of ethyl alcohol, CH_3CH_2OH, the drinking kind. In 1974, the ethyl alcohol molecule was discovered too. (There are 10^{28} fifths, calculated at 200 proof, in the molecular cloud in which it was observed.) Appendix 11 lists the interstellar molecules that are known.

There is much more to spectral-line radio astronomy than simply discovering new spectral lines. For one thing, studying the lines provides information about physical conditions in the gas clouds that emit the lines. These clouds are usually so dense that hydrogen atoms have combined into hydrogen molecules and very little 21-cm radiation is emitted. Some of the lines, for instance those of carbon monoxide (CO), have been detected in many directions in space, and maps have been made of the distribution of these molecules. Sensitive radio receivers can do this, even though the densities of these molecules in the densest clouds may be only one in every cubic meter.

Figure 21–18 One major model of the formation of interstellar molecules invokes the presence of a dust grain at intermediate stages. The mechanism by which the molecule is freed from the dust grain is not known.

DUST GRAIN

O

H

5000 YEARS LATER

OH

Figure 21–19 The 11-m (36-ft) telescope of the National Radio Astronomy Observatory, seen here through a fish-eye lens, is on Kitt Peak in Arizona (see Fig. 3–16). It is used for observations at millimeter wavelengths, and has been used to discover many interstellar molecules.

Not the least important conclusion that is reached by these spectral-line studies is that complex molecules can be readily formed by natural processes. Some of the molecules that have been discovered in space are "organic" in nature, in that they contain carbon, and we are approaching the level of complexity at which we can find simple amino acids, the building blocks of life. And we know (Section 19.1), even though we have not found amino acids themselves, that amino acids can be readily formed in a laboratory. All we need do is pass electricity through mixtures containing some of the molecules that have been discovered in space, or illuminate such a mixture with ultraviolet light. So it seems that it is much easier than we had originally thought to form the molecules from which life is made. These considerations add support to the beliefs of the many astronomers who feel that life has probably arisen spontaneously at many different locations in the universe.

SUMMARY AND OUTLINE OF CHAPTER 21

Aims: To study the interstellar medium, and to understand the techniques (especially radio astronomy and ultraviolet astronomy) used to study it

Radio astronomy (Section 21.1)
    Discovered by Jansky in 1931
    Radio waves reflected by a metal dish to a focus where an antenna collects the signal; the signal is then amplified and analyzed with electronic equipment
    Larger dishes have larger collecting areas and higher resolutions
Continuum radio astronomy (Section 21.2)
    Spectra are measured over a broad frequency range
    Radiation generated by synchrotron emission, electrons spiraling rapidly in a magnetic field, found in many non-thermal sources
Radio spectral line from interstellar hydrogen (Section 21.3)

21-cm spectral line from neutral hydrogen was discovered in 1951

Line occurs at 21 cm through a spin-flip transition

Both emission and absorption have been detected

21-cm radiation used to map the galaxy (Section 21.3a)

Distances measured using differential rotation and the Doppler effect

Molecular hydrogen (Section 21.4)

Interstellar ultraviolet absorption lines from H_2 with abundances up to 50 per cent observed from above the earth's atmosphere

Radio spectral lines from molecules (Section 21.5)

"Mysterium" lines discovered at 18 cm—later identified as OH affected by masering process

Masers first developed artificially on earth, but later discovered to exist in space

Three dozen molecules have been discovered in space, including some as massive as amino acids

QUESTIONS

1. Estimate the wavelengths for which a window screen would be a good reflector.

2. (a) List two relative advantages of radio astronomy over optical astronomy for studying our galaxy. (b) List two relative advantages of optical astronomy over radio astronomy for that purpose.

3. The angular resolution of the 100-meter telescope at a wavelength of 21 cm is approximately one arc minute. How large a telescope would you need to get the same angular resolution at a wavelength of 2 mm?

4. Why is emission from cool regions of space most likely to be detectable at radio wavelengths?

5. What is the frequency of 1 mm waves?

6. What is the wavelength of 108 MHz waves, which are those you tune in at the right end of your FM radio dial.

7. What determines whether the 21-cm line will be observed in emission or absorption?

8. If our galaxy rotated like a rigid body, would we be able to use the 21-cm line to determine distances to H I regions? Explain.

9. Why are dust grains important for the formation of interstellar molecules?

10. (a) How many interstellar molecules have 2 atoms, 3 atoms, etc?

    (b) How many interstellar molecules contain one heavy (i.e., non-H) atom, 2 heavy atoms, etc?

    (c) What fraction of the known interstellar molecules are organic?

    (d) How many interstellar molecules are chemicals of which you have heard previously?

11. Why can't we observe 21-cm radiation from interstellar molecular clouds?

12. What type of information can interstellar molecules provide?

TOPICS FOR DISCUSSION

1. What does the discovery of fairly complex molecules in space imply to you about the existence of extraterrestrial life?

2. Consider the observing schedules of astronomers who study different parts of the spectrum —for example, optical stellar astronomers work in the nighttime and optical solar astronomers work in the daytime. What are the observing schedules of x-ray astronomers, astronomers using balloons to study the infrared, solar astronomers using space telescopes, and so on? How many different kinds of specialties can you think of?

Infrared, X-Ray, and Gamma-Ray Studies of Our Galaxy

Optical and radio astronomical methods have become standard for studying our galaxy. In recent years, studies in other parts of the electromagnetic spectrum have become increasingly important.

22.1 INFRARED OBSERVATIONS

For many years the sky has been intensively studied at optical wavelengths up to about 8500 Å (0.85 micron) and to a lesser extent up to 1.1 microns, which is in the near infrared and beyond which film is not sensitive. The sky has also been studied for decades at radio wavelengths down to one or two centimeters (10,000 or 20,000 microns). But until the last few years, the sky has been studied very little at the wavelengths in between, which are the millimeter and the infrared wavelengths.

One reason that the infrared (about 1 micron to about 1 mm) was not studied is the fact that the earth's atmosphere doesn't cooperate. Most infrared radiation that hits the top of the earth's atmosphere doesn't get through; rather, it is absorbed by water vapor, carbon dioxide, or ozone. The earth's atmosphere is transparent in only a few limited regions of this broad range of the infrared spectrum.

One can observe better in the infrared from locations where there is

---

A high resolution map of the center of the Milky Way Galaxy at a wavelength of 2.2 microns. The resolution is about 1.2 arc min, and the width of the entire image is about 1.1°.

Figure 22–1 A balloon being launched to make infrared studies.

little water vapor overhead. A prime location, for example, is Mauna Kea in Hawaii, which is an especially dry site at its 4145-meter (13,600-foot) peak (that is, it is characterized by very low absorption by water vapor) even though it is surrounded by ocean. Alternatively, one can try to get above as much of the earth's atmosphere as possible by sending telescopes aloft in balloons (Fig. 22–1), or by flying as high as possible in airplanes.

Another difficulty in studying the infrared involves the detectors—it is hard to develop sufficiently sensitive instruments with which to study the infrared radiation. Remember (from Section 2.7) that $E = hc/\lambda$; since the infrared wavelengths are longer than the visible wavelengths, the energy that each infrared photon has is relatively low compared to that of a visible photon. Thus many types of detectors are simply not sensitive to infrared radiation. The longer the infrared wavelength that you want to study, the more difficult it is to find a suitable detector.

Yet another difficulty comes from the fact that ordinary heat radiation is in the infrared. The earth's atmosphere radiates conspicuously in the infrared, so the radiation coming into a telescope includes both infrared radiation from the atmosphere and from the source being observed. The telescope itself contributes its own infrared radiation. To limit the telescopic contribution of radiation, the equipment is usually bathed in liquid nitrogen, or even in liquid helium, which is colder and more expensive.

The whole part of the sky that can be observed from the top of Mount Wilson has been surveyed, and some 20,000 infrared sources were detected! Remember that there are only about 6000 stars that can be seen with the naked eye.

The brightest infrared source at a wavelength of 20 microns is associated with Eta Carinae (Color Plate 6). From observations in the infrared, it has been deduced that Eta Carinae is surrounded by a cloud of dust that is about 0.1 ly across and is at a temperature of 250 K.

Most of the infrared sources that were discovered do not coincide in space with known optical sources. The map of the radio sky doesn't look like the map of the optical sky, and the map of the infrared sky doesn't resemble either of the others. Most of these infrared-emitting objects, however, unlike the radio objects, are in our galaxy (Fig. 22–2). Following the survey program, some of the more interesting infrared objects have been individually studied much more carefully.

Perhaps the most interesting infrared source that we know is a source

Figure 22–2 Hubble's Variable Nebula, which is associated with the variable star and infrared source R Monocerotis. This was the first official photograph taken with the 5-m Hale telescope.

A

B

Figure 22–3 *(A)* A radio map of the galactic center, made at a wavelength of 6 cm with the 100-m Bonn telescope. The intense source at right is called Sgr A. *(B)* A photograph in visible light taken in the direction of the center of our galaxy, showing exactly the same region observed in the infrared that is displayed as the opening photograph to this chapter. In visible light, interstellar dust completely hides the galactic center, and we are seeing only relatively nearby stars and gas.

of very small diameter located in the direction of the center of our galaxy. (It is the photograph opening this chapter.) Since the amount of scattering by dust varies with wavelength, we can see farther through interstellar space in the infrared than we can in the visible. In this case, it seems that in the infrared we can see 8.5 kiloparsecs to the center of our galaxy. The brightest infrared source is located at the position of the radio source Sagittarius A (Fig. 22–3).

This source subtends an angle of 1 arc min, and so is about 3 parsecs (about 10 light years) across. This makes it a very small source for the prodigious amount of energy it emits: as much energy as if there were 80 million suns radiating. We don't know just what it is in the center of our galaxy that causes this radiation, but this small source does seem to give off as much as 0.1 per cent of the total radiation from our galaxy. It is interesting to know that the center of our galaxy is qualitatively different from the outer parts: the center is a different kind of place from the arms, and its greater density of stars isn't the only difference. Other galaxies have even more prominent infrared sources in their nuclei.

There are several models to account for the infrared radiation from the galactic nucleus. One model is that there are bright sources of ultraviolet or visible radiation there, such as a dense group of stars. These sources would be surrounded by shells of dust that heat up to about 50 K. One needs the equivalent of about 10 solar masses of dust, not an inconceivable amount, to make this model work. Similar processes, on a smaller scale, are known for shells surrounding hot stars.

Recent radio observations with a coast-to-coast interferometer have re-

PLEIADES ANDROMEDA GALAXY VEGA

PHOTOGRAPHIC MAGNITUDES

vealed that a powerful radio source at the nucleus of our galaxy is less than 0.001 arc sec across—at that distance, even smaller than our solar system! This indicates to some astronomers that it is a massive black hole.

22.2 HIGH-ENERGY ASTRONOMY

Thus far we have discussed how to study our galaxy not only at wavelengths in the visible region of the spectrum but also at longer wavelengths. We also study our galaxy, as of the last few years, at wavelengths shorter than those of visible radiation.

Radiation immediately below the violet part of the visible spectrum, from about 4000 Å down to 100 Å, is called *ultraviolet*. The earth's atmosphere prevents radiation shorter than 3000 Å from penetrating down to

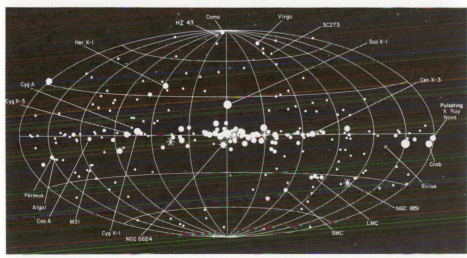

Figure 22–4 *(A)* A drawing of the Milky Way, made under the supervision of Knut Lundmark at the Lund Observatory in Sweden. 7000 stars plus the Milky Way are shown in this panorama, which is in coordinates such that the Milky Way falls along the equator. *(B)* An x-ray map of the Milky Way showing the objects observed by the Uhuru spacecraft as of mid-1976. X-ray "bursters" (see Section 4.5c) are shown with asterisks. X-ray sources observed only by other satellites are shown as open circles; these sources are detectable only at relatively long x-ray wavelengths.

Chemists studying atoms join astronomers in using angstroms as a measuring unit, although purists now claim that we should only use SI units (Système International, the set of units officially adopted by an international conference on the subject) which allow only powers of 1000 or 1/1000 meters. In this use of the metric system 1 angstrom = 0.1 nanometer. Astronomers won't give up the use of angstroms without a struggle.

the surface, so observations of shorter wavelengths must be made from rockets, balloons, or satellites above the atmosphere. We continue to call the region of the spectrum down to about 100 Å by the name "ultraviolet," but unlike the case for the "near ultraviolet" (3000 Å to 4000 Å), we must go above the atmosphere to study the shorter ultraviolet radiation. Sometimes this shorter radiation is called *XUV* or *EUV*, both of which are short for "**E**xtreme **U**ltra**v**iolet." XUV can also refer to the boundary region between x-rays and ultraviolet radiation.

Below about 100 Å, and ranging down to about 1 Å, we speak of the radiation as *x-rays*. This name survives from the discovery of x-rays in 1895 by Wilhelm Roentgen. The 1 Å length we use as a lower limit is about the diameter of a small atom.

Below about 1 angstrom, we speak of *γ-rays* (gamma rays). Again, this is just electromagnetic radiation of extremely short wavelength, and is subject to the same laws as longer wavelength radiation. The dividing lines between uv and x-rays, or between x-rays and gamma rays, are not sharp ones.

We have dealt with x-ray astronomy in several contexts before, including discussions of the sun and of black holes. Most of the x-ray sources that we detect are located throughout our galaxy (Fig. 22–4).

Some two dozen bursters are now known; only a handful are in globular clusters. None are in binary systems, though many x-ray sources that do not give off bursts of x-rays are in binaries. The x-ray binary systems seem to involve neutron stars; some scientists think that the bursters involve black holes, though this is very hard to prove.

Hundreds of x-ray sources have now been mapped, and many of them have been studied over periods of time. Some of the sources vary radically with time. The center of our galaxy is a strong and variable x-ray source; we have seen that it is also an infrared and a radio source. This again indicates that violent processes must be going on there in order to generate such powerful radiation. Although many of the x-ray sources mapped are in our galaxy, others are extragalactic (outside our galaxy). The nuclei of some other galaxies are much more active than that of the Milky Way Galaxy.

γ-ray astronomy is even harder to carry out than x-ray astronomy, and is only now getting going. Although an occasional strange object, like the Crab Pulsar, can be detected at every wavelength of the spectrum, including the x-ray and γ-ray regions, astronomers were hard put to find optical counterparts of other γ-ray sources.

Since 1975, the Cos B satellite has been orbiting; it was launched for the European Space Agency by NASA. It has detected about a dozen γ-ray sources besides the Crab and Vela pulsars. All of these sources are believed to be in our galaxy.

Just as radio astronomers prefer to work in frequency instead of wavelength units, gamma-ray astronomers also use E = hν but work in energy units. (See Section 2.7 for a definition of electron volts.) So the technical papers talk of gamma rays in millions of electron volts (abbreviated MeV; M means millions to contrast with m from milli-, which means thousandths), but we can ignore this technical point for the purposes of our general discussion. It may be useful to note that a photon with a wavelength of 1 Å has an energy of about 12,000 eV = 12 keV = 0.012 MeV.

One interesting γ-ray observation is the discovery of a general background of γ-rays concentrated along the plane of our galaxy. This diffuse background is not concentrated in particular objects, not even in the direction of the galactic center. Presumably, these γ-rays are caused by the interaction of *cosmic rays,* nuclei of atoms that travel around space at tremendous velocities, with interstellar matter.

Studies of electromagnetic radiation like x-rays and γ-rays and of rapidly-moving cosmic ray particles are part of the new field of *high-energy astrophysics.* (E = hν, and ν is very large for radiation in this part of the spectrum, so we speak of high-energy radiation.) High-energy astrophysics is an especially active field for theorists, as well as observers. The theorists study, among other things, how radiation of high energies can be generated in interactions between matter and other matter or between matter and radiation.

NASA has launched an important satellite, the High Energy Astronomy Observatory, HEAO-1, which has higher sensitivity in the x-ray spectrum than any previous spacecraft. HEAO-B, with imaging capabilities, is scheduled for launch in 1978; and HEAO-C, to study gamma rays and cosmic rays, is scheduled for launch in 1979. Discoveries are expected to concern not only the center of our galaxy but also neutron stars, black holes, and clusters of distant galaxies.

Figure 22–5 The Uhuru satellite.

SUMMARY AND OUTLINE OF CHAPTER 22

Aims: To study infrared, x-ray, and γ-ray observations of our galaxy, which are especially useful to study the nucleus of our galaxy, and to see that cosmic rays permeate the galaxy

Infrared observations (Section 22.1)
    Technological difficulties
    Radiation from the galactic nucleus indicates a lot of energy is generated in a small volume there
High-energy astronomy (Section 22.2)
    X-ray map and γ-ray map
    Radiation from the galactic nucleus indicates violent activity there
    γ-ray background radiation the result of cosmic ray interaction with matter
    HEAO satellites.

QUESTIONS

1. Why is it possible to do some infrared observations from mountain observatories while all x-ray observations must be made from space?

2. Illustrate, by sketching the appropriate Planck radiation curves, why cooling an infrared telescope from room temperature to 4 K greatly reduces the background noise.

3. Does mapping a region of the sky in the infrared more closely resemble optical or radio techniques? Explain.

4. What is the property of x-rays that makes them suitable for looking into the body?

5. Why will x-rays expose a photographic plate while infrared radiation will not?

6. What types of objects give off strong infrared radiation?

7. What types of objects give off x-rays?

8. Discuss how observations from space have added to our knowledge of our galaxy.

Part V

Galaxies and Beyond

M104, the Sombrero Galaxy in Virgo.

(*left*) The cluster of galaxies in the constellation Hercules.

Galaxies and Beyond . . .

Though the individual stars that we see are all part of the Milky Way Galaxy, discussed in the preceding three chapters, we cannot be so categorical about the locations of the conglomerations of gas and stars that can be seen through telescopes. Once they were all called "nebulae," but we now restrict the meaning of this word to gas and dust in our own galaxy. Some of the objects that were originally classed as nebulae turned out to be huge collections of gas, dust, and stars located far from our Milky Way Galaxy and of a scale comparable to that of our galaxy. These objects are galaxies in their own right, and are both fundamental units of the universe and the stepping stones that we use to extend our knowledge to tremendous distances.

In the 1770's, a French astronomer named Charles Messier was interested in discovering comets. To do so, he had to be able to recognize whenever a new fuzzy object appeared in the sky. He thus compiled a list of about 100 diffuse objects that could always be seen. To this day, these objects are commonly known by their *Messier numbers* (Appendix 9). Messier's list contains the majority of the most beautiful objects in the sky, including nebulae, star clusters, and galaxies. The list, compiled to search for comets, turns out to have this much more general importance.

Soon after, William Herschel, in England, compiled a list of 1000 nebulae and clusters, which he expanded in subsequent years to include 2500 objects. Herschel's son John continued the work, incorporating observations made in the southern hemisphere. In 1864, he published the *General Catalogue of Nebulae*. In 1888, J. L. E. Dreyer published a still more extensive catalogue, *A New General Catalogue of Nebulae and Clusters of Stars*, the *NGC*, and later published two supplementary *Index Catalogues*, *IC's*. The 100-odd non-stellar objects that have Messier numbers are known by them, or else by their numbers in Dreyer's

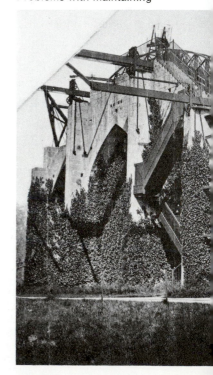

The large reflector, with a mirror 6 feet (1.8 meters) across, built by the Earl of Rosse in Ireland in 1845. Problems with maintaining

catalogue. Thus the Great Nebula in Andromeda = M31 = NGC 224. The Crab Nebula = M1 = NGC 1952.

When larger telescopes were turned to the Messier objects in later years, especially by Lord Rosse in England in about 1850, some of the objects showed traces of spiral structure, like pinwheels. They were called "spiral nebulae." But where were they located? Were they close by or relatively far away?

an accurate shape for the mirror, which was made of metal, led to the telescope's abandonment.

When such telescopes as the 0.9-meter Crossley reflector at Lick in 1898, and later the 1.5-meter and 2.5-meter reflectors on Mount Wilson, began to photograph the "spiral nebulae," they revealed many more of them. The shapes and motions of these "nebulae" were carefully studied. Some scientists thought that they were merely in our own galaxy, while others thought that they were very far away, "island universes" in their own right, so far away that the individual stars appeared blurred together. (The name "island universes" originated with the philosopher Immanuel Kant in 1755.)

The debate raged until 1924, when observations made at the Mount Wilson Observatory by Edwin Hubble proved that there were indeed other galaxies in the universe besides our own. In fact, we think of galaxies and clusters of galaxies as fundamental units in the universe. The galaxies are among the most distant objects we can study; quasars are objects on a galactic scale that are, for the most part, even farther away.

Galaxies and quasars can be studied in most parts of the spectrum. Radio astronomy, in particular, has long proved a fruitful method of study. The study across the spectrum of galaxies and quasars provides tests of physical laws at the extremes of their applications and links us to cosmological consideration of the universe on the largest scale.

Galaxies

The question of the distance to the "spiral nebulae" was settled only in 1924 by Edwin Hubble (who was shown in Fig. 3–25). He had used the Mount Wilson telescopes to observe Cepheid variables in three of the "spiral nebulae." His definitive conclusion (along the lines we described in Section 4.4a) was that the "spiral nebulae" were outside our own galaxy and of dimensions not overwhelmingly different from it, and even this distance scale has since been increased by the discovery of two types of Cepheids. Since Hubble's work there has been no doubt that the spiral forms we see in the sky are galaxies like our own. For the rest of the book we shall strictly use the term *spiral galaxies;* the incorrect, historical term, "spiral nebula," often hangs on in certain contexts, chiefly when we discuss the "Great Nebula in Andromeda," which is actually a spiral galaxy.

23.1 TYPES OF GALAXIES

Hubble went on to use the Mount Wilson telescopes to study the different types of galaxies. Actually, spiral galaxies are in the minority; there are many galaxies that have elliptical shapes and others that are irregular or abnormal in appearance. In 1925, Hubble set up a system of classification of galaxies that we shall discuss below; we normally describe a galaxy by its *Hubble type.*

23.1a Elliptical Galaxies

Most of the galaxies of which we know are elliptical in shape (Fig. 23–1). The largest of these *elliptical galaxies* may contain 10^{13} solar masses and may be 10^5 parsecs across (approximately the diameter of our own galaxy); these *giant ellipticals* are rare. Much more common are *dwarf*

Figure 23–1 A galaxy of Hubble type E0, NGC 4486, in the constellation Virgo. Globular clusters can be seen in the outer regions. This is actually a peculiar elliptical galaxy, and is a strong source of radio radiation.

An edge-on view of a galaxy of type Sb, NGC 4565, in the constellation Coma Berenices.

Figure 23–2 M31, the Great Galaxy in Andromeda, a type Sb spiral, with its accompanying elliptical galaxies NGC 205 (*top left*) and M32 (*bottom right*).

ellipticals, which may contain "only" a few million solar masses and be only 2000 parsecs across.

Elliptical galaxies range from nearly circular in shape, which Hubble called *type E0,* to very elongated, which Hubble called *type E7,* with galaxies of various amounts of apparent oblateness (a measure of the difference between the longest and shortest diameters) in between. The spiral Andromeda Galaxy, M31, shown in Figure 23–2 and in Color Plate 46, is accompanied by two elliptical companions. It is obvious on the photographs that the companions are much smaller than M31 itself.

23.1b Spiral Galaxies

Although spiral galaxies, with arms unwinding smoothly from the central regions, are a minority of all the galaxies in the universe, they form a majority in certain particular groups of galaxies. Also, since they are brighter than the more abundant small ellipticals, we tend to see the spirals as dominant in a given volume of space, while really the fainter ellipticals may make up the majority.

Sometimes the arms are tightly wound around the nucleus; Hubble called this type *Sa,* the S standing for "spiral" (Fig. 23–3). Categories of spirals with arms less and less tightly wound (that is, looser and looser) are called *types Sb* (Color Plates 49 and 50) and *Sc* (Color Plate 47). The nuclear bulge as seen from edge-on is less and less prominent as we go from Sa to Sc galaxies, as can be seen from the difference between the figures opening this chapter and this part of the book. It has been found spectroscopically that galaxies rotate in the sense that the arms trail.

Spiral galaxies can be 10,000 to 30,000 parsecs across. They contain

Our galaxy is type Sb.

NGC 1201   Type S0   NGC 2811   Type Sa   NGC 2841   Type Sb   NGC 628 M74   Type Sc

Figure 23-3 Normal spiral galaxies.

10^9 to over 10^{11} solar masses; since most stars are of less than 1 solar mass this means that spirals contain over 10^9 to over 10^{11} stars—we think our own galaxy has 10^{11} (100 billion).

In about one-third of the spirals, the arms unwind not from the nucleus but rather from a straight *bar* of stars, gas, and dust that extends to both sides of the nucleus (Fig. 23–4). These are again classified in the Hubble scheme from *a* to *c* in order of increasing openness of the arms, but with a *B* for "barred" inserted: *SBa, SBb,* and *SBc.* There is actually a complete range in the size of the bar from not visible to dominant in the appearance of a galaxy, so non-barred ("normal") spirals and barred spirals may not really be distinct types from each other.

23.1c   Irregular Galaxies

A few per cent of the galaxies show no regularity, neither spiral nor elliptical. The Magellanic Clouds, for example, are irregular galaxies; they were shown in Figure 4–13 and enlargements appear in Figure 23–5. Sometimes traces of regularity—perhaps a bar—can be seen. Irregular galaxies are classified as *Irr.*

Irregular galaxies that can be resolved into nebulae, stars, and clusters are called *type Irr I.* Other galaxies appear amorphous and cannot be resolved into nebulae, stars, and clusters. They are called *type Irr II.*

Figure 23-4   Barred spiral galaxies.

NGC 2859   Type SB0   NGC 175   Type SBab   NGC 1300   Type SBb(s)   NGC 1073   Type SBc

Figure 23–5 The Large Magellanic Cloud (*left*) and the Small Magellanic Cloud (*right*), photographed with the new British 1.2-m Schmidt telescope at Siding Spring, Australia, as part of the current project to extend the National Geographic Society–Palomar Observatory Sky Survey to the southern hemisphere. The Royal Observatory, Edinburgh, Scotland, operates the telescope as a national facility, and the exposed plates from Siding Spring are sent to Edinburgh for analysis.

23.1d Peculiar Galaxies

In some cases, as in M82 (Fig. 23–6), it appears at first as though an explosion has taken place in what might have been a regular galaxy. We now think that we are seeing light from the galaxy's nucleus scattered toward us by dust in the filaments. For other peculiar galaxies, such as those shown in Figure 23–7, we have little idea of what might have gone on.

Peculiar galaxies are classified as the corresponding Hubble type followed by (*pec*): for example, *Sa (pec)*.

Figure 23–6 M82, a most unusual galaxy that is a powerful source of radio radiation. It was once thought to be exploding, but now gentler processes are thought to cause its form and non-thermal radiation.

23.1e The Hubble Classification

Hubble drew out his scheme of classification in a *tuning-fork diagram* (Fig. 23–8). The transition from ellipticals to spirals is represented by *type S0*, which resembles spirals in having the shape of a disk but does not have spiral arms.

It has since been shown, from optical observation and from studies of the 21-cm hydrogen line, that the amount of gas between the stars in galaxies depends on the type of galaxy. Elliptical galaxies have essentially no gas or dust between their stars, while spiral galaxies have interstellar gas and dust. The relative amount of gas increases from types Sa (or SBa) to Sc (or SBc). Irregular galaxies usually have even denser interstellar media than do spirals.

At first it was thought that the arrangement of galaxies in the Hubble classification, and the differing amounts of gas in different types, might indicate that one type of galaxy evolves into another, but that is no longer thought to be the case. We believe that the differences in gas content result from differing conditions at the time of formation of the galaxies, and may be the result of processes that led some of the galaxies to become spiral and others elliptical.

23.2 CLUSTERS OF GALAXIES

Careful study of the positions of galaxies and their distances from us have revealed that most galaxies are part of clusters. These *clusters of galaxies* may have just a handful of members or may have hundreds of members; in any case, they are very large scale phenomena. (Obviously, they are on a completely different scale from that of the galactic clusters, a type of star cluster that we discussed in Section 4.5).

23.2a The Local Group

The two dozen or so galaxies nearest us form the *Local Group*. The Local Group contains a typical distribution of types of galaxies and extends

Figure 23–7 NGC 2685, a peculiar barred spiral nebula, type S0 (pec), in Ursa Major. It seems to be wrapped with helical filaments around a second axis of symmetry; the reasons for this are not understood.

Do not confuse "galactic clusters," which are clusters of stars in our galaxy (the individual objects are stars), with "clusters of galaxies," which are, as their name shows, clusters of separate galaxies (the individual objects are galaxies).

Figure 23–8 The Hubble classification of galaxies.

Figure 23-9 An artist's conception of the close companion of our galaxy that S. Christian Simonson reported to exist on the basis of 21-cm observations.

over a volume 1 megaparsec in diameter. It contains three spiral galaxies, each 15 to 50 kiloparsecs across—the Milky Way, Andromeda, and M33. There are four irregular galaxies, each 3 to 10 kiloparsecs across, including the Large and Small Magellanic Clouds. The rest of the galaxies are ellipticals, including 4 regular ellipticals, each 2 to 5 kiloparsecs across, two of which are the companions to the Andromeda Galaxy. The others are dwarf ellipticals, mostly less than 2 kiloparsecs across.

The possible detection of another new member of the Local Group was reported in 1975 by S. Christian Simonson, then of the University of Maryland. Simonson interprets maps of the 21-cm structure in the direction of the constellation Gemini to indicate that a dwarf galaxy is passing nearby and is disrupting otherwise regular spiral structure there. No optical traces of the galaxy have been detected.

If this is a new galaxy, it would be only 20 kiloparsecs or so farther out from the galactic center than is the sun (Fig. 23–9). The new galaxy has been informally called Snickers, because its mass is only 10^8 M, peanuts compared with that of the Milky Way. It would be almost touching our galaxy, and some astronomers prefer to think that the 21-cm maps show a deformed part of the spiral structure of our galaxy rather than an independent galaxy.

23.2b Farther Clusters of Galaxies

There are apparently other small groups of galaxies in the vicinity of the Local Group, each containing only a dozen or so members. The nearest cluster of many galaxies (a rich cluster as opposed to a poor cluster) can be observed in the constellation Virgo (the Virgin) and is called the Virgo Cluster. It covers a region in the sky over 6° in diameter, 12 times greater

Figure 23–10 A cluster of galaxies in the constellation Coma Berenices. A close look will reveal that many of the small dots are not round and sharp, and are thus galaxies rather than stars.

than the angular diameter of the moon. The Virgo Cluster contains thousands of galaxies of all types. It is about 2 million parsecs across, and is located about 20 million parsecs away from us.

Other rich clusters are known at greater distances, including the Coma Cluster (Fig. 23–10) in the constellation Coma Berenices (Berenice's Hair). The Coma Cluster has spherical symmetry and a concentration of galaxies toward its center, not unlike the distribution of stars in a globular cluster. Thus it is a *regular cluster* as opposed to an *irregular cluster*.

X-rays have been discovered from the three rich clusters closest to us, the Virgo, Coma, and Perseus Clusters, so it seems reasonable that rich clusters of galaxies are generally x-ray sources.

There may be as many as 10,000 galaxies in a cluster, and the density of galaxies near the center of a rich cluster may be higher than that near the Milky Way by a factor of one thousand to one million. Thousands of clusters of galaxies are known.

23.2c *The Missing Mass Problem (Optional)*

Clusters of galaxies appear to be stable configurations that have lasted for many years. But since galaxies have random velocities in various directions with respect to the center of mass of the cluster, why don't the galaxies escape? If we assume that the clusters of galaxies are actually stable, we can calculate the amount of gravity that must be present to keep the galaxies bound. Knowing the amount of gravity in turn allows us to calculate the mass. When this supposedly simple calculation is made for the Virgo Cluster, it turns out there should be fifty times more mass present than is observed. 98 per cent of the mass expected is not found. This is called the *missing mass problem.*

The missing mass problem is the discrepancy found when the mass derived from consideration of the motions of galaxies in clusters is compared with the mass that we can observe.

We know the missing mass is not simply in the form of neutral hydrogen, for if it were we would detect it through its 21-cm radiation. Now that we can survey the distribution of molecular hydrogen in our own galaxy by ultraviolet observations with the Copernicus satellite, we know that there is not enough H_2 in our own galaxy to provide the deficient amount of mass, and there is presumably not enough H_2 in the Virgo Cluster either. One leading thought is that the missing mass could be present in black holes. Another idea is that even if there were a very dense intergalactic medium, it would not radiate 21-cm emission if it were hot enough and thus almost entirely ionized. A hot gas, however, would radiate x-rays, so through x-ray observations of clusters of galaxies, we set limits on the amount of hot intergalactic gas that can be present. Thus we must continue to try to identify the sources of extragalactic x-radiation, both the component that comes from clusters of galaxies and also a diffuse component that comes isotropically (from all directions). Calculations carried out during the compilation of the Fourth Uhuru Catalogue of x-ray sources indicate that hot gas is associated with superclusters—clusters of clusters—of galaxies and that it could contain five to ten times more material than that observed in the associated galaxies. It could thus contain 10 to 50 per cent of the missing mass.

In Section 25.6, we shall see how another method of measuring the density that also does not depend on whether mass is visible or invisible has been brought to bear on calculating the density of the universe. The results of this other, very indirect, method do not seem to show the discrepancy that we call the missing mass problem.

23.3 THE EXPANSION OF THE UNIVERSE

In the decade before the problem of the location of the "spiral nebulae" was settled, Vesto M. Slipher of the Lowell Observatory took many spectra that indicated that the spirals had large redshifts. This work was to lead to a profound generalization. In 1929 Hubble announced that galaxies in all directions are moving away from us and that there is a direct proportionality between the distance of a galaxy and its redshift. Hubble, in collaboration at Mount Wilson with Milton L. Humason, went on during the 1930's to establish the relation more fully. It is known as *Hubble's law.* The redshift is presumably caused by the Doppler effect (which was described in Section 2.14), and the law is usually stated in terms of the velocity that corresponds to the measured wavelength rather than in terms of the redshift itself.

Hubble's law states that the velocity of recession of a galaxy is proportional to its distance; it is written

$$v = H_0 d,$$

A MEMBER OF A CLUSTER OF GALAXIES IN	DISTANCE IN LIGHT YEARS	REDSHIFTS
VIRGO	7×10^7	H + K 1200 KM/SEC
URSA MAJOR	9×10^8	15,000 KM/SEC
CORONA BOREALIS	1.3×10^9	22,000 KM/SEC
BOOTES	2.3×10^9	39,000 KM/SEC
HYDRA	3.6×10^9	61,000 KM/SEC

Figure 23–11 Spectra are shown at right for the galaxies at left. Distances are based on Hubble's constant = 55 km/sec/Mpc.

where v is the velocity, d is the distance, and H_o is the present-day value of the constant of proportionality, which is known as *Hubble's constant* (Figs. 23–11 and 23–12).

The best current value of Hubble's constant, measured by Allan Sandage and Gustav Tammann at the Hale Observatories, is 55 km/sec/Mpc. This is almost a factor of 10 lower than the value that Hubble originally announced, but Sandage and Tammann have been able to make use of new developments in finding the distance to distant galaxies to derive the improved value, incorporating as well earlier corrections to the distance scale.

Hubble's constant is given in units that appear strange, but it merely states that for each megaparsec (3.3×10^6 light years) of distance from the

Figure 23–12 The Hubble diagram for the galaxies shown in Fig. 23–11.

sun, the velocity increases by 55 km/sec (thus the units are 55 km/sec per Mpc). From Hubble's law, we see that a galaxy at 10 Mpc would have a redshift corresponding to 550 km/sec; at 20 Mpc the redshift of a galaxy would correspond to 1100 km/sec; and so on.

We use a subscript "o" on H_o in order to retain the letter H for describing how Hubble's constant might vary over time. Obviously, if Hubble's constant varies over time, it isn't really a constant.

The major philosophical import of Hubble's law is that objects in the universe in all directions are moving away from us; the universe is expanding. Since the time when Copernicus moved the earth out of the center of the universe (and the time when Shapley moved the earth and sun out of even the center of the Milky Way Galaxy), we have not liked to think that we could be at the center of the universe. Fortunately, Hubble's law can be accounted for without our having to be at any such favored location, as we see below.

Imagine a raisin cake (Fig. 23–13) about to go into the oven. The raisins are spaced a certain distance away from each other. Then, as the cake rises, the raisins spread apart from each other. If we were able to sit on one of those raisins, we would see our neighboring raisins move away from us at a certain speed. It is important to realize that raisins farther away from us would be moving away faster: not only would the distance from us to the neighboring raisin have increased but also the additional distance beyond the neighbor to the farther raisin would have increased. Thus no matter in what direction we looked, the raisins would be receding from us, with the velocity of recession proportional to the distance.

The next important point to realize is that it doesn't matter which raisin we sit on; all the other raisins would always seem to be receding. Of course, the raisin cake is finite in size and the universe may have no limit, but other than that the analogy is exact. The fact that all the galaxies appear to be receding from us does not put us in a unique spot in the universe; there is no center to the universe. Each observer at each location would observe the same picture.

Note also that the raisin cake—and the universe—must be expanding with respect to something. We are assuming that we are still able to measure lengths on an unchanging scale. In our analogy, the size of the raisins themselves is not changing; only the separations are changing. In the universe, the galaxies themselves and the clusters of galaxies are not expanding; only the distances between the clusters are increasing.

Note also that individual stars in our galaxy can appear to have small

Figure 23–13 From every raisin in a raisin cake, every other raisin seems to be moving away from you at a speed that depends on its distance from you. This leads to a relation like the Hubble law between the velocity and the distance. Note also that each raisin would be at the center of the expansion measured from its own position, yet the cake is expanding uniformly. For a better analogy with the universe, consider an infinite cake; clearly, there is no center to its expansion.

redshifts or blueshifts, caused either by their peculiar velocities or by the differential galactic rotation. Also, even some of the nearer galaxies have random velocities of sufficient size, or velocities less than our rotational velocity in our galaxy, so that they are approaching us. But except for these few nearby cases, all the galaxies are receding.

The major problem for setting the Hubble law on the firmest footing is finding the distances to the galaxies for which redshifts are measured. Only for the nearest galaxies can we detect Cepheid or RR Lyrae variable stars. Beyond those galaxies we use such methods as assuming that the magnitudes of supergiant stars or sizes of H II regions are more or less the same as they are in our own galaxy, and calculating, respectively, spectroscopic parallaxes or the distances that would make the observed angular dimensions of the H II regions correspond with the linear dimensions we know for those objects in the Milky Way. At still greater distances, we assume that the brightest member in a cluster of galaxies has the same absolute magnitude as all other brightest members of other clusters; sometimes to limit the difficulty that one or two members might be exceptionally bright we rather consider the third-brightest member of a cluster. (The distribution of magnitudes of galaxies in clusters has also been considered, though this is much more difficult to do.) Certainly the methods we have to use grow less and less precise as we get farther and farther away from the sun. In particular, we are seeing galaxies that emitted their light so long ago that they then were different magnitudes than similar galaxies in our neighborhood are today. Also, we are observing radiation that was emitted in the ultraviolet, even though it is now redshifted into the visible, and we do not yet know much about ultraviolet spectra of galaxies. We may thus be making a systematic error in assessing their distances. In any case, all the current evidence indicates that when we correct for such effects as best we can the Hubble relation continues to be a straight-line proportionality deep into space.

Beyond a certain range, we can no longer independently measure distances, and our only method of assessing distance is application of the Hubble law to the observed redshifts. The galaxy thus far detected with the greatest redshift, which is presumably also thus the most distant galaxy, is called 3C 123 (Fig. 23–14). It appears to be an elliptical galaxy in a compact cluster of galaxies, and is a strong source of radio waves.

We discuss the redshift formula and its applications in Section 24.2, since more quasars than galaxies have such high redshifts. The redshift of 3C 123 is so great, 64 per cent of the speed of light, that the formula used for the Doppler shift must be the version that takes the special theory of relativity into account.

Figure 23–14 The farthest known galaxy, 3C 123, a galaxy fainter than 21st magnitude with a redshift of 0.637, can barely be observed. It is 8 billion light years away. The print is a negative.

23.4 RADIO GALAXIES

In Sections 21.2 and 21.3, we saw how a radio eye has been opened on the universe, and how observations made in the radio part of the spectrum have helped us understand the structure of our own Milky Way and of objects such as supernova remnants in it. However, most of the objects that we detect in the radio sky (that is, locations from which more than the background level of radio radiation can be detected) turn out not to be located in our galaxy. The study of these *extragalactic radio sources* is the subject of this section.

23.4a Low-Resolution Observations of Radio Galaxies

The core of our galaxy, the radio source we call Sagittarius A, is one of the strongest radio sources that we can observe in our galaxy. But if the Milky Way Galaxy were at the distance of other galaxies, its radio emission would be very weak.

We are able to detect a small amount of radio emission from many spiral galaxies. Among ellipticals, some seem to have quite strong radio emission while others at equal distances seem radio quiet. The radio radiation is presumably created by such processes as synchrotron radiation (which was described in Section 21.2), but the strength of the radiation is not exceptional and we shall not further deal with such "normal" radio galaxies.

By analogy, we speak of an x-ray galaxy for an exceptionally strong x-ray emitter.

Since the earliest work on radio astronomy, it has been clear that there is a class of galaxies whose members emit quite a lot of radio radiation, many orders of magnitude (that is, many powers of ten) more than "normal" radio galaxies. These have often been called "peculiar" radio galaxies; we shall limit the use of the term *radio galaxy* to mean these relatively powerful radio sources. They often appear optically as peculiar giant elliptical galaxies.

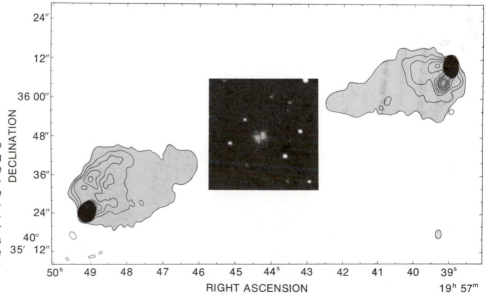

Figure 23–15 A radio map of Cygnus A, with shading and contours indicating the intensity of the radio emission. A photograph of the faint optical object or objects observable is superimposed at the proper scale. The beam size is shown by the ellipse at the lower right. An enlargement of the central region appears above.

The first radio galaxy to be detected, Cygnus A, radiates about a million times more energy in the radio region of the spectrum than does the Milky Way Galaxy. The most recent map is shown in Figure 23–15, with an optical image superimposed. As has been known for decades, Cygnus A, and dozens of other radio galaxies, emit radio radiation mostly from two zones, called *lobes*, located far to either side of the optical object that is visible. The optical object—two fuzzy blobs—has been the subject of much analysis, but its makeup is still not understood. Such *double-lobed structure* is typical of many radio galaxies.

Often there are peculiarities in the optical images that correspond to radio sources of all kinds. For example, a small jet of gas can be seen on short exposures of M87, which corresponds to the powerful radio galaxy Virgo A. Light from the jet is polarized, which confirms that the synchrotron process is at work here. M82, the galaxy that has the appearance of being in a state of explosion (Fig. 23–6), is also a radio galaxy.

23.4b Radio Interferometers

We have discussed in Section 21.2 that the resolution of single radio telescopes is very low, because of the long wavelength of radio radiation. Single radio telescopes may only be able to resolve structure a few minutes of arc or even a degree or so across, depending both on the wavelength of observation and on the size of the dish. For the past few years, the techniques of interferometry have been applied to radio astronomy, with the result that arrays of radio telescopes now in existence can map the sky with resolutions almost as high as the 1 arc sec or so that we can get with optical telescope (Fig. 23–16).

The resolution of a single dish radio telescope depends on the diameter of the telescope (let us consider, for the moment, radiation of one frequency so that we do not have to discuss the variation of resolution with frequency). If, as we see in Figure 23–17, we could somehow retain only the outer zone of the dish, the resolution would remain the same (though the collecting area would be decreased, so we would have to collect the signal for a longer time to get the same intensity).

We use a property of radiation called *interference* to compare the signals from two separated dishes, so we call the dishes plus whatever links them together an *interferometer*. Whereas for a single dish, we would call the maximum spacing of the two most distant points from which we can detect radiation by the name "diameter," for a two-dish interferometer we call it the *baseline*.

1 ARC SEC

30 ARC MIN

Figure 23–16 The Andromeda Galaxy as it would appear at different resolutions.

Figure 23–17 A large single mirror (*A*) can be thought of as a set of smaller mirrors (*B*). Since the resolution of a telescope for light of a certain wavelength depends only on the telescope's aperture, retaining only the outermost segments (*C*) matches the resolution of a full-aperture mirror. We can use a property of light called *interference* to analyze the incoming radiation. The device is then called an *interferometer*.

A DIAMETER OF MIRROR B C BASELINE

Figure 23–18 VLBI techniques with a baseline that is the diameter of the earth allow radio sources to be studied with extremely high resolution.

Figure 23–19 The lines joining the several dishes making up an interferometer represent pairs of dishes that give results equivalent to having several baselines simultaneously. The same results could be gathered with only two dishes, one of which was moved around, but it would take longer. With the three dishes shown, one can simultaneously make measurements with three different baselines.

The first interferometers involved dishes that were linked with wires in order to make sure that the length of time remains constant between the incidence of the wavefront on the first dish to its incidence on the second. In principle, however, if the dishes that make up an interferometer could be placed still farther away from each other, the resolution could be even better. A radio link between the dishes enabled an expansion of the baseline, but irregularities in the distance that the radio waves travel through the air limit this method to use within about 100 kilometers.

The breakthrough came with the invention of atomic clocks, which can keep time to an accuracy of one part in 10^{15} for the latest hydrogen masers (which is equivalent to a drift of three hundred-billionths of a second in a year). A time signal from an atomic clock can be recorded on a tape recorder on one tape band while the radio signal is put on an adjacent tape band. This signal can be compared at any later time with the signal from the other dish, synchronized accurately through comparison of the clock signals. The ability to have the signals linked so accurately freed radio astronomers of the need to have the dishes in direct contact with each other during the period of observation. Now all that is necessary is that the two telescopes observe the same object at the same period of time; the comparison of the signals can take place in a computer weeks later. With this ability, astronomers can make up an interferometer of two or more dishes very far apart, even thousands of kilometers. This technique is called *very-long-baseline interferometry (VLBI)*.

The maximum baseline that can be used in VLBI research is approximately that of the diameter of the earth (Fig. 23–18), 12,700 km. This corresponds at wavelengths of a few centimeters to resolutions as small as 0.0001 arc sec, far better than resolutions that can be gotten with optical telescopes. So radio astronomy has moved in recent years from a situation of providing inferior resolution to a situation of providing superior resolution.

VLBI techniques are difficult and time-consuming to apply. Also, when we work at high resolution, we sample only a small area of the sky at any time, and it therefore takes longer to study a region at high resolution than it does at low resolution. Therefore, VLBI techniques can be applied only to very small areas of sky. But for those few areas, chosen for their special interest, our knowledge of the structure of radio sources has been fantastically improved.

23.4c   Aperture Synthesis Techniques

By suitably arranging a set of radio telescopes across a landscape, one can simultaneously make measurements over a variety of baselines, because each two telescopes in the set can be considered to be a pair with a different baseline from each other pair (Fig. 23–19). Thus with such an arrangement one can more rapidly put together a map of a radio source than one can from two-dish interferometers. Also, the existence of several dishes instead of just two means that there is that much more collecting area.

Note that the resolution of an interferometer depends on the baseline, so that at any one time the resolution is quite good along the line in which the telescopes lie but is only the resolution of a single dish in the perpendicular direction. Fortunately, we can take advantage of the fact that the earth's rotation changes the orientation of the telescopes with respect to the stars (Fig. 23–20). Thus by observing over a 12-hour period of time, one can improve the resolution in all directions.

In effect, we have synthesized the existence of a large telescope covering an elliptical area whose longest diameter is the same as the maximum separation of the outermost telescopes. This type of interferometric technique is known as *aperture synthesis,* and was pioneered by Martin Ryle in England. The interferometers that he built at Cambridge have been responsible for many high-resolution maps, including the map of Cygnus A shown in Figure 23–15.

The most fantastic aperture synthesis radio telescope is now gradually being constructed in New Mexico. It is composed of 27 dishes, each 26 m (85 ft) in diameter, arranged in the shape of a "Y" over a flat area 27 km (17 mi) in diameter (Fig. 23–21). The control room is at the center of the "Y," and contains a large computer to analyze all the signals. This $76 million project is the major ground-based effort of U.S. astronomy. The system is prosaically called the *Very Large Array (VLA).*

When fully operational, which should be in 1981, the VLA will make pictures of a field of view a few minutes of arc across, with resolutions comparable to the 1 arc sec of optical observations from large telescopes, in

Figure 23–20 The earth's rotation carries around a straight line of dishes so that it maps out an ellipse in the sky. The figure shows the view of a line of dishes on the earth that we would see from the direction of the source being observed.

Figure 23–21 The VLA, under construction near Socorro, New Mexico.

Figure 23-22 A Westerbork synthesis map of DA 240, a giant radio galaxy with two lobes of emission. It is 34 arc min across in our sky, larger than the full moon. Its central radio component coincides with a distant galaxy observed in visible light.

about 10 hours. Though this resolution is lower than the resolution obtainable for a limited number of sources with VLBI techniques, the ability to make pictures in such a short time will be invaluable.

23.4d   Aperture Synthesis Observations

The aperture synthesis array at Westerbork in the Netherlands has discovered several giant double radio sources (Fig. 23–22), much larger than any of the double-lobed sources previously known. Some are hundreds of times larger than our own galaxy (Fig. 23–23). These are the largest single objects currently known in the universe.

Observations first with the Cambridge One Mile (1.6 km) Interferometer and then with the Westerbork array have revealed the existence of a class of galaxies with tails extending out from one side. They are called *head-tail galaxies,* and resemble tadpoles in appearance (Fig. 23–24). As these galaxies move through intergalactic space, it is thought that they expel the clouds of gas that we see as the tails.

The discrete blobs that we can see in the tails indicate that the galaxies give off puffs of ionized gas every few million years as they chug through intergalactic space. Perhaps by studying these puffs, we can learn about the main galaxies themselves as they were at earlier stages in their lives. Head-tail galaxies seem to be a common although hitherto unknown type; most are found in rich clusters of galaxies.

SUMMARY AND OUTLINE OF CHAPTER 23

Aims: To study the different types of galaxies that are observed in various parts of the spectrum, to see that galaxies are fundamental units of the universe, to study the expansion of the universe, to consider how interferometry is now allowing radio astronomy to make sig-

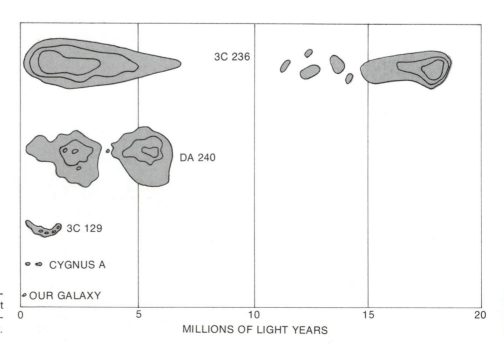

Figure 23-23 A comparison of the sizes of giant double-lobed radio galaxies with our own galaxy.

nificant advances in the study of galaxies, and to consider how and why galaxies may have formed

Observations and catalogues of non-stellar objects
    Lord Rosse's early observations of spiral forms
    Messier's catalogue, General Catalogues by the Herschels, New General Catalogue (NGC) and Index Catalogues (IC) by Dreyer
    Galaxies as "island universes"
    Galaxies were proved to be outside our Milky Way galaxy when Hubble found their distances from observations of Cepheids
Hubble classification (Section 23.1)
    Elliptical galaxies (E0–E7)
    Spiral galaxies (Sa–Sc) and barred spiral galaxies (SBa–SBc)
    Irregular galaxies (Irr I, Irr II)
    Peculiar galaxies (pec)
Clusters of galaxies (Section 23.2)
    Local Group
    Rich clusters, such as Virgo and Coma, are x-ray sources
    Missing mass problem: some gravitational effects indicate the presence in clusters of galaxies of unobserved mass
The universe is expanding (Section 23.3)
    Hubble's law, $v = H_o d$, expresses how the velocity of expansion increases with increasing distance
    Best current value for Hubble's constant is $H_o = 55$ km/sec/Mpc
    The expansion is universal, and has no center
Radio galaxies (Section 23.4)
    Some objects are powerful sources of non-thermal radiation in the radio and x-ray spectral regions
    Double-lobed shape is typical of radio galaxies
    Single-dish radio telescopes have low resolution; interferometers give resolution as high or higher than optical observations
    VLBI (very-long-baseline interferometry) uses widely separated dishes to provide the highest resolution now possible
    Aperture synthesis arrays provide quicker maps, still retain high resolution
        VLA (Very Large Array) now under construction in U.S.
    Giant radio galaxies and head-tail galaxies studied

Figure 23–24 A Westerbork radio map of the head-tail radio source 3C 129, converted to a radiophotograph in which the brightness of the image corresponds to the intensity of radio emission at a wavelength of 21 cm (near but not at the hydrogen wavelength). The resolution is about 30 arc sec. The front end of the head of the radio galaxy corresponds to the position of a faint optical galaxy, a member of a rich cluster of galaxies.

QUESTIONS

1. The sense of rotation of galaxies is determined spectroscopically (Section 23.1b). How is this done?

2. How is it possible that galaxies could exist close to our own and not have been discovered before?

3. When we say that mass is missing, why do we expect it to be there in the first place?

4. How do we know that the missing mass is not neutral or molecular hydrogen?

5. To measure the Hubble constant, you must have a means (other than the redshift) to determine the distances to galaxies. What are two methods that are used?

6. (a) At what velocity in km/sec is a galaxy 5 million parsecs away receding from us? (b) Express this velocity in miles/sec and miles/hr.

7. Does α Centauri, the nearest set of stars to us, show a redshift that follows Hubble's law? Explain.

8. What is one advantage of VLBI techniques over the VLA?

9. What is one advantage of the VLA techniques over the VLBI techniques?

24

Quasars

Quasars are enigmatic objects that, like stars, appear as points of light in the sky. But unlike the stars that we see, the quasars occur in the farthest reaches of the universe and thus may be the key to our understanding the history and structure of space. We study quasars in both the optical and radio parts of the spectrum. Although they are relatively faint in visible light, they are among the strongest radio sources in our sky, and therefore must be prodigious radiators of energy. Astronomers have so far been unable to determine how they give off so much energy. Consequently the quasars raise fundamental questions about our basic physical laws and how energy is generated in the universe.

The word *quasar* originated from QSR, a contraction of "quasi-stellar radio source." However, the most important characteristic of quasars is not that they are emitting radio radiation but that they are traveling away from earth at tremendous speeds. Even certain "quasi-stellar objects" that are not radio sources have been included among the quasars because they too have tremendous velocities. Astronomers can deduce distances by observing the quasar spectra, measuring the Doppler shifts, and applying Hubble's law (Section 23.3).

Most astronomers accept the idea that quasars are the objects most distant from us, but a small minority contends that they are in fact relatively close to us. The reasoning that leads to these opposite conclusions will be discussed later in this chapter. Even if the quasars are relatively near, they are still important to astronomy since they would show that our present methods of measuring distances to faraway objects are not always valid. Whether nearby or fantastically distant, quasars have strange properties that raise basic questions.

24.1 THE DISCOVERY OF QUASARS

Quasars are a discovery resulting from the interaction of optical astronomy and radio astronomy. When maps of the radio sky turned out to be

An enlarged portion of a photograph of the quasi-stellar radio source 3C 273, taken with the 5-m Hale telescope. The object looks like any 13th magnitude star, except for the faint narrow jet that is visible out to about 20 seconds of arc from the quasi-stellar object. The print is a negative.

Figure 24–1 The first quasar, 3C 48, photographed with the 5-m Hale telescope.

Accurate positions can be measured for radio sources either by interferometry or by lunar occultation.

Figure 24–2 This object appears stellar but is known to be not a star but a quasar.

very different in appearance from maps of the optical sky, many astronomers set out to correlate the radio objects with visible ones. The sun was easy to identify, and some of the brighter galaxies also proved to be radio sources (as discussed in Section 23.4).

Single-dish radio telescopes did not give sufficiently accurate positions of objects in the sky to allow identifications to be made, so interferometers (Section 23.4b) had to be used. At least one of the strong radio sources, 3C 48 (the 48th source in the third Cambridge catalogue), seemed to be suspiciously near a faint (16th magnitude), bluish star (Fig. 24–1). At that time, 1960, no stars had been found to emit radio waves, with the sole exception of the sun, whose weak radio radiation we can detect only because of the sun's proximity to the earth.

In Australia, the 64-meter (210-foot) radio telescope at Parkes (Color Plate 2) was used to observe the passage of the moon across the position in the sky of another bright radio source, 3C 273 (the 273rd source in the third Cambridge catalogue). Since we know the position of the moon in the sky very accurately, if it occults — hides — a radio source then we know that at the moment the signal strength decreases the source must have passed behind the advancing limb of the moon. Later, the instant the radio source emerges — for 3C 273 it was nearly an hour later — the position of the lunar limb marks another set of possible positions. Where these two sets of possible positions overlap is the actual position of the source.

There were, fortunately, three lunar occultations of 3C 273 within a few months. From the data gathered, a very accurate position was derived for it. A 13th magnitude optical object was found at that location. It is not completely starlike in appearance, for a luminous jet appears to be connected to the point nucleus (as shown in the photograph opening this chapter).

The spectra of the bluish stellar objects showed emission lines, but the lines were not of any known element. The lines had the general appearance of lines emitted by a gas of medium temperature, but the wavelengths of the lines observed were different from those that known gases emit.

The breakthrough in understanding came in 1963. At that time, Maarten Schmidt, who had used the 5-meter Hale telescope to photograph the spectrum of 3C 273, noted that its spectral lines seemed to have the same pattern as lines of hydrogen under normal terrestrial conditions — that is, they had the characteristic pattern of lines that the hydrogen spectrum has. Schmidt then made a major scientific discovery: he asked himself whether he could simply be observing a hydrogen spectrum that had been greatly shifted in wavelength by the Doppler effect. The Doppler shift required would be huge: each wavelength would have to be shifted by 16 per cent toward the red to account for the spectrum of 3C 273. This would mean that 3C 273 is receding from us at approximately 16 per cent of the speed of light. Immediately, Schmidt's colleague Jesse Greenstein recognized that the spectrum of 3C 48, another of the bluish stellar objects associated with radio sources, could be similarly explained. All the lines in the spectrum of 3C 48 were shifted by 37 per cent, a still more astounding redshift.

Besides the "quasi-stellar radio sources," another class of objects appeared. Sources in this new class resembled the quasi-stellar radio sources in being quasi-stellar and in having large redshifts but did not emit any detectable radio radiation. These were called *quasi-stellar objects* (QSO's) (Fig. 24–2). We now use the term *quasar* to include both quasi-stellar radio sources and quasi-stellar objects. Over a hundred quasars have been discovered, with Doppler shifts ranging from about 16 per cent to 353 per cent ($z = 0.16$ to $z = 3.53$; see Section 24.2). Although only a hundred spectra have been measured, astronomers estimate that there have been perhaps one million quasars in the universe. By now, however, most of them have probably lived out their lifetimes; that is, they have given off so much energy that they are no longer quasars. Perhaps about 35,000 quasars now exist, in the sense that they would be included if we could somehow take an instantaneous picture of the universe as it is at present, without consideration of the fact that because of the finite speed of light, we see distant sources by the radiation that they emitted long ago.

24.2 THE DOPPLER SHIFT IN QUASARS

Most objects in the universe show some Doppler shift with respect to the earth, for they are either approaching us or receding from us. (This is normally the only component of their motion that we can measure.) For velocities that are small compared to the velocity of light, the amount of shift in the spectrum is written simply, for rest wavelength λ, velocity v, and speed of light c, as

$$\frac{\Delta\lambda}{\lambda} = \frac{v}{c}$$ (as discussed in Section 2.14).

Astronomers often use the symbol z to stand for $\Delta\lambda/\lambda$, the amount of the redshift (Fig. 24–3).

The Doppler shifts of quasars are much greater than those of most galaxies. Even for the nearest quasar, 3C 273, $z = 0.16$ (read "a redshift of 16 per cent"), and so the spectrum has been shifted by a substantial amount. Thus Hubble's law implies that 3C 273 is as far away from us as are distant galaxies. Only a handful of galaxies are known to have greater redshifts.

Figure 24–3A The wavelengths of the visible part of the spectrum are shown at top, and the wavelengths at which spectral lines would appear after undergoing a redshift of 0.2 ($\Delta\lambda/\lambda = 0.2$) is at bottom. The lines are shifted by 0.2 times their original wavelengths, and wind up at 1.2 times their original wavelengths. The non-relativistic formula $\Delta\lambda/\lambda = v/c$ can be applied to give the approximate velocity; at still smaller redshifts this formula is even more accurate.

Figure 24–3B For a redshift of 1 ($\Delta\lambda/\lambda = 1$), as shown here, the lines are shifted (*bottom*) by an amount equal to their original wavelengths, and wind up at twice their original wavelengths. The velocity is a significant fraction of the speed of light and the simple non-relativistic formula for the Doppler shift cannot be applied. The relativistic formula allows z to be greater than 1 without the velocity exceeding the speed of light.

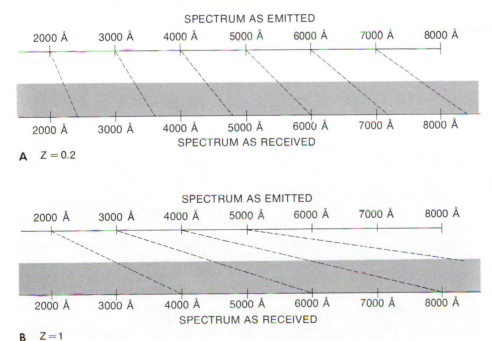

SPECTRUM AS EMITTED

2000 Å   3000 Å   4000 Å   5000 Å   6000 Å   7000 Å   8000 Å

2000 Å   3000 Å   4000 Å   5000 Å   6000 Å   7000 Å   8000 Å
SPECTRUM AS RECEIVED

A   Z = 0.2

SPECTRUM AS EMITTED

2000 Å   3000 Å   4000 Å   5000 Å   6000 Å   7000 Å   8000 Å

2000 Å   3000 Å   4000 Å   5000 Å   6000 Å   7000 Å   8000 Å
SPECTRUM AS RECEIVED

B   Z = 1

24.3   THE IMPORTANCE OF QUASARS

If we accept that the spectra of quasars are redshifted by tremendous amounts and apply Hubble's law, we realize that the quasars are the farthest known objects in the universe. At the moment, we have no independent way of measuring distances to objects so far away except by Hubble's law, and so it is important to test Hubble's law. Any deviations from the law would no doubt show up in the farthest objects, namely the quasars.

At present, Hubble's law is our only method of finding the distances to any objects with substantial redshifts. If the quasars were found not to satisfy Hubble's law, however, then doubt would be cast on all distances derived by Hubble's law, for we would never know whether we were observing an object that satisfied the law or one that did not.

If we do accept the quasars as the farthest objects, then they are billions of light years away from us and their light has taken billions of years to reach us. Thus we are looking back in time when we observe the quasars, and we can hope that they will help us understand the early phases of our universe.

24.4   THE ENERGY PROBLEM

If the quasars are as far away as this orthodox view, of which Maarten Schmidt is a leading proponent, holds, then they must be intrinsically very luminous to appear to us at their observed intensities. The optical brightness of quasars varies with time—scales of weeks or months.

If you want to pack as much energy into a quasar as they seem to radiate, you must have a certain amount of space in which to pack it. A quasar cannot be too small and still contain the 100 million solar masses that many astronomers think is necessary to generate that much energy. If something is, say, a tenth of a light year across, you would expect that it could not vary in brightness in less than a tenth of a year because one side cannot signal to the other side, so to speak, to join in the variation (Fig. 24–4). Somehow the whole object has to know at the same time to do something, and that ability is limited by the speed of light.

Figure 24–4 The figure illustrates why a large object can't fluctuate in brightness as rapidly as a smaller object. Say that each object abruptly brightens at one instant. The wave emitted from the top of the object takes somewhat longer to reach us than the wave emitted from the side of the object nearest us, just because of the additional distance it has to travel. We don't see the full effect of the variation in brightness until we have the waves from all parts of the object. This simply takes longer for larger objects than it does for smaller ones.

TAKES 1 WEEK FOR VARIATION TO BE COMPLETE

1 LIGHT WEEK

TAKES 1 YEAR FOR VARIATION TO BE COMPLETE

1 LIGHT YEAR

So the variations in intensity mean that the quasars are probably fairly small. But the mechanism that makes them vary with these very short time scales is not known.

One possible model for the energy generation in a quasar is nuclear fusion in an extremely massive body (containing thousands of solar masses) known as a *supermassive star*. There is no confirming observation that such supermassive stars exist.

A model of many exploding supernovae seemed attractive for a time because one could explain the observed sporadic variations in the intensity by variations in the number going off from one moment to another. But the notion of the immense number of supernovae boggles the mind and does not seem acceptable in terms of the amount of mass required. An alternative involves rotating bodies whose magnetic fields are at skew angles.

Many other suggestions have been made to explain the source of energy in quasars, none of which is convincing. Perhaps matter and antimatter are annihilating each other in quasars. It may be a long time before we know, or agree on, the energy sources involved. Perhaps the answer will prove to be a new method, yet unknown, that can be applied to help our energy problems here on earth.

24.5 OTHER EXPLANATIONS OF REDSHIFTS

Some scientists say that it would be too difficult for quasars to generate the amount of energy that is necessary to make them as bright as they would need to be if they were very far from us. They feel that this problem of providing enough energy is so serious that they conclude that quasars must be close to us. If the quasars are close, we must think of some other way of accounting for their great redshifts. There are at least three other ways. They are all unattractive to many astronomers because they challenge Hubble's law and thus cast doubt on our knowledge of the whole scale of the universe.

The first way also relies on velocity and assumes that quasars are close but are going away from us very rapidly. Quasars could be relatively close to us yet show high redshifts if they had been ejected from the center of a galaxy at very high velocities (Fig. 24-5). If such quasars occurred in another galaxy, we would expect to see some of them going away from us and some coming toward us. Some of them would show tremendous redshifts signifying high velocities of recession; others would have tremendous blueshifts, signifying high velocities of approach. But all the quasars have redshifts; none of them has a blueshift, so astronomers think that this idea is unlikely. But one day if we find a blueshifted quasar, much of astronomy will have to be rewritten.

Since there are no blueshifted quasars, and if the quasars exploded from the center of our galaxy or a nearby galaxy, then the explosion took place so long ago that all quasars have expanded past us and are receding in all directions. Such tremendous and perhaps repeated explosions require a source of energy so great that it would be no easier to explain than it was with the redshifts following Hubble's law.

A second kind of redshift is "gravitational." According to Einstein's theory of relativity, spectral lines are redshifted very slightly when they leave a star or any object that has a large mass and hence a large gravita-

Figure 24-5 If quasars were local, and were ejected from our galaxy, that could explain why they all have high redshifts. But a tremendous amount of energy is necessary to make the ejection. So we may not find ourselves any closer to a solution of the energy problem — where all this energy comes from.

tional pull. The sun slightly redshifts the lines in the spectrum that leave it; a white dwarf star, which is much denser than the sun, redshifts the lines much more (see Section 7.5 for a discussion of this). Several scientists tried to find a way to allow for a sufficient redshift from gravity to match the huge redshifts in quasars. But the spectra of the quasars are not the spectra of very dense stars; they resemble in part that of a thin gas. So the scientists considered that dense objects were present to provide the gravity in addition to the presence of the gas that provides the spectral lines. They were not able to work out such a system that would be free of other, unobserved consequences.

A third possibility is that the quasars are acting on principles that we do not yet understand—some new kind of physics. Some specific modifications of physical laws have even been suggested. But most scientists feel (as discussed in Section 19.5) that the basic philosophy of science forbids us from inventing new physical laws as long as we can satisfactorily use the existing ones to explain all our data.

24.6 PECULIAR GALAXIES AND QUASARS

A principal attack on the theory that quasars are at great distances has come from Halton Arp of the Hale Observatories in California. He has been making observations of "peculiar galaxies" (discussed in Section 23.1d), objects in the sky whose shapes deviate from the regularity of the shapes of ordinary galaxies. These observations have led him to conclude that quasars might be linked to galaxies, both the peculiar and the ordinary types. He contends that he has found many examples in which two quasars are located on opposite sides of a galaxy from each other. He argues that the quasars and the galaxy must therefore be linked. We know the distances to the galaxies from Hubble's law; if the quasars and galaxies are indeed physically linked, then the quasars are at the same distance from us as the galaxies. These distances are much smaller than the quasars' velocities of recession indicate.

A similar argument that Doppler distances cannot be trusted is made with Stephan's Quintet (Fig. 24–6), five apparently linked galaxies, one of which has a redshift very different from the others. Such cases can be explained without need to reject the Hubble law if the object with a discordant redshift only happens to appear in almost the same line of sight as do the other objects. Associations of a quasar with a distant goup of galaxies with redshifts that do agree with each other have been found.

Arp's data are subject to several objections, the main one being statistical. Arp feels that the associations he finds of objects of differing redshifts are genuine and not the result of chance. He points out that the probability of finding these objects so close together in the sky is very small. But most astronomers feel that the fact that a quasar and a galaxy are seen in the same directions from the earth does not prove that they are physically linked; they could be at different distances. The associations would be harder to explain in this way if we found a group of galaxies in which not just one but two discrepant redshifts were found. No such group is known. So far, most astronomers do not accept Arp's arguments and feel that quasars are in fact very far away.

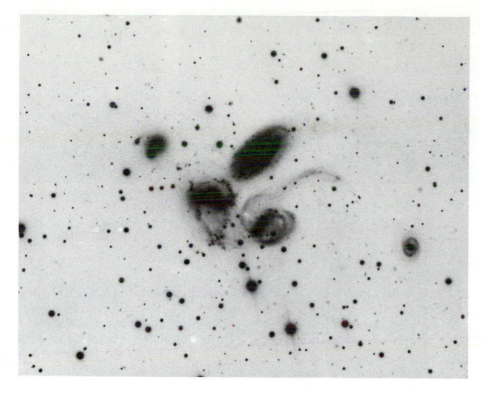

Figure 24–6 Stephan's Quintet, in a negative print of a photograph taken by Halton Arp with the 5-m telescope at the Hale Observatories.

24.7 QUASARS AND GALAXIES

Another question is whether quasars are truly a new phenomenon or just unfamiliar types of galaxies. Some spiral galaxies have especially bright nuclei while others have fainter centers. One type of galaxy, discovered by Carl Seyfert of the Mount Wilson Observatory in 1943 and named after him, has a very bright nucleus indeed (Fig. 24–7). Can the quasars that we see be only the bright nuclei of galaxies? It has been recently calculated that although we can see such structures as the nucleus and the spiral arms of close Seyferts, only the nuclei would be visible if such galaxies were very far away. Another type of galaxy, an N galaxy, also has an especially bright core. The brightest quasar (3C 273) and two Seyfert galaxies are among the few extra-galactic objects from which the Uhuru satellite has detected x-radiation.

Jerome Kristian of the Hale Observatories has studied a sample of about 25 quasars to see whether they could be galaxies with bright cores. He attempted to predict which would be close enough to us to reveal some structure around the core, which would appear as points like quasars, and which would be in between and thus might or might not show structure. He then studied the best photographs of these objects, and found that almost all objects in which one would expect to see structure did in fact show structure, almost all objects that would not show structure because they were too far away did not show structure, and that the middle group was mixed. It thus seems reasonable to believe that quasars are a class of objects somehow related to the cores of galaxies. Perhaps quasars and galaxies are the same, but are simply in different stages of evolution.

Now there is even stronger evidence that the redshift is a valid estimator of distance for quasars. A strange object named BL Lacertae

The difference between Seyfert and N galaxies is partly based on their appearance and partly based on spectroscopic observations. The distinction is partly historical, and is not always clean cut.

A

B

Figure 24–7 (A) Increasing exposure time reveals more and more galactic structure around the Seyfert galaxy NGC 4151, which appears almost stellar on the least exposed photograph. (B) The nucleus of the Seyfert galaxy photographed from an altitude of 25,000 m (80,000 ft) with the 90-cm (36-in) telescope carried aloft in 1970 by the Stratoscope II balloon.

has been known for years. At first, it was thought to be merely a variable star, but in recent years it was suspected to be stranger than that, partly because of its rapidly variable radio emission. Its spectrum showed only a featureless continuum, with no absorption or emission lines, so little could be learned about it.

A pair of Caltech astronomers had the idea of blocking out the bright central part of the star with a disk held up in the focus of their telescope, thereby obtaining the spectrum of the haze of gas that seems to surround the star. On one of the several nights that they observed, the central source was relatively faint and the spectrum of the surrounding haze of gas could be observed with an electronic device. The spectrum turns out to be a faint continuum with absorption lines that are redshifted by 7 per cent. Thus BL Lacertae is the nearest quasar!

If BL Lacertae is at the distance that corresponds to this redshift, it has several properties similar to other quasars if they are at their distances assigned from the Hubble law. This strengthens the notion that the quasars are indeed at those distances and that it is valid to use the Hubble relation. And the quasar again seems to be a special "event" in the center of a mass of gas, just as quasars may be similar to bright "events" in the centers of Seyfert galaxies or N galaxies. These observations may indicate that such "events" can happen late in the lifetime of a galaxy. Since most quasars are at large redshifts, the light we see now was emitted far back in time, so it had been thought that such events might only be linked to young galaxies.

In sum, there seems to be a strong relation between quasars and the cores of galaxies. This linkage may make the quasars seem somewhat less strange, but at the same time causes galaxies to seem more exotic.

SUMMARY AND OUTLINE OF CHAPTER 24

Aims: To understand the history and significance of quasars

Discovery of quasars (Section 24.1)
    QSR — quasi-stellar radio sources
    Interaction of optical and radio astronomy

Positions from lunar occultations or from interferometry

Huge Doppler shifts in spectra

Doppler shift in quasars (Section 24.2)

Importance of quasars (Section 24.3)

Test of Hubble's law, which is thus a test of the accuracy of the distance scale

Energy problem (Section 24.4)

Too much energy required to be accounted for

Possible explanations

Supermassive star

Immense number of supernovae

Skew magnetic field

Annihilation of matter and anti-matter

Presently unknown method

Other explanations of redshifts (Section 24.5)

Quasars ejected from cores of galaxies at very high velocities; no evidence of blueshifted quasars, however

Gravitational redshifts

Quasars operating under new physical laws?

All these alternative explanations challenge Hubble's law

Peculiar galaxies and quasars (Section 24.6)

Question of physical link between galaxies and quasars

Stephan's Quintet and other sources with discrepant redshifts might be explained as chance alignments

Quasars and galaxies (Section 24.7)

Are quasars truly a new phenomenon?

Seyfert and N galaxies have bright nuclei that may be related to quasars

BL Lacertae—the nearest quasar

QUESTIONS

1. Why was it important to know accurately the positions in the sky of radio sources for which there was at first no optical counterpart?

2. (a) You are heading toward a red traffic light so fast that it appears green. How fast are you going? (b) At $1 per mph over the speed limit of 55 mph, what would your fine be in court?

3. A quasar is receding at $1/10$ the speed of light. (a) If its distance is given by the Hubble relation, how far away is it? (b) At what wavelength would the 21-cm line appear?

4. A quasar is observed and it is found that a line whose rest wavelength is 3000 Å is observed at 4000 Å. (a) How fast is the quasar receding? (b) How far away is it if its distance is given by the Hubble relation?

5. A quasar has $z = 0.3$. What is its velocity of recession in km/sec?

6. Why does the rapid time variation in some quasars make the "energy problem" even more difficult to solve?

7. What are three differences between quasars and pulsars?

8. Briefly list the objections to each of the following "local" explanations of quasars:
   (a) They are local objects flying around at large velocities.
   (b) The redshift is gravitational.

9. If quasars were proved to be local objects, how would this help solve the "energy problem"?

10. What features of some quasars suggest that quasars may be some type of galaxy?

Cosmology

In Armagh, Ireland, in the mid-seventeenth century, Bishop Ussher declared that the universe was created at 9 a.m. on Sunday, October 23rd, in the year 4004 B.C. Nowadays we are less certain of the details of our origin.

We study the origin of the universe as part of the study of the universe as a whole. This larger field is called *cosmology*. Our discussion of astronomy has brought us to consider many different types of celestial objects and many different ways of observing them. All this observational knowledge and all these techniques, in conjunction with theoretical calculation, must be brought to bear on cosmological problems in order to understand the most fundamental questions about our universe. Still, there is a unifying theme in the study of cosmology. After all, we have only one universe, and its study is thus an investigation of just one thing.

The study of where we have come from and of what the universe is like now leads us to consider where we are going. Is the universe now in its infancy, in its prime of life, or in its old age? Will it die? It is difficult for us, who, after all, spend most of our time thinking of topics like "which TV program shall I watch tonight" or "what's for dinner," to realize that we can seriously think about the structure of space around us, and to conclude what the future of the universe will be. One must take a little time every day, as Alice was told when she was in Wonderland, to think of impossible things. By and by we become accustomed to concepts that have at first overwhelmed us. You must sit back and ponder when studying cosmology; only in time will many of the ideas that we shall discuss take shape and form in your mind.

25.1 OLBERS'S PARADOX (*Optional*)

Many of the deepest questions of cosmology can be very simply phrased. Why is the sky dark at night? Analysis of this simple observation

Albert Einstein in the Swiss Patent Office in Berne in 1905.

A

B

Figure 25–1 *(A)* If we look far enough in any direction in an infinite universe, our line of sight will hit the surface of a star. This leads to Olbers's paradox. *(B)* This painting by the Belgian surrealist René Magritte shows a situation analogous to saying that we should not see a star in any direction under the conditions of a uniform distribution of stars.

We have phrased the paradox in terms of stars, whereas we know that the stars are actually grouped into galaxies. But we can carry on the same argument with galaxies, and deduce that we must see the average surface brightness of galaxies everywhere. This is patently not what we see.

leads to profound conclusions about the universe. We certainly know that the night sky is basically dark, with light from stars and planets scattered about on a dark background. But a bit of thought will show that if there is a uniform distribution of stars in space, it shouldn't be dark anywhere in the sky. If we look in any direction at all we will eventually see a star (Fig. 25–1), so the sky should appear uniformly bright. We can make the analogy to our standing in a forest. There, we would see some trees that are closer to us and some trees that are farther away, but if the forest is big enough our line of vision will always eventually stop at the surface of a tree. If all the trees were painted white, we would see a white expanse all around us. Similarly, when looking up at the night sky we would expect the sky to have the uniform brightness of the surface of a star.

The fact that this argument doesn't work, and that the sky **is** dark at night, is called *Olbers's paradox.* Wilhelm Olbers phrased it in 1823, although the question had been discussed at least a hundred years earlier.[*] The solution of Olbers's paradox leads us into basic considerations of the structure of the universe. This solution could not have been advanced in Olbers's time because it depends on astronomical discoveries that have been made more recently.

In explaining Olbers's paradox, it is best to clear up one possible point of confusion right at the outset. Remember—the fact that the total amount of light we get from a star decreases as the star gets farther away is not relevant to this discussion. In other words, we cannot explain the paradox merely by claiming that the farthest stars are so distant that they look relatively faint and therefore the sky looks black in their directions. We must realize that since each of our lines of sight would seem to end at the surface of a star, we would see the brightness of that star per unit area of its surface and not the total amount of light. As we consider bright surfaces that are farther and farther from us, even though the total amount of light we receive gets less, the brightness of each bit of surface stays the same.

You may be familiar with an example from photography that results from the same effect of constant surface brightness. Suppose that we set the aperture and exposure time of our camera by coming up close to a subject and making measurements. When we go 15 feet back to take our picture, these settings don't change

---

[*]By Edmund Halley, Jean-Phillippe de Chéseaux, and others.

from what they were when we were up close. This is simply because the surface brightness of the objects remains the same no matter how far we are away from them.

One might think that the easiest way out of the paradox, as was realized in Olbers's time, is simply to say that there is dust or other interstellar matter absorbing the light from the distant stars and galaxies. But this doesn't solve the problem, because the dust would soon absorb so much energy that it would heat up and begin glowing. So given a long enough time, all the matter in the universe would begin glowing with the same brightness, and we would have our paradox all over again.

One solution to Olbers's paradox lies in part in the existence of the redshift, and thus in the expansion of the universe (Section 23.3). This does not mean that the answer to the paradox is just that visible light from distant galaxies is redshifted out of the visible, for at the same time ultraviolet light is continually being redshifted out of the ultraviolet into the visible. The point is rather that each quantum of light undergoes a real diminution of energy as it is redshifted. Thus we do not see the level of brightness emitted at the surface of a faraway star or galaxy because the energy that was emitted has been diminished by this redshift effect before it reaches us. Since the energy, E, of a quantum at a given wavelength, λ, corresponding to a certain frequency, ν, is $E = h\nu = hc/\lambda$, the quantum has at first an energy corresponding to its original wavelength. As its wavelength gets longer (its frequency decreases) because of the Doppler effect and Hubble's law, the formula $E = hc/\lambda$ shows that its energy diminishes. This energy is actually lost from the quantum.

There are some complications to this major point that we could mention. As we look out in space, we are looking back in time, because the light has taken a finite amount of time to travel. If we could see far enough out we would possibly see back to a time before the formation of the stars. E. R. Harrison of the University of Massachusetts recently pointed out that this is an equally valid way out of the dilemma. In fact, Harrison calculates that this explanation is a more important contribution to the solution of the paradox than is the existence of the redshift. Most of the radiation, the absence of which contributes to the existence of Olbers's paradox, would have to come from stars that were very distant and that consequently emitted their radiation a long time ago—10^{24} years. The universe is simply not that old; on the basis of Hubble's law an age of 10^{10} years or so seems appropriate.

The fact that we have to know about the expansion and the age of the universe to answer Olbers's question—why is the sky dark at night?—shows how the most straightforward questions in astronomy can lead to important conclusions. In this case, we conclude that either (1) the universe is expanding, or (2) it is too young (assuming that if it had indeed been sufficiently old, earlier generations of stars would have been present to radiate), or (3) both.

25.2 THE BIG BANG THEORY

Astronomers looking out into space have noticed that on the large scale things look about the same in all directions. That is, ignoring the presence of local effects such as our being in the plane of a particular galaxy and thus seeing a Milky Way across the sky, the universe has no direction that is special. Further, it seems that there is no change with distance either, except insofar as time and distance are linked.

This notion has been codified as the *cosmological principle*: the universe is homogeneous and isotropic throughout space. The assumption of homogeneity says that the distribution of matter doesn't vary with distance from the sun, and the assumption of isotropy says that it is about the same no matter in which direction one looks. Actually, of course, we know that we have to look in certain directions to see out of our galaxy without having our vision ended by the interstellar dust, but remember that we are ignoring inhomogeneities or lack of isotropy on this small scale.

A set of theories of the universe known as *big bang* theories satisfies the cosmological principle. Basically, these cosmologies say that once upon a time there was a great big bang that began the universe. From that moment on, the universe expanded, and as the galaxies formed they shared in the expansion.

Many students ask whether the fact that there was a big bang means that there was a center of the universe from which everything expanded. The answer is no; first of all, the big bang may have been the creation of space itself. Furthermore, the matter of this primordial cosmic egg was everywhere at once. There may be an infinite amount of matter in the universe, so it is possible that at the big bang an infinite amount of matter was compressed to an infinite density while taking up all space.

What was present before the big bang? Versions of the big bang theory (Fig. 25–2) have been put forward to deal with this question, but there is no real way to answer it. For one thing, we can say that time began at the big bang, and that it is meaningless to talk about "before" the big bang because time didn't exist. We do not now think that the universe can remain in a static condition, so it seems unlikely that the universe was always just sitting there in an infinitely compressed state, whatever "always" means. Of course, these possibilities don't answer the question of why the big bang happened. Another possibility is that there had been a prior big bang and then a recollapse, and that this most recent big bang was one in an infinite series of bangs. This version of the big bang is called the *oscillating universe*. It solves some questions, for if the universe is oscillating then we know that there were other cycles before the most recent big bang. However, the oscillating universe theory doesn't really tell us anything about a

"Big bang" is the technical as well as the popular name for these cosmologies; professionals write about the big bang in the scientific journals. The theories are based on Einstein's general theory of relativity.

Figure 25–2 To trace back the growth of our universe, we would like to know the rate at which its rate of expansion is changing. Big bang models of the universe are shown; the vertical axis represents a scale factor, R, that represents some measure of distances and how they change as a function of time, T. The universe could be open and expand forever or closed and begin to contract again. If it is closed, we do not know whether it would oscillate or whether we are in the only cycle of expansion or contraction it will ever undergo. We will see later that current evidence favors the open model.

Figure 25–3 If we could ignore the effect of gravity, then we could trace back in time very simply; the Hubble time corresponds to the inverse of the Hubble constant $(1/H_0)$. Actually, gravity has been slowing down the expansion. The vertical axis again represents some scale factor.

real origin of the universe because in this case there was no origin. The concepts discussed in this paragraph will not be easy to digest. They may take hours, years, or a lifetime to come to terms with.

If we consider Hubble's law with the current value for Hubble's constant—and the cosmologists Allan Sandage and Gustav Tammann have evidence that Hubble's constant is the same for "local" and "distant" galaxies—we can extrapolate backward in time. We simply calculate when the big bang would have had to take place for the universe to have reached its current state at its current rate of expansion. This calculation indicates that the universe is somewhere between 13 billion and 20 billion years old (Fig. 25–3). The range of values indicates in part the uncertainty in Hub-

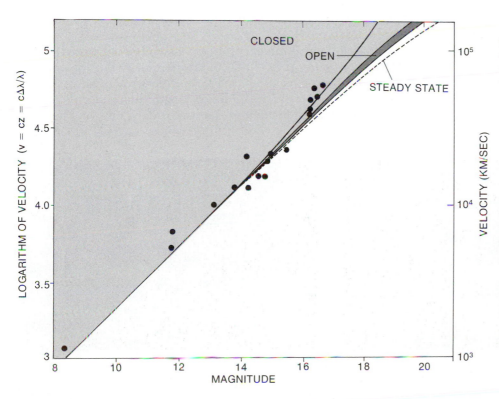

Figure 25–4 This Hubble diagram, plotted in terms of redshift and magnitude, shows faint clusters of galaxies (dots). In principle, we can determine the future of the universe from the slight deviations at upper right of the curves from a straight line. If the curve appears in any of the shaded regions, the expansion is slowing down. If it falls in the heavily shaded region, the universe is open, infinite, and will expand forever. If the curve falls farther to the left in the lightly shaded region (a possibility that is illustrated with a sample curve), then the universe is closed, finite, and will eventually begin to contract. The data are currently not sufficient to choose definitively between these possibilities. The dashed line represents the steady state universe.

ble's constant and in part the uncertainty in the calculation of how much the effect of gravity would have by now slowed down the expansion of the universe.

The universe will last more than ten billion years, so we have nothing to worry about for the immediate future, but the study of the future of the universe is an exceptionally interesting investigation. Some of the methods of tackling this question involve measuring the amount of mass in the universe and thus the amount of gravity (as we will discuss in Section 25.6). Other methods of determining our destiny involve looking at objects as distant as possible to see if any deviation from the Hubble law can be determined. This investigation of distant bodies has been going on for many years, but a deviation from the straight-line relation between velocity and distance known as Hubble's law has not been found conclusively. Sometimes one hears that a slight deviation from a straight line has been found at the outermost end of the graph, but as yet there are no definitive results (Fig. 25–4). The deviations have always been within the margin of error of the statistics.

25.3 THE STEADY STATE THEORY (*Optional*)

The cosmological principle, that the universe is homogeneous and isotropic, is very general in scope, but starting in the late 1940's, three British scientists began investigating a principle that is even more general. Hermann Bondi, Thomas Gold, and Fred Hoyle considered what they called the *perfect cosmological principle,* that the universe is not only homogeneous and isotropic in space but also unchanging in time. According to Occam's Razor (Section 19.5), we must accept the simplest theory that is in accord with all the observations, but it is a matter of personal preference whether the cosmological principle or the perfect cosmological principle is more simple.

The theory that follows from the perfect cosmological principle is called the *steady state theory* (Fig. 25–5); it has certain philosophical differences from the big bang cosmologies. For one thing, according to the steady state theory, the universe never had a beginning and will never have an end. It always looked just about the way it does now and always

Figure 25–5 In the steady state theory, as the dotted box at left expands to fill the full box at right, new matter is created to keep the density constant. In the picture, the four galaxies shown at left can all still be seen at right, but new galaxies have been added so that the number of galaxies inside the dotted box is about the same as it was before.

will look that way. Some scientists are glad to see the question of what happened before the big bang, or the need for finding a cause for the big bang, eliminated in the steady state theory.

But the steady state theory must be squared with the one major fact that we know about cosmology, that the universe is expanding. How can the universe expand continually but not change in its overall appearance? Remember that in the basic formulation of the theory we consider the matter to be smoothed out throughout all space to an average density. It follows that for the density to remain constant, new matter must be created at the same rate that the expansion would decrease the density. Thus the density remains the same. Now the matter that is created in the steady state theory is not simply matter that is being converted from energy according to Einstein's famous formula, $E = mc^2$. No, this is matter that is appearing out of nothing, and is thus equivalent to energy appearing out of nothing.

Many scientists objected very strenuously to the idea that matter could appear out of nothing. After all, the "law" that matter is conserved (i.e., cannot be created or destroyed), generalized by Einstein into a law of conservation of matter and energy, seems very basic. Nevertheless, since we know the rate at which the universe is expanding, we can calculate how much new matter would have to be created. It works out to be only one hydrogen atom per cubic centimeter of space every 10^{15} years, which is equivalent to one thousand atoms of hydrogen per year in a volume the size of the Astrodome in Houston. This is the rate necessary to maintain the overall value observed for the density of matter in space. This value is far too small for us to be able to measure on earth, or even for us to be able to hope to measure on earth in the foreseeable future. So any deviation that the steady state theory would require from the "law" of conservation of matter and energy is certainly too small to be ruled out on observational grounds.

For many years a debate raged between proponents of the big bang theories and proponents (largely the inventors) of the steady state theory. The evidence that came in—optical studies of deviations from Hubble's law in distant galaxies, or radio source counts—consistently seemed to favor big bang cosmologies over the steady state theory. But it could be fairly pointed out that none of this evidence against the steady state theory was conclusive, and in several cases either alternative explanations for the data were conceivable or the steady state theory could be modified, sometimes extensively, to be consistent with the recent discoveries. Since the time of the discovery of quasars, many astronomers have felt that the distribution of quasars—all far away from us in space, which corresponds to their having been more numerous at an earlier time—is strong evidence against the steady state cosmology. In the next section we shall discuss still stronger evidence against the steady state theory.

25.4 THE PRIMORDIAL BACKGROUND RADIATION

In 1965, a discovery of the greatest importance was made: radiation was discovered that is most readily explained as a remnant of the big bang itself.

That much is easy to state, and if we have indeed discovered radiation from the big bang itself then clearly the steady state theory is discredited. But the explanation of just how the observed radiation is interpreted to come from the big bang is fairly technical and requires an understanding of abstract concepts. However, the sheer importance of the discovery warrants our taking the time to consider this matter thoroughly.

The original discovery was made by Arno A. Penzias and Robert W. Wilson of the Bell Telephone Laboratories in New Jersey. They were testing a radio telescope and receiver system (Fig. 25–6) to try to track down all possible sources of static. The discovery, in this way, thus parallels Jansky's discovery of radio astronomy itself.

Penzias and Wilson were observing at a wavelength of 7 cm in the radio spectrum. After they had subtracted the contributions of all known sources to the static, they were left with a residual signal that they could not explain.

Actually, the Princeton group was not the first to predict that such radiation might be present. Many years earlier, similar predictions had been made, but this work was not referred to in the first 1965 papers and did not become generally known until later.

At the same time, a group at Princeton University had made the prediction that there should be radiation from the big bang that corresponded to a certain signal strength at 7 cm. These astronomers were, in fact, in the process of building their own receiver to observe at another radio wavelength. The findings of the observers from Bell Labs confirmed their prediction.

The basic prediction was that there should be radiation permeating the entire universe, and that this radiation would follow the spectrum of a black body at a certain, very low temperature (Fig. 25–7).

Figure 25–6 The large horn-shaped antenna at the Bell Telephone Laboratories' space communication station in Holmdel, New Jersey. 3° background radiation was discovered with this instrument.

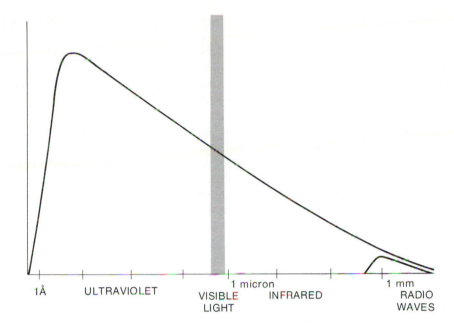

Figure 25-7 Planck curves for black bodies at different temperatures. Radiation from a 3° black body (the lower curve) peaks at very long wavelengths. On the other hand, radiation from a very hot black body (the upper curve) peaks at very short wavelengths.

In addition to the black-body nature of the radiation spectrum, the fact that the radiation was observed to be highly isotropic was other evidence that it came from the big bang. Because the big bang took place simultaneously everywhere in the universe, or even "was" the universe, we would expect that radiation from it would fill the whole universe. It would thus have the property of being isotropic to a very high degree—that is, it would be the same in any direction that we observe. It was generated all through the universe at the same time, so its remnant must seem to come from all around us now.

The leading models of the big bang consider a hot big bang, with temperatures of billions of degrees in the fractions of a second following the beginning of time. By a temperature, we mean that the amounts of radiation at different wavelengths are distributed according to the Plank curve (Section 2.5).

In the millennia right after the big bang, the universe was opaque. Photons did not travel very far before they were absorbed. But gradually the universe cooled, and after about a million years, when the universe reached 3000 K, the hydrogen ions suddenly combined with electrons to become hydrogen atoms. With this new situation, there was less continuous opacity, since hydrogen has mainly a spectrum of lines. Thus the universe suddenly became transparent. Since the matter would no longer continually absorb and reradiate all the radiation, we are left with the remnant of this original radiation, which is traveling through space forever. As the universe continues to expand, the remnant radiation retains the shape of a Planck curve but the curve corresponds to cooler and cooler temperatures. Planck curves corresponding to cooler temperatures have lower intensities of radiation and have the peaks of their radiation shifted toward the red. Because of this, the sun, at 6000 K, peaks in the yellow-green; a cool star, at 3000 K, peaks in the infrared; and the universe, much cooler yet, peaks at still longer wavelengths, equivalent to the very shortest wavelength end of the radio region of the spectrum. In a moment, we shall see just how cool the universe is.

Now, we recall that Penzias and Wilson measured a particular flux at 7 cm. This flux corresponds to what would be emitted by a black body at a temperature of only 3 K. Thus we speak of the universe's *3° background radiation,* or, sometimes, the *background radiation,* or the *primordial background radiation.** Following the original measurements at Bell Labs and at 3 cm by the Princeton group, various other groups measured values for the background radiation at different wavelengths (Fig. 25–8). Unfortunately, though, all of these measurements were obtained at wavelengths longer than that of the peak, so this left researchers with just the right-hand half of the curve defined. All the points corresponded to a temperature of approximately 3°, but still, a black-body curve does have a peak. It would have been more satisfying if some of the points measured lay on the left side of the peak graphed in Figure 25–8, or at least close enough to the peak on the right side to show that the curve in fact turns over. Points lying on the left side of the peak would prove conclusively that the radiation followed a black-body curve and was not caused by some other mechanism that could produce a straight line, or some other form that happened to mimic a 3° black-body curve in the centimeter region of the spectrum.

Though it would have been desirable to measure a few points on the short wavelength side of the black-body curve, astronomers were faced with a formidable opponent that frustrated their attempts to measure these points: the earth's atmosphere. Our atmosphere absorbs radiation of wavelengths less than about 6 mm. This is a severe limitation as the turnover in the 3° black-body curve occurs at about 2 mm, and very little radiation gets through at that wavelength.

---

*Under the new temperature notation now in use, this would be called 3 K radiation, but we shall join the astronomical community in continuing to call it 3° radiation.

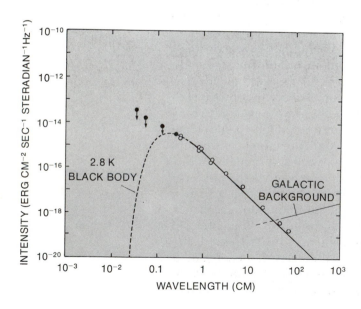

Figure 25–8 Direct measurements (dots) and upper limits from observations of the cyanogen radical (arrows) strongly indicated that the background radiation exists and is at a temperature of about 3°.

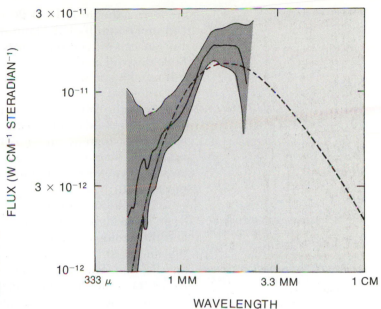

Figure 25–9 Infrared measurements from a balloon provided data that showed conclusively that the background radiation peaked and corresponded to a temperature of about 3°. The infrared data are shown with the heavy solid line, and the shaded area represents the values possible within the uncertainty of the data.

For some years, we had only an indirect method of showing that the black-body curve indeed peaked at about 2 mm. But this method did not give an accurate measure of the temperature.

In 1975, infrared observations were finally made that seem to have proved unequivocally that the radiation follows a black-body curve. Groups in the United States and England launched instruments in balloons that traveled high enough to get above most of the earth's atmosphere. The astronomers were able to measure the short wavelength side of the black-body curve down to 0.6 mm, and found that their measurements agreed with a black-body curve of 3° (Fig. 25–9).

So at present almost all astronomers consider it settled that radiation has been detected that could only have been produced in a big bang. Accepting this clearly rules out the steady state theory as a viable cosmology, and means that some version of the big bang theory must hold. It doesn't settle the question of whether the universe will expand forever, or eventually contract, or oscillate.

The importance of the discovery of the background radiation cannot be overstressed. Dennis Sciama, the British cosmologist, put it succinctly by saying that up to 1965 we carried out all our cosmological calculations knowing just one fact: that the universe expanded according to Hubble's law. After 1965, he said, we had a second fact: the existence of the background radiation. That may be a bit oversimplified, but it is essentially true.

25.5 THE CREATION OF THE ELEMENTS (*Optional*)

How did the tremendous explosion we call the big bang result in the universe we now know, with galaxies and stars and planets and people and flowers? Obviously, many complex stages of formation have taken place, and what was torn asunder at the beginning of time has now taken the form of an organized system.

We can trace back the Hubble law by following the expansion backward in time (see Fig. 25–3) and calculating how long ago all the matter we see would have been collapsed together. This calculation leads to the idea that the big bang took place some ten to twenty billion years ago—the best modern estimate is 13 to 20 billion. It is difficult to comprehend that we can meaningfully talk about the first few **seconds** of that time so long ago. But we can indeed set up sets of equations that satisfy the physical laws we have derived, and make computer simulations and calculations that we think tell us a lot about what happened in the first few moments after the origin of the universe.

According to these calculations, as the big bang took place, the universe was compressed to a fantastically high density. At first the temperature was trillions of degrees, and only the simplest kinds of matter, such as protons, neutrons, electrons, and neutrinos, were present. Perhaps their antiparticles—antiprotons, antineutrons, positrons, and antineutrinos—were present as well.

After about a hundred seconds, the temperature dropped to a billion degrees (which is low enough for a deuterium nucleus to hold together), and the protons and neutrons began to combine into heavier assemblages, the nuclei of the heavier isotopes of hydrogen and elements like helium and lithium. The study of the formation of the elements is called *nucleosynthesis*, or sometimes *nucleogenesis*.

The first nuclear amalgam to form was simply a proton and neutron together. We call this a *deuteron*; it is the nucleus of deuterium, an isotope of hydrogen (see Fig. 6–5). Then two protons and a neutron could combine to form the nucleus of a helium isotope, and then another neutron could join to form ordinary helium. Within minutes, the temperature dropped to 100 million degrees, too low for nuclear reactions to continue. Nucleosynthesis stopped, with 25 or 30 per cent of the mass of the universe in the form of helium.

Some of the earliest quantitative work on nucleosynthesis in the big bang was described in an article published under the names of Ralph Alpher, Hans Bethe, and George Gamow in 1948. Actually Alpher and Gamow did the work, and just for fun included Bethe's name in the list of authors so that the names would sound like the first three letters of the Greek alphabet. These letters seem particularly appropriate for an article about the beginning of the universe.

To test this theory, we would like to study the distribution of helium around the universe, to see if it tends to be approximately this percentage everywhere. But helium is very difficult to observe; it has few convenient spectral lines for us to study, and those are usually observable only in hot stars and gaseous nebulae, where a lot of energy is available to excite them. Furthermore, even when we can observe helium in stars we are observing only the surface layers, which do not necessarily have the same abundances of the elements as the interiors of the stars. So even though "the helium problem" has been extensively studied, one cannot as yet conclusively say that the helium is uniformly distributed around the universe, though the evidence tends to support that conclusion. Another difficulty is that some helium was surely formed in the interiors of stars.

Did all the elements form in the big bang? Only the lightest elements could be formed in the big bang. The nuclear state of mass 5 and again the state of mass 8 are not stable and don't hold together long enough to form still heavier nuclei by the addition of another proton. In a stellar interior, processes such as the triple-alpha process (Section 6.4) can get past these mass gaps. So only hydrogen (and deuterium) and helium (and traces of one isotope of lithium and one of boron) would be formed in the big bang.

The heavier elements must, then, have been formed at times after the big bang. The theory of nucleosynthesis shows that these heavier elements are built up in the interiors of ordinary stars, in the explosions of supernovae, and possible in explosions of more massive gaseous clouds.

Thus heavy elements in our bodies do not date back to the earliest part of time. They were formed more recently, in the interiors of stars or in supernovae in our own galaxy.

25.6 THE FUTURE OF THE UNIVERSE

How can we foretell how the universe will evolve in the distant future? We know that for the present it is expanding, but will that always continue? We have seen that the steady state theory now seems discredited, so let us discuss the alternatives that are predicted by different versions of big bang cosmologies.

Basically, the question is whether there is enough gravity in the universe to overcome the expansion (Fig. 25–2). If gravity is strong enough, then the expansion will gradually stop, and a contraction will begin. If gravity is not strong enough, then the rate of expansion might slow, but the universe would continue to expand forever, just as a rocket sent up from Cape Canaveral will never fall back to earth if it is launched with a high enough velocity.

To assess the amount of gravity, we must determine the average mass in a given volume of space, that is, the density. It might seem that to find the mass in a given volume we need only count up the objects in that volume: one hundred billion stars plus umpteen billion atoms of hydrogen plus so much interstellar dust, etc. But there are severe limitations to this method, for there are many kinds of mass that are invisible to us. How much matter is in black holes, for example? Until a few years ago, we couldn't measure the molecular hydrogen in space, and until a few decades ago we couldn't measure the atomic hydrogen in space either, although their abundances are now approximately known to us.

Figure 25–10 A theoretical model, worked out by Robert V. Wagoner of Stanford, that shows the changing relative abundances of the light elements in the first minutes after the big bang. Time is shown on the top axis and the corresponding temperature is shown on the bottom axis. The calculations shown here are for a certain assumed density at some early time; if a different density was present, then the curves would be different.

Moreover, we have seen in Section 23.2 that there are lines of evidence from dynamic studies of clusters of galaxies that indicate that much of the mass may be "missing," that is, invisible to us. We must find a method that doesn't depend on whether mass is visible or invisible.

One method that has recently attracted increasing interest concerns itself with the abundances of the light elements. Since these light elements were formed soon after the big bang (Fig. 25–10), they may tell us about conditions at the time of their formation.

The light elements include hydrogen, helium, lithium, beryllium, and boron (see Appendix 4).

But somehow we must distinguish between the amount of these elements that was formed in the big bang, and the amount that has been subsequently formed in stars. This consideration complicates the calculations for helium, for example, because the helium formed later in stellar interiors and then spewed out in supernovae has been added to the primordial helium formed in the big bang.

However, one isotope of hydrogen, deuterium, is free of this complication. We do not think that any is formed in stars, so all the deuterium now in existence was formed at the time of the big bang. Deuterium, "heavy hydrogen," contains one proton and one neutron in its nucleus instead of just the single proton of ordinary hydrogen. We are most familiar with deuterium as a constituent of "heavy water," HDO, which is used in atomic reactors. Deuterium is now being studied increasingly as an important constituent of the fusion process. We hope that we will be able to harness nuclear fusion, which would allow us to transform mass into energy here on earth to provide pollution-free new energy sources.

Deuterium has two properties that make it an important probe of conditions that existed in the first fifteen minutes after the big bang, when the deuterium was formed. First, the amount of deuterium, relative to the amount of hydrogen, that is formed is particularly sensitive to the density of matter at the time the deuterium formed; that is, a slight variation in the primordial density makes a much larger change in the deuterium abundance than it does in the abundance of other isotopes or elements. Second, the nuclear processes that we think take place in stars all destroy deuterium, rather than create it. So the amount of deuterium we measure today is not greater than the amount that would have existed shortly after the big bang. (Indeed, we can estimate that the factor by which the amount of deuterium is depleted is two to four.)

Why is the ratio of deuterium to (ordinary) hydrogen sensitive to density? Deuterium very easily combines with an additional neutron, briefly forming another hydrogen isotope (tritium), and the new neutron quickly decays into a proton, leaving us with an isotope of helium. If the universe was very dense in its first few minutes, then it was easy for the deuterium to meet up with neutrons, and almost all the deuterium "cooked" into helium. If, on the other hand, the density of the universe was low, then most of the deuterium that was formed still survives. The theoretical calculations can be used to indicate the density that would have been present soon after the big bang, if the ratio of deuterium to hydrogen can be determined observationally.

There have been a number of attempts to detect deuterium in interstellar space, but it was not until 1972 that any succeeded. My colleagues Diego A. Cesarsky of the Observatoire de Paris and Alan T. Moffet of Caltech, and I, observed a faint absorption feature that is probably an absorption line caused by the spin-flip transition of deuterium (see Section 21.3).

We used the 40-meter (130-foot) radio telescope at Caltech's Owens Valley Radio Observatory (Fig. 25–11). The line is so weak that we would like to have more definitive results.

At the same time as our original observations, a group of astronomers from the Bell Telephone Laboratories was discovering deuterium as one of the constituents of a "deuterated" molecule, DCN, in the cloud of gas and dust in the constellation Orion. The fact that deuterium is present is not ambiguous, for the DCN emission line is clearly present, but the interpretation of their data to give an abundance for deuterium is very complex because the processes of interstellar molecular formation are not well understood. The astronomers from Bell Labs have since mapped the strength of this DCN emission line in several molecular clouds in our galaxy.

Figure 25–11 The 40-m telescope at the Owens Valley Radio Observatory.

Since these discoveries, some of the Lyman lines of deuterium have been detected in absorption by the Princeton experiment aboard the Copernicus Orbiting Astronomical Observatory. Their observations not only show the presence of a deuterium absorption line more clearly than do the spin-flip observations of my own group but also can be interpreted with fewer complications than the molecular observations. The Copernicus satellite can study, however, only the nearest few hundred parsecs to us. The latest Copernicus results include stars up to 5 per cent of the distance to the galactic center, and establish a fraction of deuterium to hydrogen of 14 parts per million.

All the studies of deuterium thus far agree that the amount of deuterium is such that there is not, and was not, enough mass to reverse the expansion of the universe. We say that the universe is *open*. (Alternatively, if the universe were oscillating or even to fall back on itself only once, it would be *closed*.) Recent new measurements of galaxies by Sandage and Tammann at Palomar now agree with this prediction that the universe is open. Sandage and Tammann measured the *deceleration parameter* (see Fig. 25–2), a measure of how much the expansion of the universe is slowing down at the outer limits of our ability to observe, taking into account that the galaxies they are observing are being studied far back in time.

A major uncertainty is how galaxies evolve with time. Were they brighter or fainter long ago than they are now?

It is particularly interesting that all these measurements tend to suggest the same result, that the universe will not stop its expansion, and that the galaxies will continue to fly away from each other. And it is a good example of modern astronomical research to note that these results were obtained from a wide variety of methods of observation: optical observation with large telescopes on earth, radio observations using a variety of techniques, and ultraviolet observations using telescopes in space. We are making use of all the methods we know to try to foretell our future.

SUMMARY AND OUTLINE OF CHAPTER 25

Aims: To study the origin and evolution of the universe as a whole, to discuss the origin of the elements, and to consider observations that indicate the future of the universe

Olbers's paradox and its solutions (Section 25.1)
Big bang theory (Section 25.2)
    Cosmological principle: universe homogeneous and isotropic
    Hubble's constant and the "age" of the universe – 13 to 20 billion years

Steady state theory (Section 25.3)

Perfect cosmological principle: universe homogeneous, isotropic, and unchanging in time

Continuous creation of matter

Demise of the steady state theory with discovery of quasars and 3° black-body radiation

Primordial background radiation (Section 25.4)

Provides strong support for the big bang theory

Evidence for 3° black-body radiation

Direct radio measurements on one side of the peak of the black-body curve

New infrared observations provide convincing evidence for the radiation corresponding to a 3° black body because they confirm the peaked nature of the radiation

Creation of the elements (Section 25.5)

Origin of the light elements in the first minutes after the big bang

Heavier elements formed in the interiors of stars

Future of the universe (Section 25.6)

Evidence from the study of deuterium

Deuterium-to-hydrogen ratio depends sensitively on cosmic density

All deuterium observed was formed in the first minutes after the big bang

Current evidence is that universe is open, i.e., expansion will not stop

QUESTIONS

1. Hindsight has allowed solutions to be found to Olbers's paradox. For example, knowing that the universe is expanding, we can come up with a solution. However, based on the reasoning in this chapter, do you think it is possible that scientists might have used Olbers's paradox to reach the conclusion that the universe is expanding before it was determined observationally? Explain.

2. We say in the chapter that the universe is homogeneous on a large scale. What is the largest scale on which the universe does not seem to appear homogeneous?

3. List the evidence favoring each of the following theories: (a) the big bang, and (b) the steady state theory.

4. What is "perfect" about the perfect cosmological principle?

5. For a Hubble constant of 55 km/sec/Mpc, show how you calculate the Hubble time, the age of the universe ignoring the effect of gravity. (Hint: Take $1/H_0$, and simplify units so that only units of time are left.)

6. In actuality, if current interpretations are correct, the 3° background radiation is only indirectly the remnant of the big bang, but is directly the remnant of an "event" in the early universe. What event was that?

7. Apply Wien's displacement law to the solar photospheric temperature and spectral peak in order to show where the spectrum of a 3° black body peaks.

8. Why must we resort to indirect methods to find out if the universe is open or closed?

9. Why is determining the abundance of deuterium so important?

10. What are two pieces of evidence that suggest that the universe is open?

TOPICS FOR DISCUSSION

1. Which is more appealing to you: the cosmological principle or the perfect cosmological principle? Discuss why we should adopt one or the other as the basis of our cosmological theory.

2. What effect would finding out whether the universe will expand forever, will begin to contract into a final black hole, or will contract and then oscillate with cycles extending into infinite time, have on the conduct of one's life or thoughts?

26

Astronomy Now

We have seen in the preceding chapters that astronomy encompasses not only a wide range of subjects but also a wide range of techniques. No longer can a single astronomer hope to be expert in all phases of the science. We have consistently seen how new discoveries often open more questions than they answer. The advance of science seems to leave us farther and farther behind!

Up to this point, this book has been organized by type and size of object, from stars (and their planets) at the beginning up to the universe as a whole at the end. Let us take a brief tour of topics that are now under active consideration, organized, for a change, by wavelength in the electromagnetic spectrum. The discussion should give us an idea of new discoveries that are expected in the next few years.

26.1 CURRENT PLANS

The shortest wavelengths in the electromagnetic spectrum represent gamma-rays and x-rays. Of course, we are dependent on satellites above the earth's atmosphere to study these radiations efficiently. Study of this region of the spectrum, in which each individual photon has a high energy, is a NASA priority for the next few years, and we can confidently look forward to discoveries in this field that will force us to revamp our ideas of many processes in astronomy. The High-Energy Astronomy Observatories provide a capability for research in this region of the spectrum that is a significant advance. X-ray data will be used for further mapping of the x-ray sky and for further detailed studies of particularly interesting or suspicious objects. The Vela series of satellites, which mostly look for gamma-ray bursts in order to detect explosions of hydrogen bombs, are equally ready to report gamma-ray bursts from outer space. In the meantime, theoreticians continue to make progress in understanding the high-

Two of the first VLA antennas to be completed.

energy processes that dominate emission in this region of the spectrum. And most people want to know whether we find more (or any) black holes. We should also see whether the current tendency to invoke the presence of black holes to explain puzzling phenomena is a panacea or a fad.

In the ultraviolet, the Copernicus satellite continues to study the stars and interstellar space. Two of its recent observations, of deuterium in interstellar space and of the spectrum of a white dwarf, illustrate the unique and important research that can be carried out with this Orbiting Astronomical Observatory. The European TD-1 continues its ultraviolet studies of the stars with high spectral resolution.

The Space Telescope, currently scheduled for 1983, should expand our horizons considerably.

In visible light, a number of new giant telescopes are just beginning their operation or are about to be completed. Our ability to observe the optical sky will be increased manyfold. Electronic devices to receive and record the photons that are focused by the huge mirrors increase the capability of even the older telescopes. But we are still faced with a limitation in the size of telescopes with single mirrors. The completion of the multi-mirror telescopes now under construction will tell us whether this kind of instrument will allow us to study fainter objects than we have previously been able to study. If we can, we will be able to observe regions more distant from us, which could not help but improve the observations on which cosmological theories are based. Also in the optical spectrum, we can expect improvements in our knowledge of the outer planets. The images that will be sent to us from Saturn in 1979 and possibly images of Uranus from a subsequent spacecraft should increase manyfold our understanding of the planets, their atmospheres, and their origins.

The infrared is a burgeoning area of research. We are seeing stars in formation and so are getting to understand better the processes of stellar birth. We can study dust wherever it is in space—in the solar system or between the stars. New telescopes devoted exclusively to infrared studies are becoming operational, and electronic instrumentation used to study this part of the spectrum is improving rapidly.

In radio astronomy, there will continue to be tremendous strides in making high-resolution pictures of astronomical objects. The VLA (Very Large Array) will shortly become operational and will enable large-scale mapping of the sky in a manner never before possible (Fig. 23–21).

This is but the beginning of a long list of topics of active research. Different astronomers put high priority on very different areas. I suppose that we are fortunate that we don't all want to work on the same thing at the same time! There is surely no shortage of research topics available for astronomers.

Perhaps the most significant gains will come when observations from several different parts of the spectrum are put together. The sun, for example, is under careful scrutiny from wavelengths down below 0.01 angstroms up to wavelengths above 3 meters, a factor of more than a trillion (10^{12}).

In a textbook of astronomy, it is relatively easy to get across a sense of progress in observational results compared to transmitting a sense of the importance of theoreticians in the whole enterprise. The computer, especially, has enhanced the ability of theoreticians to make detailed models of what is occurring in a stellar atmosphere, in a stellar interior, or in a disk of material around a black hole. But computers aren't everything. The ability of a scientist to sit quietly with a pencil and paper, confidently calculating how bodies 10^{26} times the astronomer's size will act, remains at the heart of modern research.

"No one has yet programmed a computer to be of two minds about a hard problem, or to burst out laughing. . . ."
Lewis Thomas,
THE LIVES OF A CELL
(New York: Viking Press, 1974)

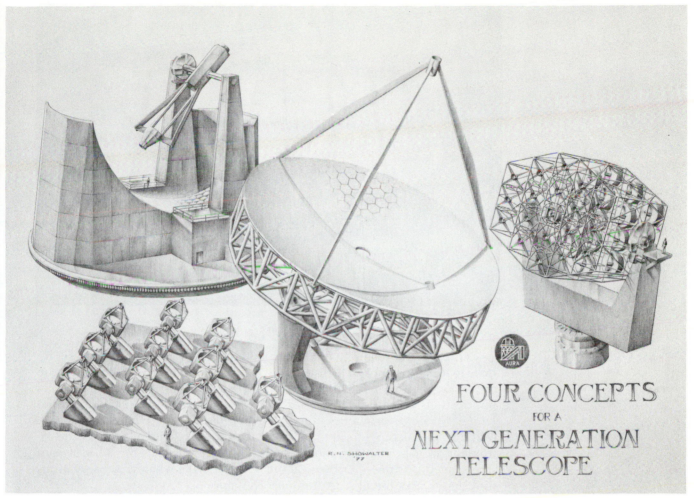

FOUR CONCEPTS FOR A NEXT GENERATION TELESCOPE

R. N. SHOWALTER '77

Figure 26-1 A study group at the Kitt Peak National Observatory is working on the Next Generation Telescope (NGT). Whether it will be a mosaic of small mirrors, a collection of medium-sized telescopes, or something else entirely must be decided. The equivalent aperture will be about 25 meters (1000 inches).

26.2   THE SCIENTIFIC ENTERPRISE (OPTIONAL)

Science of all kinds, including astronomy, grows increasingly dependent on large instrumentation. It becomes more and more difficult—and is really nigh unto impossible—to carry out significant research as a lone individual. Even theoreticians are often dependent on the availability of large computers, and their calculations may be limited in accuracy mainly by the amount of computer time that is available to them.

All this points to the importance of funding to provide the equipment and the financial backing for the scientists' time and effort. Although many smaller sources of funds exist, the role of government funding has increased to the point where it is clearly the dominant source. A government supports scientific research for a variety of reasons. Its interest in supporting science for the sake of knowledge itself is one of them, just as governments often support museums of art, or opera companies.

But this is surely not the major reason why governments support science. Usually there is a desire for practical results. Astronomers do not ordinarily undertake their investigations for practical ends, but over the past centuries a number of basic ideas and processes of practical importance have been discovered in an astronomical context. We can confidently expect that today's astronomical research will pay off similarly in the long run.

Figure 26–2 The control room of the 100-meter radio telescope near Bonn of the Max Planck Institute for Radio Astronomy. This instrument is the largest fully-steerable radio telescope in the world.

Often, astronomers benefit from apparatus or projects that were set up mainly for other purposes. The whole Apollo program to land astronauts on the moon by the end of the decade of the 1960's had important political implications and implications for national pride and goals; these led the United States government to spend on the project sums previously unparallelled in the history of scientific support. But for whatever reasons the astronauts were sent to the moon, or for whatever reasons Skylab was sent into orbit, the astronomers were at the ready to make whatever observations they could with the new techniques available.

In the United States, the National Science Foundation is the organization that is the primary source of support for basic research in astronomy. The NSF might spend 30 million dollars or so on astronomy in a given year, a tiny fraction of a per cent of the national budget.

Astronomy has also certainly benefited by the interest of NASA in astronomical problems. The cost of space experiments is many times the cost of ground-based observing, so one must be prepared to think on another scale of expenses when one considers using a satellite to make observations. Of course, most of the costs of experimenting in space are not really from the astronomical experiments but are rather engineering costs. And the tremendous amounts of money that are spent do not just go up into outer space; the money is spent here on earth, largely to pay contractors and subcontractors to build, test, and maintain the equipment.

26.3 GRANTS

How does a scientist, say a professor at a university, go about getting a grant from the National Science Foundation? First, he or she would have to

Figure 26–3 A mock proposal, including most of the kinds of descriptions and budgeting stages that a real proposal would have. Reprinted from the *Journal of Irreproducible Results* and *Theology Today.*

HOPKINS OBSERVATORY
WILLIAMS COLLEGE
WILLIAMSTOWN, MASS. 01267
To: Universal Creation Foundation
REQUEST FOR SUPPLEMENT TO U.C.F. GRANT
#000-00-00000-001
"CREATION OF THE UNIVERSE"
This report is intended only for the
internal uses of the contractor.
Period: Present to Last Judgment
Principal Creator: Creator
Proposal Writers and Contract Monitors:
Jay M. Pasachoff and Spencer R. Weart

BACKGROUND

Under a previous grant (U.C.F. Grant #000-00-00000-001), the Universe was created. It was expected that this project would have lasting benefits and considerable spinoffs, and this has indeed been the case. Darkness and light, good and evil, and Swiss Army knives were only a few of the useful concepts developed in the course of the Creation. It was estimated that the project would be completed within four days (not including a mandated Day of Rest, with full pay), and the 50% overrun on this estimate is entirely reasonable, given the unusual difficulties encountered. Infinite funding for this project was requested from the Foundation and granted. Unfortunately, this has not proved sufficient. Certain faults in the original creation have become apparent, which it will be necessary to correct by means of miracles. Let it not be said, however, that we are merely correcting past errors; the final state of the Universe, if this supplemental request is granted, will have many useful features not included in the original proposal.

PROGRESS TO DATE

Interim progress reports have already been submitted ("The Bible," "The Koran," "The Handbook of Chemistry and Physics," etc.). The millennial report is currently in preparation, and a variety of publishers for the text (tentatively entitled, "Oh, Genesis!") will be created. The Gideon Society has applied for the distribution rights. Full credit will be given to the Foundation.

Materials for the Universe and for the Creation of Man were created out of the Void at no charge to the grant. A substantial savings was generated when it was found that materials for Woman could be created out of Man, since the establishment of Anti-Vivisection Societies was held until Phase Three. Given the limitations of current eschatological technology, it can scarcely be denied that the Contractor has done His work at a most reasonable price.

SUPPLEMENT

We cannot overlook a certain tone of dissatisfaction with the Creation which has been expressed by the Foundation, not to mention by certain of the Created. Let us state outright that this was to be expected, in view of the completely unprecedented nature of the project. The need for a supplement is to be ascribed solely to inflation (not to be confused with expansion of the Universe, which was anticipated). Union requests for the accrual of Days of Rest at the rate of one additional Day per week per millenium ($Dw^{-1}m^{-1}$) must also be met. Concerning the problem of Sin, we can assure you that extensive experimentation is under way. Considerable experience is being accumulated and we expect a breakthrough before long. When we are satiated with Sin, we shall go on to consider Universal Peace.

We cannot deny — in view of the cleverness of the Foundation's auditors — that the bulk of the supplemental funds will go to pay off old bills. Nevertheless we do not anticipate the need for future budget requests, barring unforeseen circumstances. If this project is continued successfully, additional Universe — anti-Universe pairs can be created without increasing the baryon number, and we would keep them out of the light cone of the Original. By the simple grant of an additional Infinity of funds (and note that this proposal is merely for Aleph Null), the officers of the Foundation will be able to present their Board of Directors with the accomplished Creation of one or more successful Universes, instead of the current incomplete one.

We will attempt to minimize the additional delays that may temporarily exist during the changeover from fossil fuels to fusion for some minor locations in the Universe. For the time being, elements with odd masses (hydrogen, lithium, etc.) will be created only on odd days of the month and those with even masses (helium, beryllium, etc.) on the even days, except, of course, for Sundays. The 31st of each month will be devoted to creation of trans-uranic elements. The Universe has been depleted of deuterium; new creation of this will take place only on February 29th in leap years.

PROSPECTIVE BUDGET

Remedial miracles on fish of the sea	∞
Remedial miracles on fowl of the air	∞
Creeping things that creep upon the earth, etc.	∞
Hydrogen	n/c (created)
Heavier elements	n/c (nucleosynthesis)
(Note: The carbon will be reclaimed and ecologically recycled)	
Mountains (Sinai, Ararat, Palomar)	∞
Extra quasars, neutron stars	∞
Black holes (no-return containers)	∞
Miscellaneous, secretarial, office supplies, etc.	∞
Telephone installation (Princess model, white, one-time charge, tax included)	$16.50

SALARIES

Creator (1/4 time)	at His own expense
Archangels	
Gabriel	1 trumpet (Phase 5)
Beelzebub	misc. extra brimstone (low sulfur)
Others	assorted halos
Prophets	
Moses	stone tablets (to replace breakage)
Geniuses	finite

N.B. Due to the Foundation's regulations and changes in Exchange Rates, we have not yet been able to reimburse Euclid (drachmae), Leonardo (lire), Newton (pounds sterling), Descartes (francs), or Catfish Hunter (dollars). Future geniuses will be remunerated indirectly via the Alfred Nobel Foundation.

Graduate students (2 at 2/5 time)	reflected glory

MONITORING EQUIPMENT & MISC.

1 5-meter telescope (maintenance)	finite
Misc. other instruments, particle accelerators, etc.	large but finite
Travel to meetings	∞
Pollution control equipment	+40%
Total	$\infty + 40\% = \infty$
Overhead (51.97%)	∞
Total funds requested	∞

Starting date requested:
Immediate. Pending receipt of supplemental funds, layoffs are anticipated to reach the 19.5% level in the ranks of angels this quarter.

Figure 26–3 *See opposite page for legend*

have a clear idea of the project under consideration, which would normally require some preliminary research. This preliminary research must be supported in some other way, perhaps by funds from the university. Eventually the project would be developed to a point where a proposal for a grant could be written (Fig. 26–3).

The proposal might be some two dozen pages in length. It would contain a clear exposition of the proposed project, its importance, and why it should be undertaken in this specific manner. The scientist's ability to undertake the project would have to be justified; his or her research ability as demonstrated by prior published papers would be an important factor in assessing this. If staff or students will be included in the project, their tasks and abilities must be explained. The budget is an important part of the proposal.

At the astronomy section of the NSF, the program managers would send out copies of the proposal to perhaps half a dozen scientists across the country who are knowledgeable in the specific field of research under discussion. These scientists are asked to act as anonymous *referees* of the proposal in a process known as *peer review*.

The program managers try to assess the referees' reports and assign ratings and priorities to each proposal. Then they try to divide their limited funds appropriately. There are always many more acceptable proposals than can be funded.

The process of review can take six months or more, so prospective proposers have to decide long in advance what they want to do. One day, successful proposers receive in the mail copies of letters announcing the awarding of grants. The projects can then get under way.

26.4 THE NEW ASTRONOMY

In the decade of the seventies, astronomy seems to be maintaining its momentum as one of the most active sciences. It is exploding with new results.

The discussions in this book have necessarily been limited to a fraction of the topics with which contemporary astronomers concern themselves. Yet it is to be hoped that the reader has become aware of topics and ongoing investigations that will provide a basis for the astronomy of the 1980's and afterward.

Close to earth, the planets will continue to be explored, and we hope to gain a fundamental understanding of all the planets in the solar system. We are better prepared now to study comets than we have ever been, and the reappearance of Halley's comet in 1985 or of another bright comet will lead to much excellent data.

The sun remains the center of active investigation, not only for its fundamental properties as the nearest example to earth of a star but also because of its effect on our environment in space. From neutrinos and particles in the solar wind to the entire electromagnetic spectrum, we can study the sun in more ways than we can study any other object in the universe.

The stars and galaxies are also under scrutiny across the electromagnetic spectrum. They are all much fainter than the sun, of course, and so are correspondingly more difficult to study. But we seem to be approaching the

Figure 26–4 A close-up view of the Martian satellite Phobos, taken by the Viking 1 orbiter from a distance of only 120 km. The picture covers an area 3 by 3.5 km. Craters and grooves can be seen.

point where we begin to understand the fundamental processes that govern stars and galaxies (although I would hate to read these words from the perspective of a hundred years hence, when it will be seen that we really now know nothing at all). Will quasars turn out to be something completely new?

We are also making progress on our understanding of cosmology, a subject that has been at the center of human thought for over two thousand years but about which we have a lot to learn. The detection of the primordial radiation has brought us tidings of the big bang itself, and we hope that within a few years careful measurements of this radiation in different directions will tell us about the inhomogeneous nature of the universe in the first few instants of time some 13 billion years ago.

Astronomy has come a long way since early humans first noticed the objects in the sky and their motions. The pace of progress is dramatically increasing, and it is exciting to see the discoveries made in such quick succession. Still, we have at least as far to go.

Figure 26–5 The spiral galaxy M81, photographed with the new 4-meter telescope at the Kitt Peak National Observatory.

SUMMARY AND OUTLINE OF CHAPTER 26

Aims: To summarize some of the future developments that we may expect in astronomical research, and to discuss the funding and publication of astronomical research

Some next steps (Sections 26.1 and 26.4)
The scientific enterprise (Section 26.2)
    Importance of funding
    National Science Foundation (NSF) oversees most funding for astronomy in the U.S.
Grants (Section 26.3)
    How a proposal is prepared and submitted
    Peer review

TOPICS FOR DISCUSSION

1. What are the astronomical projects **you** think are the most important for the next decade?

2. What is the value of astronomy to **you**?

Appendices

Appendix 1
Greek Alphabet

Upper Case	Lower Case	Pronunciation
A	α	alpha
B	β	beta
Γ	γ	gamma
Δ	δ	delta
E	ϵ	epsilon
Z	ζ	zeta
H	η	eta
Θ	θ	theta
I	ι	iota
K	κ	kappa
Λ	λ	lambda
M	μ	mu
N	ν	nu
Ξ	ξ	xi
O	o	omicron
Π	π	pi
P	ρ	rho
Σ	σ	sigma
T	τ	tau
Υ	υ	upsilon
Φ	ϕ	phi
X	χ	chi
Ψ	ψ	psi
Ω	ω	omega

Appendix 2

Measurement Systems

Metric units

Basic units

length	meter (m)
volume	liter (l)
mass	gram (gm or g)
time	second (sec or s)

Other metric units

1 micron (μ) = 10^{-6} meter

1 Ångstrom (Å or A) = 10^{-10} meter

$= 10^{-8}$ cm

Prefixes for use with basic units of metric system

Prefix	Symbol	Power		Equivalent
tera	T	10^{12} =	1,000,000,000,000	Trillion
giga	G	10^{9} =	1,000,000,000	Billion
mega	M	10^{6} =	1,000,000	Million
kilo	k	10^{3} =	1,000	Thousand
hecto	h	10^{2} =	100	Hundred
deca	da	10^{1} =	10	Ten
– – –	–	10^{0} =	1	One
deci	d	10^{-1} =	.1	Tenth
centi	c	10^{-2} =	.01	Hundredth
milli	m	10^{-3} =	.001	Thousandth
micro	μ	10^{-6} =	.000001	Millionth
nano	n	10^{-9} =	.000000001	Billionth
pico	p	10^{-12} =	.000000000001	Trillionth
femto	f	10^{-15} =	.000000000000001	
atto	a	10^{-18} =	.000000000000000001	

Examples: 1000 meters = 1 kilometer = 1 km

10^{6} hertz = 1 megahertz = 1 MHz

10^{-3} sec = 1 millisecond = 1 msec

Some other units used in astronomy

Energy:	1 joule (J) = 10^{7} ergs
	1 electron volt (eV) = 1.60207×10^{-12} ergs
Power:	1 watt (W) = 1 joule per sec (J/sec)
Frequency:	hertz (Hz) = cycles per sec (sec^{-1})
	1 megahertz (MHz) = 10^{6} Hz

Conversion factors

1 cm = 0.3937 in	1 in = 25.400 mm
1 m = 1.0936 yd	= 2.54 cm
1 km = 0.6214 mi	1 ft = 0.3048 m
= 5/8 mi	1 yd = 0.9144 m
1 kg = 2.2046 lb	1 mi = 1.6093 km
1 gm = 0.0353 oz	= 8/5 km
	1 lb = 0.4536 kg

Appendix 3

Basic Constants

Physical constants

Speed of light*	c	$= 299\ 792\ 458$ m/sec
Constant of gravitation*	G	$= 6.672 \times 10^{-11}$ m^3/kg·sec^2
Planck's constant	h	$= 6.6262 \times 10^{-27}$ erg·sec
Boltzmann's constant	k	$= 1.3806 \times 10^{-16}$ erg/kelvin
Stefan-Boltzmann constant	σ	$= 5.66956 \times 10^{-5}$ erg/cm^2·deg^4·sec
Wien displacement constant	$\lambda_{max}T$	$= 0.289789$ cm·K $= 28.9789 \times 10^6$ Å·K
Mass of hydrogen atom	m_H	$= 1.6735 \times 10^{-24}$ gm
Mass of neutron	m_n	$= 1.6749 \times 10^{-24}$ gm
Mass of proton	m_p	$= 1.6726 \times 10^{-24}$ gm
Mass of electron	m_e	$= 9.1096 \times 10^{-28}$ gm
Rydberg's constant	R	$= 1.09677 \times 10^5$/cm

Mathematical constants

$\pi = 3.1415926536$
$e = 2.7182818285$

Astronomical constants

Astronomical unit*	1 A.U.	$= 1.495\ 978\ 70 \times 10^{11}$ m
Solar parallax*	$\pi_\odot$	$= 8.794148$ arc sec
Parsec	pc	$= 206\ 264.806$ A.U.
		$= 3.261633$ light years
		$= 3.085678 \times 10^{18}$ cm
Light year	ly	$= 9.460530 \times 10^{17}$ cm
		$= 6.324 \times 10^4$ A.U.
Tropical year (1900)* — equinox to equinox		$= 365.24219878$ ephemeris days
1 Julian century*		$= 36525$ days
1 day*		$= 86400$ sec
Sidereal year		$= 365.256366$ ephemeris days
		$= 3.155815 \times 10^7$ sec
Mass of sun*	$M_\odot$	$= 1.9891 \times 10^{33}$ gm
Radius of sun*	$R_\odot$	$= 696000$ km
Luminosity of sun	$L_\odot$	$= 3.827 \times 10^{33}$ erg/sec
Mass of earth*	$M_\oplus$	$= 5.9742 \times 10^{27}$ gm
Equatorial radius of earth*	$R_\oplus$	$= 6\ 378.140$ km
Mean distance center of earth to center of moon		$= 384\ 403$ km
Radius of moon*	R_M	$= 1\ 738$ km
Mass of moon*	M_M	$= 7.35 \times 10^{25}$ gm
Solar constant	S	$= 135.3$ mW/cm^2
Direction of galactic center (precessed for 1950)	α	$= 17^h42.4^m$
	δ	$= -28°\ 55'$

---

*Adopted as "IAU (1976) system of astronomical constants" at the General Assembly of the International Astronomical Union.

Appendix 4

Elements and Solar Abundances*

Atomic Number	Element	Name	Atomic Weight	Solar Abundance (Relative to Hydrogen = 10^{12})
1	H	hydrogen	1.01	10^{12}
2	He	helium	4.00	6.3×10^{10}
3	Li	lithium	6.94	10×10^{1}
4	Be	beryllium	9.01	1.4×10^{1}
5	B	boron	10.81	1.3×10^{2}
6	C	carbon	12.01	4.2×10^{8}
7	N	nitrogen	14.01	8.7×10^{7}
8	O	oxygen	16.00	6.9×10^{8}
9	F	fluorine	19.00	3.6×10^{4}
10	Ne	neon	20.18	3.7×10^{7}
11	Na	sodium	22.99	1.9×10^{6}
12	Mg	magnesium	24.31	4.0×10^{7}
13	Al	aluminum	26.98	3.3×10^{6}
14	Si	silicon	28.09	4.5×10^{7}
15	P	phosphorus	30.97	3.2×10^{5}
16	S	sulphur	32.06	1.6×10^{7}
17	Cl	chlorine	35.45	3.2×10^{5}
18	Ar	argon	39.95	1.0×10^{6}
19	K	potassium	39.10	1.4×10^{5}
20	Ca	calcium	40.08	2.2×10^{6}
21	Sc	scandium	44.96	1.1×10^{3}
22	Ti	titanium	47.90	1.1×10^{5}
23	V	vanadium	50.94	1.0×10^{4}
24	Cr	chromium	52.00	5.1×10^{5}
25	Mn	manganese	54.94	2.6×10^{5}
26	Fe	iron	55.85	3.2×10^{7}
27	Co	cobalt	58.93	7.9×10^{4}
28	Ni	nickel	58.71	1.9×10^{6}
29	Cu	copper	63.55	1.1×10^{4}
30	Zn	zinc	65.37	2.8×10^{4}
31	Ga	gallium	69.72	6.3×10^{2}
32	Ge	germanium	72.59	3.2×10^{3}
33	As	arsenic	74.92	
34	Se	selenium	78.96	
35	Br	bromine	79.90	
36	Kr	krypton	83.80	
37	Rb	rubidium	85.47	4.0×10^{2}
38	Sr	strontium	87.62	7.9×10^{2}
39	Y	yttrium	88.91	1.3×10^{2}
40	Zr	zirconium	91.22	5.6×10^{2}
41	Nb	niobium	92.91	7.9×10^{1}
42	Mo	molybdenum	95.94	1.4×10^{2}
43	Tc	technetium	98.91	
44	Ru	ruthenium	101.07	6.8×10^{1}
45	Rh	rhodium	102.91	2.5×10^{1}
46	Pd	palladium	106.4	3.2×10^{1}
47	Ag	silver	107.87	7.1
48	Cd	cadmium	112.40	7.1×10^{1}
49	In	indium	114.82	4.5×10^{1}
50	Sn	tin	118.69	1.0×10^{2}
51	Sb	antimony	121.75	1.0×10^{1}
52	Te	tellurium	127.60	
53	I	iodine	126.90	
54	Xe	xenon	131.30	
55	Cs	caesium	132.91	$<7.9 \times 10^{1}$

Appendix 4 continued on the opposite page

Appendix 4 *Continued*

Atomic Number	Element	Name	Atomic Weight	Solar Abundance (Relative to Hydrogen = 10^{12})
56	Ba	barium	137.34	1.2×10^2
57	La	lanthanum	138.91	1.3×10^1
58	Ce	cerium	140.12	3.5×10^1
59	Pr	praseodymium	140.91	4.6
60	Nd	neodymium	144.24	1.7×10^1
61	Pm	promethium	146	
62	Sm	samarium	150.4	5.2
63	Eu	europium	151.96	5.0
64	Gd	gadolinium	157.25	1.3×10^1
65	Tb	terbium	158.93	
66	Dy	dysprosium	162.50	1.1×10^1
67	Ho	holmium	164.93	
68	Er	erbium	167.26	5.8
69	Tm	thulium	168.93	1.8
70	Yb	ytterbium	170.04	7.9
71	Lu	lutetium	174.97	5.8
72	Hf	hafnium	178.49	6.3
73	Ta	tantalum	180.95	
74	W	tungsten	183.85	5.0×10^1
75	Re	rhenium	186.2	≤ 0.5
76	Os	osmium	190.2	5.0
77	Ir	iridium	192.2	7.1
78	Pt	platinum	195.09	5.6×10^1
79	Au	gold	196.97	5.6
80	Hg	mercury	200.59	$< 1.3 \times 10^2$
81	Tl	thallium	204.37	7.9
82	Pb	lead	207.19	8.5×10^1
83	Bi	bismuth	208.98	$< 7.9 \times 10^1$
84	Po	polonium	210	
85	At	astatine	210	
86	Rn	radon	222	
87	Fr	francium	223	
88	Ra	radium	226.03	
89	Ac	actinium	227	
90	Th	thorium	232.04	1.6
91	Pa	protactinium	230.04	
92	U	uranium	238.03	< 4.0
93	Np	neptunium	237.05	
94	Pu	plutonium	242	
95	Am	americium	242	
96	Cm	curium	245	
97	Bk	berkelium	248	
98	Cf	californium	252	
99	Es	einsteinium	253	
100	Fm	fermium	257	
101	Md	mendelevium	257	
102	No	nobelium	255	
103	Lr	lawrencium	256	
104	Rf	rutherfordium	261	
105	Ha	hahnium	262	
106		(Reported 1974)	263	

°The solar abundance for hydrogen has been arbitrarily set at 10^{12}.

Atomic weights are averages for terrestrial abundances.

Solar abundances from John E. Ross and Lawrence H. Aller, *Science*, *191*, 1223, 1976. Gaps indicate that the element has not been observed on the sun.

Appendix 5

The Planets

Appendix 5a. *Intrinsic and Rotational Properties*

Name	Equatorial Radius km	Equatorial Radius ÷Earth's	Mass ÷Earth's	Mean Density (gm/cm³)	Oblate-ness	Surface Gravity (Earth = 1)	Sidereal Rotation Period	Inclination of Equator to Orbit	Apparent Magnitude at Opposition
Mercury	2,439	0.3824	0.0553	5.44	0.0	0.378	59^d	<28°	−1.8
Venus	6,052	0.9489	0.8150	5.24	0.0	0.894	244.3^d R	177°	−4.3
Earth	6,378.140	1	1	5.497	0.0034	1	23^{h}56^{m}04.1^s	23°27′	−
Mars	3,397.2	0.5326	0.1074	3.9	0.009	0.379	24^{h}37^{m}22.6^s	23°59′	−2.01
Jupiter	71,398	11.194	317.89	1.3	0.063	2.54	9^{h}50^m to >9^{h}55^m	3°05′	−2.55
Saturn	60,000	9.41	95.17	0.7	0.098	1.07	10^{h}14^m to >10^{h}38^m	26°44′	+0.67
Uranus	27,900	4.4	14.56	1.0	0.01	0.8	12^h R	97°55′	+5.52
Neptune	24,300	3.8	17.24	1.7	0.02	1.2	15^{h}48^m	28°48′	+7.84
Pluto	2,500	0.4	0.11?	?	?	?	6^{d}9^{h}17^m	?	+14.9

R signifies retrograde rotation.

The masses and diameters are the values recommended by the International Astronomical Union in 1976, except for a 1977 value for the radius of Uranus. Densities and surface gravities were calculated from these values.

Appendix 5b. *Orbital Properties*

Name	Semimajor Axis A.U.	Semimajor Axis 10⁶ km	Sidereal Period Years	Sidereal Period Days	Synodic Period (days)	Eccentricity	Inclination to Ecliptic
Mercury	0.3871	57.9	0.24084	87.96	115.9	0.2056	7°00′26″
Venus	0.7233	108.2	0.61515	224.68	584.0	0.0068	3°23′40″
Earth	1	149.6	1.00004	365.26	−	0.0167	0°00′14″
Mars	1.5237	227.9	1.8808	686.95	779.9	0.0934	1°51′09″
Jupiter	5.2028	778.3	11.862	4337	398.9	0.0483	1°18′29″
Saturn	9.5388	1427.0	29.456	10,760	378.1	0.0560	2°29′17″
Uranus	19.1914	2871.0	84.07	30,700	369.7	0.0461	0°48′26″
Neptune	30.0611	4497.1	164.81	60,200	367.5	0.0100	1°46′27″
Pluto	39.5294	5913.5	248.53	90,780	366.7	0.2484	17°09′03″

Mean elements of planetary orbits for 1980, referred to the mean ecliptic and equinox of 1950 (Seidelmann, P. K., L. E. Doggett, and M. R. DeLuccia, Astronomical Journal 79, 57, 1974). Periods are calculated from them.

Appendix 6

Planetary Satellites

Satellite	Semimajor Axis of Orbit (km)	Sidereal Period (days)	Orbital Eccentricity	Orbital Inclination (°)	Radius (km)	Mass ÷ Mass of Planet	Discoverer	Apparent Magnitude at Opposition
SATELLITE OF THE EARTH								
The Moon	385,000	27.322	0.055	18–29	1738	0.01230002°	—	−12.7
SATELLITES OF MARS								
Phobos	9,380	0.3189	0.018	1.0	14	2.7×10^{-8}	Hall (1877)	11.6
Deimos	23,500	1.262	0.002	1.3	8	4.8×10^{-9}	Hall (1877)	12.7
SATELLITES OF JUPITER								
V Amalthea	181,000	0.4982	0.003	0.4	80	18×10^{-10}	Barnard (1892)	13.0
I Io	422,000	1.769	0.000	0	1830	4.70×10^{-5}°	Galileo (1610)	5.0
II Europa	671,000	3.551	0.000	0	1550	2.56×10^{-5}°	Galileo (1610)	5.3
III Ganymede	1,070,000	7.155	0.001	0	2640	7.84×10^{-5}°	Galileo (1610)	4.6
IV Callisto	1,880,000	16.69	0.01	0	2500	5.6×10^{-5}°	Galileo (1610)	5.6
XIII Leda	11,110,000	239	0.147	26.7	8	5×10^{-13}	Kowal (1974)	20
VI Himalia	11,500,000	250.6	0.158	27.6	60	8.5×10^{-10}	Perrine (1904)	14.7
VII Elara	11,700,000	259.7	0.207	24.8	20	0.35×10^{-10}	Perrine (1905)	16.0
X Lysithea	11,900,000	263.6	0.130	29.0	7	0.010×10^{-10}	Nicholson (1938)	18.8
XII Ananke	21,200,000	631.1 R	0.169	147	6	0.007×10^{-10}	Nicholson (1951)	18.3
XI Carme	22,600,000	692.5 R	0.207	164	7	0.020×10^{-10}	Nicholson (1938)	18.6
VIII Pasiphae	23,500,000	738.9 R	0.378	145	6	0.077×10^{-10}	Melotte (1908)	18.1
IX Sinope	23,700,000	758 R	0.275	153	7	0.015×10^{-10}	Nicholson (1914)	18.8
XIV							Kowal (1975)	20
SATELLITES OF SATURN								
10 Janus	169,500	0.749	0.0	0.0	100	3×10^{-8}	Dollfus (1966)	14
1 Mimas	186,000	0.942	0.020	1.5	200	6.59×10^{-8}	W. Herschel (1789)	12.1
2 Enceladus	238,000	1.370	0.004	0.0	300	1.48×10^{-7}	W. Herschel (1789)	11.8
3 Tethys	295,000	1.888	0.000	1.1	500	1.10×10^{-6}	Cassini (1684)	10.3
4 Dione	377,000	2.737	0.002	0.0	400	2.04×10^{-6}	Cassini (1684)	10.4
5 Rhea	527,000	4.518	0.001	0.4	750	3.2×10^{-6}	Cassini (1672)	9.7
6 Titan	1,220,000	15.95	0.029	0.3	2900	2.41×10^{-4}°	Huygens (1655)	8.4
7 Hyperion	1,480,000	21.28	0.104	0.4	200	2×10^{-7}	Bond (1848)	14.2
8 Iapetus	3,560,000	79.33	0.028	14.7	750	3.94×10^{-6}	Cassini (1671)	11.0
9 Phoebe	13,000,000	550.5 R	0.163	150	100	5.2×10^{-8}	W. Pickering (1898)	16.5
SATELLITES OF URANUS								
5 Miranda	130,000	1.414	0.017	3.4	200	1×10^{-6}	Kuiper (1948)	16.5
1 Ariel	191,000	2.520 R	0.003	0	700	15×10^{-6}	Lassell (1851)	14.4
2 Umbriel	260,000	4.144 R	0.004	0	500	6×10^{-6}	Lassell (1851)	15.3
3 Titania	436,000	8.706 R	0.002	0	900	50×10^{-6}	W. Herschel (1787)	14.0
4 Oberon	583,000	13.46 R	0.001	0	800	29×10^{-6}	W. Herschel (1787)	14.2
SATELLITES OF NEPTUNE								
Triton	354,000	5.877 R	0.000	160.0	1900	2×10^{-3}°	Lassell (1846)	13.6
Nereid	5,570,000	359.4	0.76	27.4	300	10^{-6}	Kuiper (1949)	18.7

°Values recommended by the International Union in 1976.

Most other values from Wilkins, G. A., and A. T. Sinclair (Proc. Royal Society London A *336*, 85, 1974), with diameters for 5 of Saturn's satellites from Veverka, J., J. Elliot, and J. Goguen (Sky and Telescope, December 1975, p. 356.)

Apparent magnitudes at discovery given for Jupiter XIII and XIV.

R signifies retrograde revolution.

Appendix 7

The Brightest Stars

Star	Name	Position 1980 R.A.	Dec.	Apparent Magnitude (V)	Spectral Type		Absolute Magnitude	Distance (ly)	Proper Motion (sec)
1. α CMa A	Sirius	06 44.2	−16 42	−1.47	A1	V	+1.45	8.7	1.324
2. α Car	Canopus	06 23.5	−52 41	−0.72	F0	Ib-II	−3.1	98	0.025
3. α Boo	Arcturus	14 14.8	+19 17	−0.06	K2	IIIp	−0.3	36	2.284
4. α Cen A	Rigil Kentaurus	14 38.4	−60 46	0.01	G2	V	+4.39	4.3	3.676
5. α Lyr	Vega	18 36.2	+38 46	0.04	A0	V	+0.5	26.5	0.345
6. α Aur	Capella	05 15.2	+45 59	0.05	G8 III:+F		−0.6	45	0.435
7. β Ori A	Rigel	05 13.6	−08 13	0.14v	B8	Ia	−7.1	900	0.001
8. α CMi A	Procyon	07 38.2	+05 17	0.37	F5	IV-V	+2.7	11.3	1.250
9. α Ori	Betelgeuse	05 54.0	+07 24	0.41v	M2	Iab	−5.6	520	0.028
10. α Eri	Achernar	01 37.0	−57 20	0.51	B3	Vp	−2.3	118	0.098
11. β Cen AB	Hadar	14 02.4	−60 16	0.63v	B1	III	−5.2	490	0.035
12. α Aql	Altair	19 49.8	+08 49	0.77	A7	IV-V	+2.2	16.5	0.658
13. α Tau A	Aldebaran	04 34.8	+16 28	0.86v	K5	III	−0.7	68	0.202
14. α Vir	Spica	13 24.1	−11 03	0.91v	B1	V	−3.3	220	0.054
15. α Sco A	Antares	16 28.2	−26 23	0.92v	M1	Ib+B	−5.1	520	0.029
16. α PsA	Fomalhaut	22 56.5	−29 44	1.15	A3	V	+2.0	22.6	0.367
17. β Gem	Pollux	07 44.1	+28 05	1.16	K0	III	+1.0	35	0.625
18. α Cyg	Deneb	20 40.7	+45 12	1.26	A2	Ia	−7.1	1600	0.003
19. β Cru	Beta Crucis	12 46.6	−59 35	1.28v	B0.5	III	−4.6	490	0.049
20. α Leo A	Regulus	10 07.3	+12 04	1.36	B7	V	−0.7	84	0.248
21. α Cru A	Acrux	12 25.4	−62 59	1.39	B0.5	IV	−3.9	370	0.042
22. ε CMa A	Adhara	06 57.8	−28 57	1.48	B2	II	−5.1	680	0.004
23. λ Sco	Shaula	17 32.3	−37 05	1.60v	B1	V	−3.3	310	0.031
24. γ Ori	Bellatrix	05 24.0	+06 20	1.64	B2	III	−4.2	470	0.015
25. β Tau	Elnath	05 25.0	+28 36	1.65	B7	III	−3.2	300	0.178

Based on a table compiled by Donald A. MacRae in the *Observer's Handbook 1976* of the Royal Astronomical Society of Canada.

Appendix 8
The Nearest Stars

Name	1980 α	1980 δ	Parallax π	Distance	Spectral Type	Proper Motion μ	Visual Magnitude m	Luminosity
	h m	° ′	″	ly		″		
Sun					G2		−26.8	1.0
α Cen A	14 38	−60 46	0.760	4.3	G2	3.68	0.1	1.3
B					K5		1.5	0.36
C	14 28	−62 36			M5e		11.0	0.00006
Barnard's	17 56	+04 36	.552	5.9	M5	10.30	9.5	0.00044
Wolf 359	10 56	+07 10	.431	7.6	M6e	4.84	13.5	0.00002
Lalande 21185	11 03	+36 07	.402	8.1	M2	4.78	7.5	0.0052
Sirius A	6 44	−16 42	.377	8.6	A1	1.32	−1.5	23.
B					wd		7.2	0.008
Luyten 726-8A	1 37	−18 04	.365	8.9	M6e	3.35	12.5	0.00006
B (UV Ceti)					M6e		13.0	0.00004
Ross 154	18 49	−23 50	.345	9.4	M5e	0.74	10.6	0.0004
Ross 248	23 40	+44 04	.317	10.3	M6e	1.82	12.2	0.00011
ε Eri	03 32	−09 32	.305	10.7	K2	0.97	3.7	0.30
Luyten 789-6	22 38	−15 28	.302	10.8	M6	3.27	12.2	0.00012
Ross 128	11 47	+00 58	.301	10.8	M5	1.40	11.1	0.00033
61 Cyg A	21 06	+38 38	.292	11.2	K5	5.22	5.2	0.083
B					K7		6.0	0.040
ε Ind	22 03	−56 52	.291	11.2	K5	4.67	4.7	0.13
Procyon A	07 39	+05 17	.287	11.4	F5	1.25	0.3	7.6
B					wd		10.8	0.0005
Σ 2398 A	18 42	+59 36	.284	11.5	M3.5	2.29	8.9	0.0028
B					M4		9.7	0.0013
Groombridge 34 A	00 18	+43 54	.282	11.6	M1	2.91	8.1	0.0058
B					M6		11.0	0.00040
Lacaille 9352	23 05	−35 59	.279	11.7	M2	6.87	7.4	0.012
τ Ceti	01 43	−16 03	.273	11.9	G8	1.92	3.5	0.44
BD+5°1668	07 27	+05 27	.266	12.2	M4	3.73	9.8	0.0014
L725-32	01 11	−17 06	.262	12.4	M5e	1.31	11.5	0.0003
Lacaille 8760	21 16	−38 58	.260	12.5	M1	3.46	6.7	0.025
Kapteyn's	05 11	−44 59	.256	12.7	M0	8.79	8.8	0.0040
Kruger 60 A	22 27	+57 36	.254	12.8	M4	0.87	9.7	0.0017
B					M6		11.2	0.00044
Ross 614 A	06 28	−02 48	.249	13.1	M5e	0.97	11.3	0.0004
B					?		14.8	0.00002
BD−12°4523	16 30	−12 36	.249	13.1	M5	1.18	10.0	0.0013
van Maanen's	00 48	+05 19	.234	13.9	wdF	2.98	12.4	0.00017
Wolf 424 A	12 33	+09 09	.229	14.2	M6e	1.87	12.6	0.00014
B					M6e		12.6	0.00014
CD−37°15492	00 04	−37 27	.225	14.5	M3	6.09	8.6	0.0058
G158 27	00 06	−07 38	.224	14.6		2.1	13.8	0.00005
Groombridge 1618	10 10	+49 33	.217	15.0	M0	1.45	6.6	0.040
CD−46°11540	17 28	−46 53	.216	15.1	M4	1.15	9.4	0.0030
CD−49°13515	21 32	−49 11	.214	15.2	M3	0.78	8.7	0.0058
CD−44°11909	17 37	−44 17	.213	15.3	M5	1.14	11.2	0.00063
Luyten 1159-16	01 59	+13 00	.212	15.4	(M7)	2.08	12.3	0.00023
Lalande 25372	13 44	+15 01	.208	15.7	M3.5	2.30	8.5	0.0076
AOe 17415-6	17 37	+68 22	.207	15.7	M3.5	1.31	9.1	0.0044
CC 658	11 44	−64 42	.206	15.8	wd	2.69	11.0	0.0008
Ross 780	22 52	−14 22	.206	15.8	M5	1.17	10.2	0.0016
40 Eri A	04 14	−07 41	.205	15.9	K0	4.08	4.4	0.33
B					wdA		9.9	0.0027
C					M4e		11.2	0.00063

Based on a table compiled by Alan H. Batten in the *Observer's Handbook 1976* of the Royal Astronomical Society of Canada.

Appendix 9
Messier Catalogue

M	NGC	α 1980	δ	m_v	Description
		h m	° '		
1	1952	5 33.3	+22 01	11.3	Crab Nebula in Taurus
2	7089	21 32.4	−00 54	6.3	Globular cluster in Aquarius
3	5272	13 41.3	+28 29	6.2	Globular cluster in Canes Venatici
4	6121	16 22.4	−26 27	6.1	Globular cluster in Scorpio
5	5904	15 17.5	+02 11	6	Globular cluster in Serpens
6	6405	17 38.9	−32 11	6	Open cluster in Scorpio
7	6475	17 52.6	−34 48	5	Open cluster in Scorpio
8	6523	18 02.4	−24 23		Lagoon Nebula in Sagittarius
9	6333	17 18.1	−18 30	7.6	Globular cluster in Ophiuchus
10	6254	16 56.0	−04 05	6.4	Globular cluster in Ophiuchus
11	6705	18 50.0	−06 18	7	Open cluster in Scutum
12	6218	16 46.1	−01 55	6.7	Globular cluster in Ophiuchus
13	6205	16 41.0	+36 30	5.8	Globular cluster in Hercules
14	6402	17 36.5	−03 14	7.8	Globular cluster in Ophiuchus
15	7078	21 29.1	+12 05	6.3	Globular cluster in Pegasus
16	6611	18 17.8	−13 48	7	Open cluster in Serpens
17	6618	18 19.7	−16 12	7	Omega Nebula in Sagittarius
18	6613	18 18.8	−17 09	7	Open cluster in Sagittarius
19	6273	17 01.3	−26 14	6.9	Globular cluster in Ophiuchus
20	6514	18 01.2	−23 02		Trifid Nebula in Sagittarius
21	6531	18 03.4	−22 30	7	Open cluster in Sagittarius
22	6656	18 35.2	−23 55	5.2	Globular cluster in Sagittarius
23	6494	17 55.7	−19 00	6	Open cluster in Sagittarius
24	6603	18 17.3	−18 27	6	Open cluster in Sagittarius
25	IC4725	18 30.5	−19 16	6	Open cluster in Sagittarius
26	6694	18 44.1	−09 25	9	Open cluster in Scutum
27	6853	19 58.8	+22 40	8.2	Dumbbell Nebula; planetary nebula in Vulpecula
28	6626	18 23.2	−24 52	7.1	Globular cluster in Sagittarius
29	6913	20 23.3	+38 27	8	Open cluster in Cygnus
30	7099	21 39.2	−23 15	7.6	Globular cluster in Capricornus
31	224	0 41.6	+41 09	3.7	Andromeda Galaxy (Sb)
32	221	0 41.6	+40 45	8.5	Elliptical galaxy in Andromeda; companion to M31
33	598	1 32.8	+30 33	5.9	Spiral galaxy (Sc) in Triangulum
34	1039	2 40.7	+42 43	6	Open cluster in Perseus
35	2168	6 07.6	+24 21	6	Open cluster in Gemini
36	1960	5 35.0	+34 05	6	Open cluster in Auriga
37	2099	5 51.5	+32 33	6	Open cluster in Auriga
38	1912	5 27.3	+35 48	6	Open cluster in Auriga
39	7092	21 31.5	+48 21	6	Open cluster in Cygnus
40	–	–	–		Double star in Ursa Major
41	2287	6 46.2	−20 43	6	Open cluster in Canis Major
42	1976	5 34.4	−05 24		Orion Nebula
43	1982	5 34.6	−05 18		Orion Nebula; smaller part
44	2632	8 38.8	+20 04	4	Praesepe; open cluster in Cancer
45	–	3 46.3	+24 03	2	The Pleiades; open cluster in Taurus
46	2437	7 40.9	−14 46	7	Open cluster in Puppis
47	2422	7 35.6	−14 27	5	Open cluster in Puppis
48	2548	8 12.5	−05 43	6	Open cluster in Hydra
49	4472	12 28.8	+08 07	8.9	Elliptical galaxy in Virgo
50	2323	7 02.0	−08 19	7	Open cluster in Monoceros

Appendix 9 continued on the opposite page

M NGC	α 1980 h m	δ °	m$_v$	Description
51 5194	13 29.0	+47 18	8.4	Whirlpool Galaxy; spiral galaxy (Sc) in Canes Venatici
52 7654	23 23.3	+61 29	7	Open cluster in Cassiopeia
53 5024	13 12.0	+18 17	7.7	Globular cluster in Coma Berenices
54 6715	18 53.8	−30 30	7.7	Globular cluster in Sagittarius
55 6809	19 38.7	−31 00	6.1	Globular cluster in Sagittarius
56 6779	19 15.8	+30 08	8.3	Globular cluster in Lyra
57 6720	18 52.9	+33 01	9.0	Ring Nebula; planetary nebula in Lyra
58 4579	12 36.7	+11 56	9.9	Spiral galaxy (SBb) in Virgo
59 4621	12 41.0	+11 47	10.3	Elliptical galaxy in Virgo
60 4649	12 42.6	+11 41	9.3	Elliptical galaxy in Virgo
61 4303	12 20.8	+04 36	9.7	Spiral galaxy (Sc) in Virgo
62 6266	16 59.9	−30 05	7.2	Globular cluster in Scorpio
63 5055	13 14.8	+42 08	8.8	Spiral galaxy (Sb) in Canes Venatici
64 4826	12 55.7	+21 48	8.7	Spiral galaxy (Sb) in Coma Verenices
65 3623	11 17.8	+13 13	9.6	Spiral galaxy (Sa) in Leo
66 3627	11 19.1	+13 07	9.2	Spiral galaxy (Sb) in Leo; companion to M65
67 2682	8 50.0	+11 54	7	Open cluster in Cancer
68 4590	12 38.3	−26 38	8	Globular cluster in Hydra
69 6637	18 30.1	−32 23	7.7	Globular cluster in Sagittarius
70 6681	18 42.0	−32 18	8.2	Globular cluster in Sagittarius
71 6838	19 52.8	+18 44	6.9	Globular cluster in Sagitta
72 6981	20 52.3	−12 39	9.2	Globular cluster in Aquarius
73 6994	20 57.8	−12 44		Open cluster in Aquarius
74 628	1 35.6	+15 41	9.5	Spiral galaxy (Sc) in Pisces
75 6864	20 04.9	−21 59	8.3	Globular cluster in Sagittarius
76 650	1 40.9	+51 28	11.4	Planetary nebula in Perseus
77 1068	2 41.6	−00 04	9.1	Spiral galaxy (Sb) in Cetus
78 2068	5 45.8	+00 02		Small emission nebula in Orion
79 1904	5 23.3	−24 32	7.3	Globular cluster in Lepus
80 6093	16 15.8	−22 56	7.2	Globular cluster in Scorpio
81 3031	9 54.2	+69 09	6.9	Spiral galaxy (Sb) in Ursa Major
82 3034	9 54.4	+69 47	8.7	Irregular galaxy (Irr) in Ursa Major
83 5236	13 35.9	−29 46	7.5	Spiral galaxy (Sc) in Hydra
84 4374	12 24.1	+13 00	9.8	Elliptical galaxy in Virgo
85 4382	12 24.3	+18 18	9.5	Elliptical galaxy (SO) in Coma Berenices
86 4406	12 25.1	+13 03	9.8	Elliptical galaxy in Virgo
87 4486	12 29.7	+12 30	9.3	Elliptical galaxy (Ep) in Virgo
88 4501	12 30.9	+14 32	9.7	Spiral galaxy (Sb) in Coma Berenices
89 4552	12 34.6	+12 40	10.3	Elliptical galaxy in Virgo
90 4569	12 35.8	+13 16	9.7	Spiral galaxy (Sb) in Virgo
91 −	−	−		M58 ?
92 6341	17 16.5	+43 10	6.3	Globular cluster in Hercules
93 2447	7 43.6	−23 49	6	Open cluster in Puppis
94 4736	12 50.1	+41 14	8.1	Spiral galaxy (Sb) in Canes Venatici
95 3351	10 42.8	+11 49	9.9	Barred spiral galaxy (SBb) in Leo
96 3368	10 45.6	+11 56	9.4	Spiral galaxy (Sa) in Leo
97 3587	11 13.7	+55 08	11.1	Owl Nebula; planetary nebula in Ursa Major
98 4192	12 12.7	+15 01	10.4	Spiral galaxy (Sb) in Coma Berenices
99 4254	12 17.8	+14 32	9.9	Spiral galaxy (Sc) in Coma Berenices
100 4321	12 21.9	+15 56	9.6	Spiral galaxy (Sc) in Coma Berenices

Appendix 9 continued on the following page

Appendix 9 *Continued*

M NGC	α 1980	δ	m$_v$	Description
	h m	° ′		
101 5457	14 02.5	+54 27	8.1	Spiral galaxy (Sc) in Ursa Major
102 –	–	–		M101 ?
103 581	1 31.9	+60 35	7	Open cluster in Cassiopeia
104 4594	12 39.0	−11 35	8	Sombrero Nebula; spiral galaxy (Sa) in Virgo
105 3379	10 46.8	+12 51	9.5	Elliptical galaxy in Leo
106 4258	12 18.0	+47 25	9	Spiral galaxy in (Sb) Canes Venatici
107 6171	16 31.8	−13 01	9	Globular cluster in Ophiuchus
108 3556	11 10.5	+55 47	10.5	Spiral galaxy (Sb) in Ursa Major
109 3992	11 56.6	+53 29	10.6	Barred spiral galaxy (SBc) in Ursa Major

Based in part on a table in the *Observer's Handbook 1976* of the Royal Astronomical Society of Canada.

Appendix 10

The Constellations

Latin Name	Genitive	Abbreviation	Translation
Andromeda	Andromedae	And	Andromeda*
Antlia	Antliae	Ant	Pump
Apus	Apodis	Aps	Bird of Paradise
Aquarius	Aquarii	Aqr	Water Bearer
Aquila	Aquilae	Aql	Eagle
Ara	Arae	Ara	Altar
Aries	Arietis	Ari	Ram
Auriga	Aurigae	Aur	Charioteer
Boötes	Boötis	Boo	Herdsman
Caelum	Caeli	Cae	Chisel
Camelopardalis	Camelopardalis	Cam	Giraffe
Cancer	Cancri	Cnc	Crab
Canes Venatici	Canum Venaticorum	CVn	Hunting Dogs
Canis Major	Canis Majoris	CMa	Big Dog
Canis Minor	Canis Minoris	CMi	Little Dog
Capricornus	Capricorni	Cap	Goat
Carina	Carinae	Car	Ship's Keel**
Cassiopeia	Cassiopeiae	Cas	Cassiopeia*
Centaurus	Centauri	Cen	Centaur*
Cepheus	Cephei	Cep	Cepheus*
Cetus	Ceti	Cet	Whale
Chamaeleon	Chamaeleonis	Cha	Chameleon
Circinus	Circini	Cir	Compass
Columba	Columbae	Col	Dove
Coma Berenices	Comae Berenices	Com	Berenice's Hair*
Corona Australis	Coronae Australis	CrA	Southern Crown
Corona Borealis	Coronae Borealis	CrB	Northern Crown
Corvus	Corvi	Crv	Crow
Crater	Crateris	Crt	Cup
Crux	Crucis	Cru	Southern Cross
Cygnus	Cygni	Cyg	Swan

Appendix 10 continued on the opposite page

Appendix 10 *Continued*

Latin Name	Genitive	Abbreviation	Translation
Delphinus	Delphini	Del	Dolphin
Dorado	Doradus	Dor	Swordfish
Draco	Draconis	Dra	Dragon
Equuleus	Equulei	Equ	Little Horse
Eridanus	Eridani	Eri	River Eridanus*
Fornax	Fornacis	For	Furnace
Gemini	Geminorum	Gem	Twins
Grus	Gruis	Gru	Crane
Hercules	Herculis	Her	Hercules*
Horologium	Horologii	Hor	Clock
Hydra	Hydrae	Hya	Hydra* (water monster)
Hydrus	Hydri	Hyi	Sea serpent
Indus	Indi	Ind	Indian
Lacerta	Lacertae	Lac	Lizard
Leo	Leonis	Leo	Lion
Leo Minor	Leonis Minoris	LMi	Little Lion
Lepus	Leporis	Lep	Hare
Libra	Librae	Lib	Scales
Lupus	Lupi	Lup	Wolf
Lynx	Lyncis	Lyn	Lynx
Lyra	Lyrae	Lyr	Harp
Mensa	Mensae	Men	Table (mountain)
Microscopium	Microscopii	Mic	Microscope
Monoceros	Monocerotis	Mon	Unicorn
Musca	Muscae	Mus	Fly
Norma	Normae	Nor	Level (square)
Octans	Octantis	Oct	Octant
Ophiuchus	Ophiuchi	Oph	Ophiuchus* (serpent bearer)
Orion	Orionis	Ori	Orion*
Pavo	Pavonis	Pav	Peacock
Pegasus	Pegasi	Peg	Pegasus* (winged horse)
Perseus	Persei	Per	Perseus*
Phoenix	Phoenicis	Phe	Phoenix
Pictor	Pictoris	Pic	Easel
Pisces	Piscium	Psc	Fish
Piscis Austrinus	Piscis Austrini	PsA	Southern Fish
Puppis	Puppis	Pup	Ship's Stern**
Pyxis	Pyxidis	Pyx	Ship's Compass**
Reticulum	Reticuli	Ret	Net
Sagitta	Sagittae	Sge	Arrow
Sagittarius	Sagittarii	Sgr	Archer
Scorpius	Scorpii	Sco	Scorpion
Sculptor	Sculptoris	Scl	Sculptor
Scutum	Scuti	Sct	Shield
Serpens	Serpentis	Ser	Serpent
Sextans	Sextantis	Sex	Sextant
Taurus	Tauri	Tau	Bull
Telescopium	Telescopii	Tel	Telescope
Triangulum	Trianguli	Tri	Triangle
Triangulum Australe	Trianguli Australis	TrA	Southern Triangle
Tucana	Tucanae	Tuc	Toucan
Ursa Major	Ursae Majoris	UMa	Big Bear
Ursa Minor	Ursae Minoris	UMi	Little Bear
Vela	Velorum	Vel	Ship's Sails**
Virgo	Virginis	Vir	Virgin
Volans	Volantis	Vol	Flying Fish
Vulpecula	Vulpeculae	Vul	Little Fox

*Proper names.
**Formerly formed the constellation Argo Navis, the Argonauts' ship.

Appendix 11

Interstellar Molecules

Name of Molecule	Chemical Symbol	Year of Discovery	Part of Spectrum	First Wavelength Observed		Telescope Used for Discovery
methylidyne	CH	1937	visible	4300 Å	2 5-m	Mt. Wilson
cyanogen radical	CN	1940	visible	3875 Å	2.5-m	Mt. Wilson
methylidyne ion	CH^+	1941	visible	4232 Å	2.5-m	Mt. Wilson
hydroxyl radical	OH	1963	radio	18 cm	26-m	Lincoln Lab
ammonia	NH_3	1968	radio	1.3 cm	6-m	Hat Creek
water	H_2O	1968	radio	1.4 cm	6-m	Hat Creek
formaldehyde	H_2CO	1969	radio	6.2 cm	43-m	NRAO/Green Bank
carbon monoxide	CO	1970	radio	2.6 mm	11-m	NRAO/Kitt Peak
hydrogen cyanide	HCN	1970	radio	3.4 mm	11-m	NRAO/Kitt Peak
cyanoacetylene	HC_3N	1970	radio	3.3 cm	43-m	NRAO/Green Bank
hydrogen	H_2	1970	ultraviolet	1013-1108 Å		NRL rocket
methyl alcohol	CH_3OH	1970	radio	36 cm	43-m	NRAO/Green Bank
formic acid	HCOOH	1970	radio	18 cm	43-m	NRAO/Green Bank
"X-ogen"	HCO^+	1970	radio	3.4 mm	11-m	NRAO/Kitt Peak
formamide	$HCONH_2$	1971	radio	6.5 cm	43-m	NRAO/Green Bank
carbon monosulfide	CS	1971	radio	2.0 mm	11-m	NRAO/Kitt Peak
silicon monoxide	SiO	1971	radio	2.3 mm	11-m	NRAO/Kitt Peak
carbonyl sulfide	OCS	1971	radio	2.7 mm	11-m	NRAO/Kitt Peak
methyl cyanide, acetonitrile	CH_3CN	1971	radio	2.7 mm	11-m	NRAO/Kitt Peak
isocyanic acid	HNCO	1971	radio	3.4 mm	11-m	NRAO/Kitt Peak
methylacetylene	CH_3C_2H	1971	radio	3.5 mm	11-m	NRAO/Kitt Peak
acetaldehyde	CH_3CHO	1971	radio	28 cm	43-m	NRAO/Green Bank
thioformaldehyde	H_2CS	1971	radio	9.5 cm	64-m	Parkes
hydrogen isocyanide	HNC	1971	radio	3.3 mm	11-m	NRAO/Kitt Peak
hydrogen sulfide	H_2S	1972	radio	1.8 mm	11-m	NRAO/Kitt Peak
methanimine	H_2CNH	1972	radio	5.7 cm	64-m	Parkes
sulfur monoxide	SO	1973	radio	3.0 mm	11-m	NRAO/Kitt Peak
(no name)	N_2H^+	1974	radio	3.2 mm	11-m	NRAO/Kitt Peak
ethynyl radical	C_2H	1974	radio	3.4 mm	11-m	NRAO/Kitt Peak
methylamine	CH_3NH_2	1974	radio	3.5, 4.1 mm	11-m, 6-m	NRAO and Tokyo
dimethyl ether	$(CH_3)_2O$	1974	radio	9.6 mm	11-m	NRAO/Kitt Peak
ethyl alcohol	CH_3CH_2OH	1974	radio	2.9-3.5 mm	11-m	NRAO/Kitt Peak
sulfur dioxide	SO_2	1975	radio	3.6 mm	11-m	NRAO/Kitt Peak
silicon sulfide	SiS	1975	radio	2.8, 3.3 mm	11-m	NRAO/Kitt Peak
acrylonitrile	H_2CCHCN	1975	radio	22 cm	64-m	Parkes
methyl formate	$HCOOCH_3$	1975	radio	18 cm	64-m	Parkes
nitrogen sulfide radical	NS	1975	radio	2.6 mm	5-m	Texas
cyanamide	NH_2CN	1975	radio	3.7 mm	11-m	NRAO
cyanodiacetylene	HC_5N	1976	radio	3.0 cm	46-m	Algonquin
formyl radical	HCO	1976	radio	3.5 mm	11-m	NRAO/Kitt Peak
acetylene	C_2H_2	1976	infrared	2.4 μ	4-m	KPNO
cyanohexatriyne	HC_7N	1977	radio	–	46-m	Algonquin
cyanoethynyl radical	C_3N	1977	radio	3.4 mm	11-m	NRAO/Kitt Peak
ketene	H_2C_2O	1977	radio	3.0 mm	11-m	NRAO/Kitt Peak

Notes: Only the first wavelength or wavelengths observed are listed. Discoveries of forms including isotopes not included. Mt. Wilson is one of the Hale Observatories. Hat Creek is the site of the University of California's radio observatory. NRAO has most of its telescopes at Green Bank, West Virginia, its millimeter-wave telescope on Kitt Peak in Arizona, and the VLA in New Mexico. The Naval Research Laboratory (NRL) is in Washington, D.C. The Australian National Radio Astronomy Observatory is at Parkes, N.S.W. The millimeter telescope at Fort Davis, Texas, is operated by the University of Texas. The Herzberg Institute's radio telescope is at Algonquin Park, Canada. The Kitt Peak National Observatory (KPNO) is in Arizona.

Glossary

The first phrase following the word or term is the *definition* or *defining property*. Further explanation or examples may follow in a separate sentence.

absolute magnitude — The magnitude that a star would appear to have if it were at a distance of ten parsecs from us.

absolute zero — The lowest point on a temperature scale; in particular, for kinetic temperature, the point at which all classical motion of particles stops. In the centigrade scale it is −273°C.

absorption line — Wavelengths (or frequencies) at which the intensity of radiation is less than it is at neighboring wavelengths (or frequencies).

accretion — The addition of matter by a body. For example, the accretion of mass by a star is the addition of mass to a star, perhaps of interstellar matter or of matter drawn off a companion star.

achromatic — Non-color-dependent; a more limited technical definition is an optical system that is constructed so that two wavelengths are focused at the same point, as distinguished from an apochromat, in which three colors are focused at the same point.

active sun — The group of solar phenomena that vary with time, or in a regular pattern over the solar surface. Examples are sunspots, flares, and prominences.

albedo — The fraction of light reflected by a body.

alpha particle — A helium nucleus; consists of two protons and two neutrons.

alt-azimuth — A two-axis telescope mounting in which motion around one of the axes, which is vertical, provides motion in azimuth, and motion around the perpendicular axis provides up and down (altitude) motion. Continuous motion along both axes is necessary to track the stars, but with modern minicomputers, this is no longer a serious limitation for large telescopes.

altitude — (a) Height above the surface of a planet, or (b) for a telescope mounting, elevation in angular measure above the horizon.

amino acid — A type of molecule containing the group NH_2 (the amino group). Amino acids are fundamental building blocks of life.

angstrom — A unit of length equal to 10^{-8} cm.

angular momentum — An intrinsic property of a system corresponding to the amount of its revolution or spin. The amount of angular momentum of a body orbiting around a point is the mass of the orbiting body times its (linear) velocity of revolution times its distance from the point. The amount of angular momentum of a spinning sphere is the moment of inertia, an intrinsic property of the distribution of mass, times the angular velocity of spin.

angular velocity — The rate at which a body rotates or revolves expressed as the angle covered in a given time (for example, in degrees per hour).

annular eclipse — A type of solar eclipse in which a ring (annulus) of solar photosphere remains visible around the moon.

anorthosite — A type of rock, common in the lunar highlands though rare on earth.

antimatter — A type of matter in which each particle (antiproton, antineutron, etc.)

is opposite in charge and certain other properties to a corresponding particle (proton, neutron, etc.) of the same mass of the ordinary type of matter from which the solar system is made. Particles of antimatter are known to exist, but it is not known why matter is dominant in this region of the universe or whether regions exist in which antimatter is common.

apastron — For an orbit around a star, the farthest point from that star.

aperture — The diameter of the lens or mirror that defines the amount of light focused by an optical system.

aphelion — For an orbit around the sun, the farthest point from the sun.

apparent magnitude — The brightness of a star as seen by an observer, given in a specific system in which five magnitudes corresponds to a brightness ratio of one hundred times and the scale is fixed by correspondence with a historical background.

apsides (**sing.** *apsis*) — The points at the ends of the major axis of an elliptical orbit. The *line of apsides* is the line that coincides with the major axis of the orbit.

association — A physical grouping of stars; in particular we talk of O and B associations or T associations.

asteroid — A "minor planet"; a non-luminous, atmosphere-less chunk of rock smaller than planet-size.

asteroid belt — A region of the solar system, between the orbits of Mars and Jupiter, in which most of the asteroids orbit.

astrology — A non-scientific system based on superstition, that purports to explain or predict human actions by study of celestial positions.

astrometric binary — A system of two stars in which the existence of one star can be deduced by study of its gravitational effect on the proper motion of the other star.

astrometry — The branch of astronomy that involves the detailed measurement of the positions of stars and other celestial bodies.

astronomical unit — The average distance from the earth to the sun.

atom — The smallest possible unit of a chemical element. When an atom is subdivided the parts no longer have properties of any chemical element. An atom consists of a nucleus with orbiting electrons.

atomic clock — A system that uses atomic properties to provide a measure of time.

atomic number — The number of protons in an atom.

atomic weight — The number of protons and neutrons in an atom, averaged over the abundances of the different isotopes.

aurora — Glowing lights visible in the sky, resulting from processes in the earth's upper atmosphere.

autumnal equinox — One of the two locations in the sky where the ecliptic crosses the celestial equator that the sun passes each year when moving from northern to southern declinations.

azimuth — The angular distance, around the horizon from the northern direction, usually expressed in angular measure from 0° for an object in the northern direction, to 180° for an object in the southern direction, around to 360°. Together, the altitude and the azimuth define the direction to an object.

Baily's beads — Beads of light visible around the rim of the moon at the beginning and end of a total solar eclipse. They result from the solar photosphere shining through valleys at the edge of the moon.

Balmer lines — Spectral absorption or emission lines resulting from transitions to or from the second energy level (first excited level) of hydrogen.

barred spiral galaxy — A type of spiral galaxy in which the arms unwind from a straight structure in the plane of the galaxy called a "bar."

basalt — A type of rock resulting from the cooling of lava.

beta particle — An electron or positron outside an atom.

big bang theory — A cosmological model, based on Einstein's general theory of relatively, in which the universe was once compressed to infinite density and has been expanding ever since.

binary star — Two stars revolving around each other.

black body — A hypothetical object that, if it existed, would absorb all radiation that hit it and would emit radiation that exactly followed Planck's law.

black dwarf — A non-radiating ball of gas that results either when a white dwarf radiates all its energy or when gas contracts gravitationally but contains too little mass to begin nuclear fusion.

black hole — A region of space from which, according to the general theory of relativity, neither radiation nor matter can escape.

Bode's law — A numerical scheme, known for 200 years, that gives the radii of the orbits of the seven innermost planets and the radius of the asteroid belt.

Bohr atom — Niels Bohr's model of the hydrogen atom, in which the energy levels are depicted as concentric circles of radii that increase as (level number)^2.

bolometer — A device for measuring the total amount of radiation from an object.

bolometric magnitude — The magnitude of a celestial object corrected to take account of the radiation in parts of the spectrum other than the visible.

bound-free transition — An atomic transition in which an electron starts bound to the atom and winds up free from it.

breccia — A type of rock made up of fragments of several types of rocks. Breccias are common on the moon.

brightness — The amount of energy received per unit area per unit of time in a given spectral region.

burster — One of the sources of bursts of x-rays.

calorie — A measure of energy in the form of heat, originally corresponding to the amount of heat required to raise the temperature of 1 gram of water by 1° C.

carbon cycle — A chain of nuclear reactions, involving carbon at its intermediate stages, that transforms four hydrogen atoms into one helium atom with a resulting release in energy. The carbon cycle is only important in stars hotter than the sun.

Cassegrainian (or Cassegrain) — A type of reflecting telescope in which the light focused by the primary mirror is intercepted short of its focal point and refocused and reflected by a secondary mirror through a hole in the center of the primary mirror.

Cassini's division — The major division in the rings of Saturn.

catalyst — A substance that participates in a reaction but that is left over in its original form at the end of the reaction.

celestial equator — The intersection of the celestial sphere with the plane that passes through the earth's equator.

celestial poles — The intersection of the celestial sphere with the axis of rotation of the earth.

celestial sphere — The hypothetical sphere centered at the center of the earth to which it appears that the stars are affixed.

center of mass — The "average" location of mass; the point in a body or system of bodies at which we may consider all the mass to be located for the purpose of calculating the gravitational effect of that mass or its mean motion when a force is applied.

Cepheid variable — A type of supergiant star that oscillates in brightness in a manner similar to the star δ Cephei. The periods of Cepheid variables, which are between 1 and 100 days, are linked to the absolute magnitude of the stars by known relationships; this allows the distances to Cepheids to be found.

charm — An arbitrary name that corresponds to a property that distinguishes certain elementary particles, including types of quarks, from each other.

chemical element — See element.

chondrite — A type of stony meteorite that contains small crystalline spherical particles called chondrules.

chromatic aberration — A defect of lens systems in which different colors are focused at different points.

chromosphere — The part of the atmosphere of the sun (or another star) between the photosphere and the corona. It is probably entirely composed of spicules and probably roughly corresponds to the region in which mechanical energy is deposited.

circumpolar stars — For a given observing location, stars that are close enough to the celestial pole that they never set or close enough to the opposite celestial pole that they never rise.

classical — When discussing atoms, not taking account of quantum mechanical effects.

closed universe — A type of big bang universe in which the universe will eventually contract. A closed universe has finite volume.

cluster — (a) Of stars, a physical grouping of many stars in space; (b) of galaxies, a physical group of at least a few galaxies in space.

cluster variable — A star that varies in brightness with a period of 0.1 to one day, similar to the star RR Lyrae. They are found in globular clusters. All cluster variables have approximately the same brightness, which allows their distance to be readily found.

CNO cycle — Same as carbon cycle.

color — (a) Of an object, a visual property that depends on wavelength; (b) an arbitrary name assigned to a property that distinguishes three kinds of quarks.

color index — The difference, expressed in magnitudes, of the brightness of a star or other celestial object measured at two different wavelengths. The color index is a measure of temperature.

coherent radiation — Radiation in which the phases of waves at different locations in a cross-section of radiation have a definite relation to each other; in non-coherent radiation, the phases are random. Only coherent radiation shows interference.

color-magnitude diagram — A Hertzsprung-Russell diagram in which the temperature on the horizontal axis is expressed in terms of color index.

coma — (a) Of a comet, the region surrounding the head; (b) of an optical system, an off-axis aberration in which the images of points appear with comet-like asymmetries.

comet — A type of object in the solar system, in very elongated orbit around the sun, that when relatively near to the sun shows a coma and a tail.

comparison spectrum — A spectrum of known elements on earth usually photographed on the same photographic plate as a stellar spectrum in order to provide a known set of wavelengths or zero Doppler shift.

conjunction — The closest apparent approach in the sky of two celestial objects. When only one body is named it is understood that the second body is the sun.

conservation law — A statement that the total amount of some property (angular momentum, energy, etc.) of a body or set of bodies does not change.

constellation — One of the 88 areas into which the sky has been divided for convenience in referring to the stars or other objects therein. Many of the constellations were delineated, not necessarily in their current shape, long ago, and names and pictures from mythology attached.

continental drift — The informal name for the theory of plate tectonics that describes the surface of the earth as a set of shifting regions called plates.

continuous spectrum — A spectrum with radiation at all wavelengths but with neither absorption nor emission lines.

continuum — The continuous spectrum that we would measure from a body if no spectral lines were present.

convection — The method of energy transport in which the rising motion of masses from below carries energy upward in a gravitational field. Boiling is an example.

convergent point — The point in the sky toward which the members of a star cluster appear to be converging or from which they appear to be diverging. Referred to in the moving cluster method of determining distances.

corona — The outermost region of the sun (or of other stars), characterized by high temperatures of millions of degrees.

coronagraph — A type of telescope with which the corona can be seen at times other than that of a total solar eclipse.

cosmic rays — Nuclear particles or nuclei arriving at the earth with high velocity.

cosmogony — The study of the origin of the universe, usually applied in particular to the origin of the solar system.

cosmological constant — A constant arbitrarily added by Einstein to an equation in his general theory of relativity in order to provide a solution in which the universe did not expand. It was only subsequently discovered that the universe did expand after all.

cosmological principle — The principle that on the whole the universe looks the same in all directions.

cosmology — The study of the universe as a whole.

coudé focus — A focal point of large telescopes in which the light is reflected by a series of mirrors so that it comes to a point at the end of a polar axis. The image does not move even when the telescope moves, which permits the mounting of heavy equipment.

deceleration parameter (q_0) — A particular measure of the rate at which the expansion of the universe is slowing down.

declination — Celestial latitude, measured in degrees north or south of the celestial equator.

deferent—In the Ptolemaic system of the universe, the larger circle, centered at the earth, on which the centers of the epicycles revolve.

degenerate matter—Matter whose properties are controlled, and which is prevented from further contraction, by quantum mechanical laws as expressed by the Pauli Exclusion Principle.

density—Mass divided by volume.

density-wave theory—The explanation of spiral structure of galaxies as the effect of a wave of compression that rotates around the center of the galaxy and causes the formation of stars in the compressed region.

deuterium—An isotope of hydrogen that contains one proton and one neutron.

diamond ring effect—The last Baily's bead glowing brightly at the beginning of the total phase of a solar eclipse, or its counterpart at the end of totality.

differential rotation—Rotation of a body in which different parts have different angular velocities (and thus different periods of rotation).

differentiation—For a planet, the existence of layers of different structure or composition.

diffraction—A phenomenon affecting light as it passes any obstacle, spreading it out in a complicated fashion.

diffraction grating—A very closely ruled series of lines that, through their diffraction of light, provide a spectrum of radiation that falls on it.

disk—(a) Of a galaxy, the disk-like flat portion, as opposed to the nucleus or the halo; (b) of a star or planet, the two-dimensional projection of its surface.

dispersion—(a) Of light, the effect that different colors are bent by different amounts when passing from one substance to another; (b) of the pulses of a pulsar, the effect that a given pulse, which leaves the pulsar at one instant, arrives at the earth at different times depending on the different wavelength or frequency at which it is observed. Both of these effects arise because light of different wavelengths travels at different speeds except in a vacuum.

D lines—A pair of lines from sodium that appear in the yellow part of the spectrum.

DNA—Deoxyribonucleic acid, a long chain of molecules that contains the genetic information of life.

Doppler effect—A change in wavelength that results when a source of waves and the observer are moving relative to each other.

dwarf—A main-sequence star.

dynamo—A device that generates electricity through the effect of motion in the presence of a magnetic field. The solar dynamo explains sunspots and the solar activity cycle.

earthshine—Sunlight illuminating the moon after having been reflected by the earth.

eccentricity—A measure of the flatness of an ellipse, defined as half the distance between the foci divided by the semi-major axis.

eclipse—The hiding of one celestial body by another. The hiding of a star by a body of larger angular size is usually called an occultation.

eclipsing binary star—A binary star in which one member periodically hides the other.

ecliptic—The path followed by the sun across the celestial sphere in the course of a year.

ecliptic plane—The plane of the earth's orbit around the sun.

electromagnetic force—One of the four fundamental forces of nature, the force that gives rise to electromagnetic radiation.

electromagnetic radiation—Radiation resulting from changing electric and magnetic fields.

electric field—A force field set up by an electric charge.

electron—A particle of one negative charge, 1/1830 the mass of a proton, that is not affected by the strong force. In an atom, the electron is normally thought of as orbiting the nucleus.

electron temperature—The kinetic temperature of the electrons present.

electron volt (eV)—The energy necessary to raise an electron through a potential of one volt.

element—A kind of atom characterized by a certain number of protons in its nucleus. All atoms of a given element have similar chemical properties.

elementary particle—One of the constituents of an atomic nucleus freed from the nucleus.

ellipse—A curve with the property that the sum of the distances from any point on the curve to two given points, called the foci, is constant.

elliptical galaxy—A type of galaxy characterized by elliptical appearance. Elliptical galaxies have no interstellar matter.

emission line—Wavelengths (or frequencies) at which the intensity of radiation is greater than it is at neighboring wavelengths (or frequencies).

emission nebula—A glowing cloud of interstellar gas.

energy—A fundamental quantity usually defined in terms of the ability of a system to do something that is technically called "work," that is, the ability to move an object by application of force where the work is the force times the displacement.

energy level—A state corresponding to an amount of energy that an atom is allowed to have by the laws of quantum mechanics.

ephemeris—A listing of astronomical positions and other data that change with time. From the same root as *ephemeral*.

epicycle—In the Ptolemaic theory, a small circle, riding on a larger circle called the deferent, on which a planet moves. The epicycle is used to account for retrograde motion.

equator—(a) Of the earth, a great circle on the earth, midway between the poles; (b) celestial, the projection of the earth's equator onto the celestial sphere.

equatorial mount—A type of telescope mounting in which one axis, called the polar axis, points toward the celestial pole and the other axis is perpendicular. Motion around only the polar axis is sufficient to completely counterbalance the effect of the earth's rotation.

equinox—An intersection of the ecliptic and the celestial equator. The center of the sun is geometrically above and below the horizon for equal lengths of time on the two days of the year when the sun passes the equinoxes; if the sun were a point and atmospheric refraction were absent, then day and night would be of equal length on those days.

erg—A unit of energy in the metric system, corresponding to the work done by a force of one dyne (the force that is required to accelerate one gram by one cm/sec²) producing a displacement of one centimeter.

ergosphere—A region surrounding a rotating black hole (or other system satisfying Kerr's solution) from which work can be extracted.

Euclidean space—A space with zero curvature; a space where the sum of the angles of a triangle is 180°.

event horizon—The sphere around a black hole from within which nothing can escape; the place at which the exit cones close.

excitation—The raising of atoms to higher energy states than the lowest possible.

exclusion principle—The quantum mechanical rule that certain types of elementary particles cannot exist in completely identical states.

exit cone—The cone that, for each point within the photon sphere of a black hole, defines the directions of rays of radiation that escape.

exobiology—The study of life located elsewhere than on earth.

extinction—The dimming of starlight by scattering and absorption as the light traverses interstellar space.

extragalactic—Exterior to the Milky Way Galaxy.

filament—A feature of the solar surface seen in Hα as a thin, dark wavy line. A filament is a prominence projected on the solar disk.

filtergram—A photograph taken through a filter that passes only a very narrow band of wavelengths; usually applied to solar photographs.

fireball—An exceptionally bright meteor.

fission, nuclear—The splitting of an atomic nucleus.

flare—A rapid eruption of material from the surface of the sun or other star.

flash spectrum—The solar chromospheric spectrum seen in the few seconds before or after totality at a solar eclipse.

fluorescence—The transformation of photons of relatively high energy to photons of lower energy through interactions with atoms, and the resulting radiation.

flux—The amount of something (such as energy) passing through a surface per unit time.

focal length—The distance from a lens or mirror to the point at which rays from an object at infinity are focused.

focus—(a) A point to which radiation is made to converge; (b) of an ellipse, one of the two points the sum of the distances to which remains constant.

force – In physics, something that can or does cause a change of momentum, measured by the rate of change of momentum with time.

force field – A way of describing phenomena that result from action at a distance, that is, even though objects are not touching.

Fraunhofer lines – The absorption lines of a solar or other stellar spectrum.

frequency – The rate at which waves pass a given point.

full moon – The phase of the moon when the side facing the earth is fully illuminated by sunlight.

fusion – The amalgamation of nuclei into heavier nuclei.

galactic cluster – An asymmetric type of collection of stars that shared a common origin.

galaxy – A collection of stars, gas, and dust, typically containing millions of solar masses, that is a basic building block of the universe. When we speak of *the galaxy* or *our galaxy,* we mean the Milky Way Galaxy.

Galilean satellites – The four brightest satellites of Jupiter.

gamma rays – Electromagnetic radiation with wavelengths shorter than approximately 1 angstrom.

giant planets – Jupiter, Saturn, Uranus, and Neptune.

giant star – A type of star brighter than main sequence stars of the same spectral type.

gibbous moon – The phases between half moon and full moon.

globular cluster – A spherically symmetric type of collection of stars that shared a common origin.

gluon – The as yet undiscovered particle that would carry the color force (and thus the strong – nuclear – force).

granule – One of the bright features, approximately one arc sec in diameter, that cover the face of the quiet sun.

grating – A surface ruled with closely spaced lines that, through diffraction, breaks up light into its spectrum.

gravitation – One of the four fundamental forces of nature, the force by which two masses attract each other.

gravitationally – Controlled by the force of gravity.

gravitational radius – The radius that, according to Schwarzschild's solutions to Einstein's equations of the general theory of relativity, corresponds to the event horizon of a black hole.

gravitational redshift – A redshift of light caused by the presence of mass according to the general theory of relativity.

gravitational waves – As yet undiscovered waves that many scientists consider to be a consequence of the general theory of relativity.

great circle – The intersection of a plane that passes through the center of a sphere with the surface of that sphere; the largest possible circle that can be drawn on the surface of a sphere.

greenhouse effect – The effect by which the amosphere of a planet heats up above its equilibrium temperature because it is transparent to incoming visible radiation but opaque to the infrared radiation that is emitted by the surface of the planet.

Gregorian – A type of reflecting telescope in which the light focused by the primary mirror is reflected by a secondary mirror after it passes its prime focus through a hole in the center of the primary mirror. The long focal length of a Gregorian telescope makes it unwieldy and thus it is little used.

Gregorian calendar – The calendar in current use, with normal years that are 365 days long, with leap years every fourth year except for years that are divisible by 100 but not by 400.

H_0 – The Hubble constant.

H I region – An interstellar region of neutral hydrogen.

H II region – An interstellar region of ionized hydrogen.

H line – The spectral line of ionized calcium at 3968 angstroms.

Hα – The first line of the Balmer series of hydrogen, at 6563 angstroms.

half-life – The length of time for half a set of particles to decay through radioactivity or instability.

halo – Of a galaxy, the region of the galaxy that extends far above and below the plane of the galaxy, as opposed to the disk or the nucleus.

head – Of a comet, the nucleus and coma together.

heliocentric — Sun-centered; using the sun rather than the earth as the point to which we refer. A heliocentric measurement, for example, omits the effect of the Doppler shift caused by the earth's orbital motion.

helium flash — The rapid onset of fusion of helium into carbon through the triple-alpha process that takes place in most red giant stars.

hertz — The measure of frequency, with units of sec^{-1}; formerly called cycles per second.

Hertzsprung gap — A region above the main sequence in a Hertzsprung-Russell diagram through which stars evolve rapidly and thus in which few stars are found.

Hertzsprung-Russell diagram — A graph of temperature or equivalent vs. luminosity or equivalent for a group of stars.

horizontal branch — A part of the Hertzsprung-Russell diagram of a globular cluster, corresponding to stars that are all at approximately zero absolute magnitude and have evolved past the red giant branch and are moving leftward on the diagram.

hour angle — Of a celestial object as seen from a particular location, the difference between the local sidereal time and the right ascension (H.A. = L.S.T. − R.A.).

H-R diagram — Hertzsprung-Russell diagram.

Hubble constant (H_0) — The constant of proportionality in Hubble's law linking the velocity of recession of a distant object and its distance from us. The best current value is 55 km/sec/Mpc.

Hubble law — The linear relation between the velocity of recession of a distant object and its distance from us, $v = H_0 d$.

hyperfine level — A subdivision of an energy level caused by such relatively minor effects as changes resulting from the angular momentum of the spinning and orbiting electron and the spinning nucleus.

IC — Index Catalogue, one of the supplements to Dreyer's New General Catalogue.

image tube — An electronic device that receives incident radiation and intensifies it or converts it to a wavelength at which photographic plates are sensitive.

inclination — Of an orbit, the angle of the plane of the orbit with respect to the ecliptic plane.

Index Catalogue — See IC.

inferior conjunction — The closest alignment of an inferior planet as it passes nearly between the earth and the sun.

inferior planet — A planet whose orbit around the sun is within the earth's, namely, Mercury and Venus.

infrared — Radiation beyond the red, from about 7000 angstroms up to 1 mm.

interference — The property of radiation, explainable by the wave theory, in which waves that are in phase can add (constructive interference) and waves that are out of phase can subtract (destructive interference); for light, this gives alternate light and dark bands.

interferometer — A device that uses the property of interference to measure such properties of objects as their positions or structure.

International Date Line — A crooked imaginary line on the earth's surface, roughly corresponding to 180° longitude, at which, when crossed from east to west, time jumps by one day.

interplanetary medium — Gas and dust between the planets.

interstellar medium — Gas and dust between the stars.

interstellar reddening — The phenomenon by which the extinction of blue light by interstellar matter is greater than the extinction of red light.

inverse Compton scattering — A method of converting photons from lower to higher energy through interaction with electrons that are moving with velocities close to the speed of light.

ion — An atom that has gained or lost one or more electrons, usually referring to the case where one or more electrons are lost. See also *negative ion*.

ionization temperature — A temperature defined by the number and types of ions present.

irregular galaxy — A type of galaxy that lacks defined shape or symmetry.

isotope — A form of a chemical element with a different number of neutrons.

isotropic — Being the same in all directions.

Jeans length — The minimum diameter of gas that is unstable to gravitational collapse.

Jovian planet — Same as giant planet.

Julian calendar — The predecessor to the Gregorian calendar, the calendar with 365 day years and leap years every fourth year without exception.

Julian day — The number of days since noon on January 1, 4713 B.C. Used for keeping track of variable stars or other astronomical events.

K line — The spectral line of ionized calcium at 3933 angstroms.

Keplerian — Following Kepler's law.

kiloparsec — One thousand parsecs.

kinetic temperature — The temperature defined by the velocities of particles in a gas.

laser — An acronym for "light amplification by stimulated emission of radiation," a device by which certain energy levels are populated by more electrons than normal, resulting in an especially intense emission of light at a certain frequency when the electrons drop to a lower energy level.

latitude — Number of degrees north or south of the equator measured from the center of the earth.

leap year — A year in which a 366th day is added.

libration — The effect by which we can see slightly more than half the lunar surface even though the moon basically has one half that always faces us.

light — Electromagnetic radiation between about 3000 and 7000 angstroms in wavelength; sometimes used more generally for other wavelengths of radiation in the ultraviolet or infrared.

light curve — The graph of the magnitude of an object vs. time.

light year — The distance that light travels in a year.

limb — The edge of a star or planet.

limb darkening — The decreasing brightness of the disk of the sun or another star as one looks from the center of the disk closer and closer to the limb.

line profile — The graph of the intensity of radiation vs. wavelength for a spectral line.

Local Group — The two dozen or so galaxies, including the Milky Way Galaxy, that form a subcluster.

local standard of rest — The system in which the average velocity of nearby stars is zero. We usually refer measurements of velocity of distant objects to the local standard of rest.

logarithmic — Increasing by multiples of ten. This is as opposed to linear, in which increases are additive.

longitude — The angular distance around a body measured along the equator from some particular point; for a point not on the equator, it is the angular distance along the equator to a great circle that passes through the poles and through the point.

luminosity — The total amount of energy given off by an object per unit time.

luminosity class — Regions of the H–R diagram that define classes of luminosity of objects of the same spectral type.

lunar eclipse — The passage of the moon into the earth's shadow.

Lyman alpha — The spectral line that corresponds to a transition between the lowest two major energy levels of a hydrogen atom.

Lyman lines — The spectral lines that correspond to transitions to or from the lowest major energy level of a hydrogen atom.

Magellanic Clouds — Two small irregular galaxies, the closest neighbors to the Milky Way Galaxy, visible in the southern sky.

magnetic field — A force field set up of a type that affects magnets in a certain way, attracting one part of the magnet while repelling another part.

magnetohydrodynamics — The study of the interaction of moving matter and a magnetic field (magneto-, from the magnetic field, and hydro-, which means water, from the historical use of the term hydrodynamics to mean the study of water).

magnetosphere — A region of magnetic field around a planet.

magnitude — A factor of $\sqrt[5]{100} = 2.511887\ldots$ in brightness. See absolute magnitude and apparent magnitude; an *order of magnitude* is a power of ten.

main sequence — A band on a Hertzsprung-Russell diagram in which stars fall during the main, hydrogen-burning, phase of their lifetimes. The main sequence runs diagonally from hot, bright stars at the upper left to cool, faint stars at the lower right.

major axis—The longest diameter of an ellipse; the line from one side of an ellipse to the other that passes through the foci. Also, the length of that line.

mantle—The shell of rock separating the core of a differentiated planet from its thin surface crust. The earth's mantle starts about 40 km below the surface and continues nearly half-way toward the center.

mare—One of the smooth areas on the moon or on some of the other planets.

mascon—A concentration of mass under the surface of the moon, discovered from its gravitational effect on spacecraft orbiting the moon.

maser—An acronym for "microwave amplification by stimulated emission of radiation," a device by which certain energy levels are more populated than normal, resulting in an especially dense emission of radio radiation at a certain frequency when the system drops to a lower energy level.

mass—A measure of the inherent amount of matter in a body.

mass function—A complicated term involving the masses of both components of a binary star system. When the spectrum of only one member of the pair is visible, one can derive only the mass function and not the individual masses.

mass-luminosity relation—A well defined relation between the mass and luminosity for main sequence stars.

megaparsec (Mpc)—A million parsecs.

meridian—The great circle on the celestial sphere that passes through the celestial poles and the observer's zenith.

Messier object—One of the objects in Messier's catalogue of nonstellar objects or in the additions to Messier's list.

metal—(a) For stellar abundances, any element higher in atomic number than 2, that is, heavier than helium. (b) For a planet or solid, matter that is a good conductor of heat and electricity.

meteor—A track of light in the sky from rock or dust burning up in the earth's atmosphere.

meteorite—An interplanetary chunk of rock after it impacts on a planet or moon, especially on the earth.

meteoroid—An interplanetary chunk of rock smaller than an asteroid.

micrometeorite—A tiny meteorite. The micrometeorites that hit the earth's surface are often sufficiently slowed down that they can reach the ground without being vaporized.

micron—The former name, still in use by astronomers, for a micrometer (10^{-6} meter).

Milky Way—The band of light across the sky from the stars and gas in the plane of the Milky Way Galaxy.

minor axis—The shortest diameter of an ellipse; the line from one side of an ellipse to the other that passes midway between the foci and is perpendicular to the major axis. Also, the length of that line.

Mira variable—A long-period variable star similar to Mira (omicron Ceti).

molecule—A collection of atoms bound together that is the smallest collection that exhibits a certain set of chemical properties.

momentum—A measure of the tendency that a moving body has to keep moving. The momentum in a given direction (the "linear momentum") is equal to the mass of the body times its component of velocity in that direction. See also angular momentum.

nebular hypothesis—The particular nebular theory for the formation of the solar system advanced by Laplace.

nebular theories—The theories that the sun and the planets formed out of a cloud of gas and dust.

neutrino—Spinning, neutral elementary particle. A neutrino has no rest mass and always travels at the speed of light.

neutron—A massive, neutral elementary particle, one of the fundamental constituents of an atom. Neutrons and protons are similar in mass, 1830 times more massive than electrons.

neutron star—A star that has collapsed to the point where it is supported against gravity by neutron degeneracy.

New General Catalogue—The common name for "A New General Catalogue of Nebulae and Clusters of Stars," put together by J. L. E. Dreyer in 1888.

new moon—The phase of the moon when the side of the moon facing the earth is the side that is not illuminated by sunlight.

Newtonian telescope—A type of reflecting telescope where the beam reflected by the primary mirror is reflected by a flat secondary mirror so that the focus falls to the side of the telescope tube.

N galaxy—A type of galaxy with a blue nucleus that dominates the galaxy in that it provides most of the radiation. The emission lines from N galaxies are generally broader than emission lines from Seyferts. N galaxies are probably a type of elliptical galaxy. Historically, N galaxies and Seyfert galaxies were defined by different astronomers on the basis of different information, and the difference is not always clearcut.

node—A point of intersection between two great circles. Eclipses of the Sun and Moon occur when these bodies are simultaneously near the nodes of their paths in the sky.

nonthermal radiation—Radiation that cannot be characterized by a single number (the temperature). Normally, we derive this number from Planck's law, so that radiation that does not follow Planck's law is called nonthermal.

nova—A star that suddenly increases in brightness. Most novae are thought to result in binary systems when matter from the giant component falls on the white dwarf component.

nucleosynthesis—The formation of the elements.

nucleus—(a) Of an atom, the core of an atom, which has a positive charge, contains most of the mass, and takes up only a small part of the volume; (b) of a comet, the chunks of matter, taking up a volume no more than a few kilometers across, at the center of the head of a comet; (c) of a galaxy, the innermost regions of a spiral galaxy; it does not show spiral structure and is visible from the sky as a bulge in the otherwise flat disk of the galaxy.

O and B association—A group of O and B stars close together in space. The members of an O and B association were formed at roughly the same time.

objective—The principal lens or mirror of an optical system.

oblate—Having an equatorial diameter greater than the polar diameter.

occultation—The hiding of one astronomical body by another, such as the occultation of a star by the moon.

Olbers's paradox—The observation that the sky is dark at night contrasted to a simple argument that shows that the sky should be uniformly bright.

Oort comet cloud—The hundreds of millions of incipient comets surrounding the solar system in a sphere of perhaps 50,000 A.U. in radius.

opacity—The lack of complete transparency of a gas.

open cluster—A galactic cluster, an asymmetric type of star cluster.

open universe—The version of big bang cosmology in which the universe will expand forever. An open universe has infinite volume.

opposition—The passage of a planet through the point most directly opposite the sun on the other side of the earth.

optical depth—The number of factors of the transcendental number (2.71828. . .) that radiation is dimmed in passage through a gas.

optical double—A pair of stars that appear extremely close together in the sky even though they are at different distances from us and are not physically linked.

organic—Containing carbon in its molecular structure.

Ozma—One of two projects involving the search for radio signals from extraterrestrial civilizations on nearby stars.

parallax—(a) When used by itself, the word "parallax" refers to trigonometric parallax, half the angle through which a star appears to be displaced when the earth moves from one side of the sun to the other, that is, through 2 A.U. The parallax of a star is inversely proportional to the distance to the star from the sun. (b) Some of the other ways of measuring distances, usually in those cases referred to with an adjective, as in spectroscopic parallax.

parsec—The distance from which one astronomical unit would subtend one second of arc. A parsec works out to be approximately 3.26 light years.

Pauli Exclusion Principle—See exclusion principle.

peculiar velocity—The velocity of a star with respect to the local standard of rest.

penumbra—(a) For an eclipse, the part of the shadow from which the sun or other radiating body is only partially occulted; (b) of a sunspot, the outer region of the sunspot, not as dark as the central umbra.

perfect cosmological principle—The assumption that on a large scale the universe is homogeneous and isotropic in space and unchanging in time.

periastron—The near point of the orbit of a body to the star around which it is orbiting.

perihelion—The near point to the sun of the orbit of a body orbiting the sun.

phase—(a) Of a planet, the varying shape of the lighted part of a planet or moon as seen from some vantage point; (b) the relation of the maxima or minima of a set of waves. For example, when light waves are in phase (coherent), all the waves have their maxima simultaneously cross a plane that is perpendicular to the direction in which the waves are travelling. Colloquially speaking, waves that are in phase go up and down together, that is, they are in step.

photometry—The measurement of the amount of light.

photomultiplier—An electronic device that through a series of internal stages multiplies the small current that is given off when light is incident on the device so that a relatively large current results.

photon—A packet of energy that can be thought of as a particle of light travelling at the speed of light. A photon of energy E is equivalent to an electromagnetic wave of wavelength hc/E, where h is Planck's constant and c is the speed of light in a vacuum.

photosphere—The region of a star from which most of its light is radiated.

plage—The part of a solar active region that appears bright when viewed in Hα.

Planck's constant—The constant of proportionality between the frequency of an electromagnetic wave and the energy of an equivalent photon. $E = h\nu$, or, since $\lambda\nu = c$, $E = hc/\lambda$.

Planck's law—The formula that predicts, for gas at a certain temperature, how much radiation there would be at any given wavelength.

planet—A celestial body of substantial size (more than about 1000 km across), basically non-radiating and of insufficient mass for nuclear reactions ever to begin, ordinarily in orbit around a star.

planetesimal—One of the small bodies into which the primeval solar nebula condensed and from which the planets formed.

plasma—An electrically neutral gas composed of approximately equal numbers of ions and electrons.

plate tectonics—See continental drift.

polar axis—The axis of an equatorial telescope mounting that is parallel to the earth's axis of rotation.

polarization—The arrangement of electromagnetic waves so that all the planes in which the waves are oscillating are parallel to each other.

Population I—The class of stars set up by Walter Baade to describe the younger stars typical of the spiral arms. These stars have relatively high abundances of metals.

Population II—The class of stars set up by Walter Baade to describe the older stars typical of the galactic halo. These stars have very low abundances of metals.

positron—The antiparticle corresponding to an electron. Its charge is +1.

precession—The slowly changing position of stars in the sky with respect to earth-based coordinates, resulting from the slowly varying orientation (the precession) of the earth's axis. The apparent precession of the stars is the result of the actual precession of the earth's axis.

primary cosmic rays—The cosmic rays arriving at the top of the earth's atmosphere.

prime focus—The location at which the main lens or mirror of a telescope focuses an image without being reflected or refocused by another mirror or other optical element.

primordial background radiation—Radiation detected in millimeter and submillimeter wavelength regions that is coming from all directions in space and interpreted to be the remnant of the big bang. Also known as 3° background radiation, background radiation, or the remnant of the primeval fireball.

prolate—Having the diameter along the axis of rotation longer than the equatorial diameter.

prominence—Solar gas protruding over the limb, visible to the naked eye only at eclipses but also observed outside of eclipses by its emission line spectrum. Prominences are at approximately the same temperature as the chromosphere.

proper motion—Motion across the sky with respect to a framework of galaxies or fixed stars, usually measured in seconds of arc per century.

proto-—A prefix from the Greek for "before." When used in conjunction with the name of a celestial body, means the state of the body just before it is considered to have formed.

proton — Massive elementary particle with positive charge 1, one of the fundamental constituents of an atom. Protons and neutrons are similar in mass, 1830 times more massive than electrons.

proton-proton chain — A set of nuclear reactions by which four hydrogen nuclei combine one after the other to form one helium nucleus with a resulting release of energy.

pulsar — A celestial object that gives off pulses of radio waves. A very few objects are known to pulse gamma rays, x-rays, or visible light.

q_0 — The deceleration parameter, a cosmological parameter that describes the rate at which the expansion of the universe is slowing up.

quantum mechanics — The branch of 20th century physics that describes atoms and radiation; the theory involves bundles of energy known as quanta.

quark — One of the subatomic particles from which many modern theoreticians believe such elementary particles as protons and neutrons are composed. The various kinds of quarks have positive or negative charges of 1/3 or 2/3.

quasar — One of the very large redshift objects that is almost stellar (point-like) in appearance. The word quasar derives from QSR, which stands for quasi-stellar radio source, although radio-quiet quasars are now also known.

quiet sun — The collection of solar phenomena that do not vary with the solar activity cycle.

radar — The acronym for **r**adio **d**etection **a**nd **r**anging; an active rather than passive radio technique in which radio signals are transmitted and their reflections received and studied.

radial velocity — The velocity of an object along a line (the radius) joining the object and the observer; the component of velocity toward or away from the observer.

radian — The unit of angular measure, defined as the ratio of a length of arc intercepted by two radii to the length of the radius. π radians = 180°.

radiant — The point in the sky from which all the meteors in a meteor shower appear to be coming.

radiation — Electromagnetic radiation.

radioactive — Having the property of spontaneously changing into another isotope or element.

radio galaxy — A galaxy that emits radio radiation many orders of magnitude stronger than radio radiation from normal galaxies.

radio telescope — An antenna or set of antennas, often together with a focusing reflecting dish, that is used to detect radio radiation from space.

radio waves — Electromagnetic radiation with wavelengths longer than about one millimeter.

ray — (a) A light ray, a wave of electromagnetic radiation; (b) on the surface of a moon or planet, a streak of material that is relatively light in shade, presumable representing material ejected when a crater was formed.

recombination — The addition of an electron to an ion, usually resulting in radiation (recombination lines) when the electron subsequently jumps down to lower energy states.

recurrent nova — An object that undergoes the nova phenomenon more than once.

red giant — A post-main-sequence stage of the lifetime of a star; the star becomes relatively bright and relatively cool.

reddening — The phenomenon by which the extinction of blue light by interstellar matter is greater than the extinction of red light so that the redder part of the continuous spectrum is enhanced.

redshift — The shift of a spectrum, usually of spectral lines in particular, to longer wavelengths.

reflecting telescope — A type of telescope that uses a mirror or mirrors to form the primary image.

reflection nebula — Interstellar gas and dust that we see because it is reflecting light from a nearby star or stars.

refracting telescope — A type of telescope in which the image is formed by a lens or lenses.

refraction — The bending of electromagnetic radiation as it passes from one medium to another or between parts of a medium that has varying properties. The *index of refraction* of a substance is the ratio of the speed of light in a vacuum to that in the substance.

relativistic — Having a velocity that is such a large fraction of the speed of light that the special theory of relativity must be applied.

relativity — Either of the theories of relativity worked out by Albert Einstein. The special theory of relativity (1905) is a theory of relative motion. The general theory of relativity (1916) is a theory of gravitation.

resolution — The ability of an optical system to distinguish fine detail.

retrograde motion — The apparent motion of the planets when they appear to move backwards with respect to the stars from the direction that they move ordinarily.

retrograde rotation — The rotation of a moon or planet opposite to the dominant direction of orbiting and rotating of the sun and planets.

revolution — The orbiting of one body around another.

right ascension — Celestial longitude, measured eastward along the celestial equator in hours of time from the vernal equinox.

Roche's limit — The sphere for each mass inside of which blobs of gas cannot agglomerate by gravitational interaction without being torn apart by tidal forces.

rotation — Spin on an axis.

Russell-Vogt theorem — The theorem that the evolution of a star is completely determined by its mass and chemical composition.

RR Lyrae star — A short period "cluster" variable. All RR Lyrae stars have approximately equal absolute magnitude and so are used to determine the distances to globular clusters.

saros — The 18 year $11^{1/3}$ day (sometimes 18 year $10^{1/3}$ day or 18 year $12^{1/3}$ day) period at which similar eclipses repeat.

Schmidt camera — A type of telescope that uses a spherical mirror and a thin lens to provide photographs of a wide field.

Schwarzschild radius — The radius that, according to Schwarzschild's solutions to Einstein's equations of the general theory of relativity, corresponds to the event horizon of a black hole.

secondary cosmic rays — High energy particles that are generated in the earth's atmosphere by primary cosmic rays.

seeing — The steadiness of the earth's atmosphere as it affects the resolution that can be obtained in astronomical observations. Good seeing corresponds to a steady atmosphere, and bad seeing corresponds to an unsteady atmosphere.

seismology — The study of waves propagating through a body and the resulting deduction of the internal properties of the body. The prefix "seismo-" comes from the Greek word for earthquake.

semimajor axis — Half the major axis, that is, for an ellipse, half the longest diameter.

Seyfert galaxy — A type of spiral galaxy that has a bright nucleus and whose spectrum shows emission lines. Historically, N galaxies and Seyfert galaxies were defined by different astronomers on the basis of different information, and the difference between them is not always clearcut.

sidereal — With respect to the stars.

sidereal time — The hour angle of the vernal equinox. Equivalent to the right ascension of objects on the observer's meridian.

singularity — A point in space where quantities become exactly zero or become infinitely large; a singularity is present in a black hole.

solar activity cycle — The 11- or 22-year cycle with which such solar activity as sunspots, flares, and prominences varies.

solar constant — The total amount of energy that would hit each square centimeter of the top of the earth's atmosphere at the earth's average distance from the sun.

solar time — A system of time-keeping with respect to the sun such that the sun is overhead of a given location at noon.

solar wind — An outflow of particles from the sun representing the expansion of the corona.

solid angle — A three-dimensional angle.

solstice — The point on the celestial sphere of northernmost or southernmost declination of the sun in the course of a year.

space velocity — The velocity of a star with respect to the sun.

spectral type — One of the categories O, B, A, F, G, K, M, into which stars can be classified from study of their spectral lines, or extensions of this system. The sequence of spectral types corresponds to a sequence of temperature.

spectrograph — A device used to photograph a spectrum.

spectroheliograph — A device that makes an image of the sun at a particular wavelength by scanning a spectrograph slit across a solar image and simultaneously scanning a second slit that defines the wavelength being observed.

spectroscope—A device for looking at a spectrum.

spectroscopic binary—A type of binary star that is known to have more than one component because of the changing Doppler shifts of the spectral lines that are observed.

spectroscopic parallax—The distance to a star derived from comparison of its apparent magnitude with its absolute magnitude deduced from study of its position on an H–R diagram determined by observation of its spectrum (spectral type and luminosity class).

spectrum—A display of electromagnetic radiation spread out by wavelength or frequency.

spicule—A small jet of gas at the edge of the quiet sun, approximately 1000 km in diameter and 10,000 km high with a lifetime of about 15 minutes.

spiral galaxy—A class of galaxy characterized by arms that unwind like a pinwheel.

sporadic meteor—A meteor that is not associated with a shower.

star—A self-luminous ball of gas that shines or has shone because of nuclear reactions in its interior.

steady state theory—The cosmological theory based on the perfect cosmological principle.

Stefan-Boltzmann Law—The radiation law that states that the energy emitted by a black body varies with the fourth power of the temperature.

stones—A stony type of meteorite, including the chondrites.

stratosphere—One of the upper layers of the atmosphere of a planet, above the weather. The earth's stratosphere ranges from about 20 to 50 km in altitude.

steradian—The unit of solid angular measure, defined as the ratio of the surface area of that section of a sphere intercepted by a solid angle to the square of the radius. A full sphere subtends 4π steradians from its center.

strong force—The nuclear force, the strongest of the four fundamental forces of nature.

subtend—The angle that an object appears to take up in your field of view; actually, the angle between lines drawn from opposite sides of the object to your eye. For example, the full moon subtends $1/2°$.

sunspot—A relatively dark area of the solar surface. Sunspots appear dark because they are relatively cool; they represent regions of extremely high magnetic field.

sunspot cycle—The 11-year cycle of variation of the number of sunspots visible on the sun.

supergiant—A post-main-sequence phase of evolution of stars of more than about 4 solar masses. Supergiants fall in the extreme upper right of the H–R diagram.

supernova—The explosion of a star with the resulting release of tremendous amounts of radiation and presumable high-energy particles as well.

synchrotron radiation—A type of non-thermal radiation emitted by electrons spiralling at relativistic velocities in a magnetic field.

synodic—Measured with respect to an alignment of astronomical bodies other than or in addition to the sun or the stars (usually the moon or a planet). For example, a synodic month depends on the positions of the sun, earth, and moon. The synodic period of Mars depends on the relative positions of the earth and Mars as they orbit the sun.

syzygy—An alignment of three celestial bodies. Sometimes applied more specifically to an alignment of the sun, earth, and moon.

T association—A grouping of several T Tauri stars, presumably formed out of the same cloud of interstellar dust and gas.

tektites—Small glassy objects found scattered around the southern part of the southern hemisphere of the earth.

terminator—The line between nighttime and daytime on a moon or planet; the edge of the part of a moon or planet that is lighted by the sun.

terrestrial planets—Mercury, Venus, Earth, and Mars.

thermal radiation—Radiation whose distribution of intensity over wavelength can be characterized by a single number (the temperature). Black-body radiation, which follows Planck's law, is an example of thermal radiation.

tidal force—A force caused by the differential effect of the gravity from one body being greater on the near side of a second body than on the far side.

transit—The passage of one celestial body in front of another celestial body. When a planet is in transit, we understand that it is passing in front of the sun.

trigonometric parallax — See parallax.

triple-alpha process — A chain of fusion processes by which three helium nuclei (alpha particles) combine to form a carbon nucleus.

Trojan asteroids — A group of asteroids that precede or follow Jupiter in its orbit by 60°.

tropical year — The length of time between two successive vernal equinoxes.

T Tauri star — A type of irregularly varying star, like T Tauri, whose spectrum shows broad and very intense emission lines. T Tauri stars have presumably not yet reached the main sequence and are thus very young.

UBV system — A system of photometry that uses three standard filters to define wavelength regions in the ultraviolet, blue, and green-yellow (visual) regions of the spectrum.

U.L.E. — Ultra-Low Expansion, Corning Glass's successor to Pyrex as a material out of which telescope mirrors are made. U.L.E. expands or contracts very little when the temperature changes, and thus a mirror made out of it holds its shape as the temperature varies without distorting the image.

ultraviolet — The region of the spectrum between about 100 and 4000 angstroms; also used in the restricted sense of ultraviolet radiation that reaches the ground, namely, that between about 3000 and 4000 angstroms.

umbra — (a) Of a sunspot, the dark central region; (b) of an eclipse shadow, the part from which the sun cannot be seen at all.

uvby — A system of photometry that uses four standard filters to define wavelength regions in the ultraviolet, blue, and yellow regions of the spectrum.

Van Allen belts — Regions of high energy particles trapped by the magnetic field of the earth.

variable star — A star whose brightness changes over time.

velocity — The measure of the speed and direction of an object; the distance travelled by an object per unit of time.

vernal equinox — The intersection of the ecliptic and the celestial equator that the sun passes each year when moving from southern to northern declinations.

visual binary — A binary star that can be seen through a telescope to be double.

VLA — The Very Large Array, a set of radio telescopes being built by the National Radio Astronomy Observatory in New Mexico to make aperture synthesis measurements.

VLBI — Very-long-baseline interferometry, the technique of using simultaneous measurements made with radio telescopes at widely separated locations to obtain extremely high resolution.

wavelength — The distance over which a wave goes through a complete oscillation.

weak force — One of the four fundamental forces of nature, weaker than the strong force and the electromagnetic force. It is important only in the decay of certain elementary particles.

weight — The force determined by the gravitational pull on a mass.

white dwarf — The final stage of the evolution of a star of between 0.07 and 1.4 solar masses; a star supported by electron degeneracy. White dwarfs are found to the lower left of the main sequence of the H–R diagram.

Wien's displacement law — The expression of the inverse relationship of the temperature of a black body and the wavelength of the peak of its emission.

Wolf-Rayet star — A type of O star whose spectrum shows very broad emission lines.

W Virginis star — A type of II Cepheid, one of the fainter class of Cepheid variable stars characteristic of Cepheids in globular clusters.

x-rays — Electromagnetic radiation between about 1 and 100 angstroms.

year — The period of revolution of a planet around its central star; more particularly, the earth's period of revolution around the sun.

ZAMS — See zero-age main sequence.

Zeeman effect — The splitting of certain spectral lines in the presence of a magnetic field.

zenith — The point in the sky directly overhead an observer.

zero-age main sequence — The curve on an H–R diagram determined by the location of stars at the time they begin nuclear fusion.

zodiac — The band of constellations through which the sun, moon, and planets move in the course of the year.

zodiacal light — A glow in the nighttime sky near the ecliptic representing sunlight reflected by interplanetary dust.

Selected Readings

(A more detailed list of readings, including articles from magazines and journals, appears in the Student Study Guide.)

Non-Technical Books on Astronomy (in approximately the order of the text)

General Books

Herbert Friedman, *The Amazing Universe* (Washington: National Geographic Society, 1975). *National Geographic's* survey of astronomy, written by a pioneer in space science and profusely illustrated in color.

Donald H. Menzel, *Astronomy* (New York: Random House, Inc., 1970). A well-illustrated large-format popular description of the universe.

Isaac Asimov, *The Universe* (New York: Avon Books, 1966). A popular survey.

Kees Boecke, *Cosmic View: The Universe in 40 Jumps* (New York: John Day Co., Inc., 1957). Views of objects in the universe increasing in scale by factors of 10, ranging up to clusters of galaxies and down to the nucleus of an atom.

Stars and Galaxies

Lawrence H. Aller, *Atoms, Stars, and Nebulae*, rev. ed. (Cambridge, Mass.: Harvard University Press, 1971). Readable discussions of the material in Chapters 2 and 4.

Bart J. Bok and Priscilla F. Bok, *The Milky Way*, 4th ed. (Cambridge, Mass.: Harvard University Press, 1974). A readable and well-illustrated survey of the Milky Way; includes good discussions of star clusters and H-R diagrams.

John C. Brandt, *The Sun and Stars* (New York: McGraw-Hill Book Co., 1966). A general discussion.

Donald H. Menzel, *Our Sun*, rev. ed. (Cambridge, Mass.: Harvard University Press, 1959). In need of updating, but still the best available popular discussion of the sun.

Robert Jastrow, *Red Giants and White Dwarfs* (New York: Harper & Row Pubs., Inc., 1967; New American Library, 1969). "The evolution of stars, planets and life."

Harry L. Shipman, *Black Holes, Quasars, and the Universe* (Boston: Houghton Mifflin Co., 1976). A careful presentation to a non-specialist audience.

Frederick Golden, *Quasars, Pulsars, and Black Holes* (New York: Charles Scribner's Sons, 1976). A popular discussion.

William J. Kaufmann III, *Relativity and Cosmology*, 2nd ed. (New York: Harper & Row Pubs., Inc., 1977). A brief and easy treatment of black holes, curved space, quasars, and other strange concepts.

The Solar System and Life in the Universe

Fred L. Whipple, *Earth, Moon, and Planets*, 3rd ed. (Cambridge, Mass.: Harvard University Press, 1968). A non-technical discussion of the solar system.

Richard S. Lewis, *The Voyages of Apollo* (New York: Quadrangle, 1974). A general discussion of the moon-landing program written for non-scientists.

Walter Sullivan, *Continents in Motion* (New York: McGraw-Hill Book Co., 1974). A popular discussion of the earth and plate tectonics.

Carl Sagan, *The Cosmic Connection: An Extraterrestrial Perspective* (Dell Publishing Co., Inc., 1975). Anecdotes and reminiscences about the search for extraterrestrial life.

Carl Sagan, *The Dragons of Eden* (New York: Random House, 1977). "Speculations on the Evolution of Human Intelligence."

Ronald Bracewell, *The Galactic Club—Intelligent Life in Outer Space* (San Francisco: W. H. Freeman & Co., 1974). On extraterrestrial life.

Donald H. Menzel and Ernest H. Taves, *The UFO Enigma* (New York: Doubleday, 1977).

Galaxies and Cosmology

Richard Berendzen, Richard Hart, and Daniel Seeley, *Man Discovers the Galaxies* (New York: Neale Watson Academic Publications, Inc., 1976). A historical view of studies of the Milky Way and other galaxies.

Charles A. Whitney, *The Discovery of Our Galaxy* (New York, Alfred A. Knopf, Inc., 1971). A readable history.

Harlow Shapley, *Galaxies*, 3rd ed., revised by Paul W. Hodge (Cambridge, Mass.: Harvard University Press, 1972). A non-technical study of galaxies, written by the master.

Simon Mitton, *Exploring the Galaxies* (New York: Charles Scribner's Sons, 1976). A readable discussion of galaxies and quasars and their study.

Gerrit Verschuur, *The Invisible Universe* (New York: Springer-Verlag New York, Inc., 1974). A popular survey of radio astronomy.

George Gamow, *One, Two, Three . . . Infinity* (New York: Bantam Books, Inc., 1971). A reprinting of a wonderful description of the structure of space that has introduced many a contemporary astronomer to his or her profession.

Evry Schatzman, *The Structure of the Universe* (New York: McGraw-Hill Book Co., 1968). A popular discussion.

Fred Hoyle, *Highlights in Astronomy* (San Francisco: W. H. Freeman & Co., 1975). A well-illustrated personal tour through the heavens.

Lloyd Motz, *The Universe: Its Beginning and End* (New York: Charles Scribner's Sons, 1975). A non-technical survey of the universe, stars, and the solar system.

Timothy Ferris, *The Red Limit, The Search for the Edge of The Universe* (New York: William Morrow & Co., 1977). "About the major discoveries in astronomy and cosmology in our time, and about the men and women who made these discoveries."

Stephen Weinberg, *The First Three Minutes* (New York: Basic Books, 1977). The first minutes after the big bang, described for a general audience.

Collections of Articles

Owen Gingerich, ed., *Frontiers in Astronomy, New Frontiers in Astronomy* (San Francisco: W. H. Freeman & Co., 1971 and 1976, respectively). Two collections of reprints from Scientific American.

John C. Brandt and Stephen P. Maran, *The New Astronomy and Space Science Reader* (San Francisco: W. H. Freeman & Co., 1977). A varied collection of articles.

The Solar System (San Francisco: W. H. Freeman & Co., 1975). A reprint of the special Scientific American issue of September 1975.

Owen Gingerich, ed., *Cosmology + 1* (San Francisco, W. H. Freeman & Co., 1977). Ten Scientific American reprints on cosmology, quasars, and black holes, plus one on the search for extraterrestrial intelligence.

More Advanced Books

Eugene H. Avrett, ed., *Frontiers in Astrophysics* (Cambridge, Mass.: Harvard University Press, 1976). A series of chapters, each written by an expert, on contemporary research.

Allan Sandage, *The Hubble Atlas* (Washington, D.C.: Carnegie Institution of Washington, 1961). Beautiful photographs of galaxies and their descriptions.

Elske v. P. Smith and Kenneth C. Jacobs, *Introductory Astronomy and Astrophysics* (Philadelphia, W. B. Saunders Co., 1973). A mathematical introductory textbook.

Monthly Non-Technical Magazines on Astronomy

Sky and Telescope, 49-50-51 Bay State Road, Cambridge, MA 02138.
Astronomy, 411 East Mason Street, Milwaukee, WI 53202.
Mercury, Astronomical Society of the Pacific, 1244 Noriega Street, San Francisco, CA 94122.
The Griffith Observer, Griffith Observatory, 2800 East Observatory Road, Los Angeles, CA 90027.

Magazines and Annuals Carrying Articles on Astronomy

Science News, 1719 N Street, N.W. Washington, DC 20036. Published weekly.
Scientific American, 415 Madison Avenue, New York, NY 10017.
National Geographic, Washington, DC 20036.
Natural History, Membership Services, Box 6000, Des Moines, IA 50340.
Science Year (Chicago, IL, Field Enterprises Educational Corp.). The World Book Science Annual.
Smithsonian, 900 Jefferson Drive, Washington, D.C. 20560.

Observing Reference Books

Donald H. Menzel, *A Field Guide to the Stars and Planets* (Boston: Houghton Mifflin Co., 1975).
Charles A. Whitney, *Whitney's Star Finder* (New York, Alfred A. Knopf, Inc., 1974).
Arthur P. Norton, *Norton's Star Atlas and Reference Handbook,* 16th ed., revised by the successors of the late Mr. Norton (Cambridge, Mass.: Sky Publishing Corp., 1973).
The Observer's Handbook (yearly), Royal Astronomical Society of Canada, 252 College Street, Toronto M5T 1R7, Canada.
The American Ephemeris and Nautical Almanac (yearly), U.S. Government Printing Office, Washington, DC 20402.
Hans Vehrenberg, *Atlas of Deep Sky Splendors* (Cambridge, Mass.: Sky Publishing Corp., 1971). Photographs, descriptions, and finding charts for hundreds of beautiful objects.

General Reference Books

C. W. Allen, *Astrophysical Quantities,* 3rd ed. (London: The Athlone Press of the University of London, 1973). Tables and lists of almost every conceivable kind.
Kenneth R. Lang, *Astrophysical Formulae* (New York: Springer-Verlag New York, Inc., 1974). Formulas of all kinds, plus many tables.
Van Nostrand's Scientific Encyclopedia, 5th ed. (New York: Van Nostrand Reinhold Co., 1976). A one-volume encyclopedia.
Jeanne Hopkins, *Glossary of Astronomy and Astrophysics* (Chicago: University of Chicago Press, 1976). A technical glossary.

For Information about Amateur Societies

American Association of Variable Star Observers (AAVSO), 187 Garden Street, Cambridge, MA 02138.
Association of Lunar and Planetary Observers (ALPO), 8930 Raven Drive, Waco, TX 76710.

Careers in Astronomy

The Executive Officer, American Astronomical Society, 211 FitzRandolph Road, Princeton, N.J. 08450. A free booklet is available on request.
Space for Women, derived from *The Earth in the Cosmos: Space for Women,* a Symposium for Women on Careers in Astronomy, Astrophysics, and Earth and Planetary Sciences. Sponsored by Center for Astrophysics, Harvard University, Radcliffe College, and Smithsonian Institution. For free copies, write to: Center for Astrophysics, 60 Garden Street, Cambridge, Mass. 02138.

Illustration Acknowledgments

COLOR PLATES

Plate 1   Copyright by The Association of Universities for Research in Astronomy, Inc. The Kitt Peak National Observatory;

Plates 2, 7, 8, and 9   Jay M. Pasachoff;

Plate 3   Bausch and Lomb; Deutsches Museum, Munich (Fraunhofer);

Plates 4, 6, and 48   Copyright by The Association of Universities for Research in Astronomy, Inc. The Cerro Tololo Inter-American Observatory;

Plates 5, 19, 20, 21, 23, 24, 25, 42, 43, 44, 45, 46, 47, and 49   Copyright California Institute of Technology and Carnegie Institution of Washington;

Plate 10   William C. Atkinson;

Plates 11 and 12   Naval Research Laboratory/NASA, courtesy J. David Bohlin;

Plates 13, 14, and 15   High Altitude Observatory/NASA, courtesy Robert M. MacQueen and Charles L. Ross;

Plates 16, 17, and 18   Harvard College Observatory/NASA, courtesy Peter V. Foukal, Edward J. Schmahl, and Edmond M. Reeves;

Plate 22   Richard J. Borken and Saul A. Rappaport;

Plates 26, 27, 28, 29, 30, 31, 32, 35, 36, 37, and 38   NASA;

Plate 33   International Planetary Patrol photograph furnished by Lowell Observatory;

Plate 34   Gustav Lamprecht;

Plate 39   NASA/Lunar and Planetary Laboratory, University of Arizona;

Plate 40   Lunar and Planetary Laboratory, University of Arizona;

Plate 41   Martin Grossmann;

Plate 50   Netherlands Foundation for Radio Astronomy, National Radio Astronomy Observatory, Arnold H. Rots and William W. Shane. Optical image from the Hale Observatories;

Front Cover — Martin Grossmann.

Back Cover — Copyright California Institute of Technology and Carnegie Institution of Washington.

BLACK-AND-WHITE ILLUSTRATIONS AND MARGINAL NOTES

End Papers — From the atlases of Bayer and Hevelius, by permission of The Houghton Library, Harvard University.

PART 1 — Opener   Lick Observatory;

Figs. 1-2 and 1-3   Jay M. Pasachoff;

Fig. 1-4   Skyviews Survey Inc.;

Figs. 1-5 and 1-6   NASA;

Fig. 1-12   Harvard College Observatory;

Figs. 1-13 and 1-14   Hale Observatories.

PART II—Opener (left) Hale Observatories, *(right)* Big Bear Solar Observatory;

PART II—Introduction NASA and TRW Systems Group;

CHAPTER 2—Opener By permission of The Houghton Library, Harvard University;

Fig. 2–1 From THE LITTLE PRINCE by Antoine de Saint-Exupéry, copyright 1943, by Harcourt Brace Jovanovich, Inc. in America and William Heinemann Ltd. in England; copyright 1971, by Consuelo de Saint-Exupéry. Reproduced by permission of the publishers.

Fig. 2–7 From Leo Goldberg, "Ultraviolet Astronomy," *Scientific American,* June 1969, and *Frontiers in Astronomy,* reprinted courtesy of *Scientific American;*

Fig. 2–13 © National Geographic Society—Palomar Observatory Sky Survey. Reproduced by permission from the Hale Observatories;

Fig. 2–14 American Institute of Physics, Meggers Gallery of Nobel Laureates;

Fig. 2–15 Harvard College Observatory;

Fig. 2–16 The Kitt Peak National Observatory;

Fig. 2–25 American Institute of Physics, Niels Bohr Library, Margrethe Bohr Collection;

Fig. 2–27 After Bok and Bok, *The Milky Way,* courtesy Harvard University Press;

Fig. 2–28 Hale Observatories;

Fig. 2–30 © 1971 by *Playboy Magazine;*

Fig. 2–37 Peter van de Kamp, Sproul Observatory.

CHAPTER 3—Opener The Kitt Peak National Observatory;

Fig. 3–4 Lick Observatory;

Fig. 3–5 American Institute of Physics, Niels Bohr Library;

Fig. 3–10B Copyright Royal Greenwich Observatory;

Figs. 3–12 and 3–13 Hale Observatories;

Fig. 3–14A Dale P. Cruikshank, University of Hawaii;

Fig. 3–14B Novisti from Sovfoto;

Fig. 3–15 NASA;

Fig. 3–16 The Kitt Peak National Observatory;

Fig. 3–17 The Cerro Tololo Inter-American Observatory;

Fig. 3–18 The Kitt Peak National Observatory and The Cerro Tololo Inter-American Observatory/Arthur A. Hoag;

Fig. 3–20 Center for Astrophysics [of the Harvard College Observatory and the Smithsonian Astrophysical Observatory];

Fig. 3–21 Hale Observatories;

Fig. 3–23 © National Geographic Society—Palomar Observatory Sky Survey. Reproduced by permission from the Hale Observatories;

Fig. 3–25 *(left)* Jay M. Pasachoff, *(middle)* Harvard College Observatory, *(right)* Hale Observatories;

Fig. 3–26 Thomas B. McCord, Massachusetts Institute of Technology;

Fig. 3–27 Hale Observatories;

Fig. 3–28 Giovanni Fazio, Center for Astrophysics;

Fig. 3–30A Herbert Gursky, Center for Astrophysics;

Fig. 3–30B Gordon P. Garmire, California Institute of Technology;

Figs. 3–31 and 3–32 Hale Observatories;

Fig. 3–33 Jay M. Pasachoff;

Fig. 3–38 Emil Schulthess, Black Star;

Fig. 3–39 Dale P. Cruikshank, University of Hawaii;

Fig. 3–40 Yerkes Observatory;

Fig. 3–43 © United Features Syndicate, Inc.

CHAPTER 4—Opener Hale Observatories;

Fig. 4–2 Hale Observatories;

Fig. 4–3 Lick Observatory;

Fig. 4–6 Peter van de Kamp;

Fig. 4–7 Based on data from D. L. Harris III, K. Aa. Strand, and C. E. Worley in *Basic Astronomical Data,* K. Aa. Strand, ed., University of Chicago Press, copyright 1963 by The University of Chicago;

Fig. 4–8 R. Hanbury Brown;

Fig. 4–9 Peter van de Kamp and Sarah Lee Lippincott, from *Vistas in Astronomy,* 19, 231 (1975), courtesy Pergamon Press;

Fig. 4–10   After Bok and Bok, *The Milky Way*, courtesy Harvard University Press;
Fig. 4–12   Harvard College Observatory;
Fig. 4–13   Shigetsugu Fujinami;
Fig. 4–14   After Bok and Bok, *The Milky Way*, courtesy Harvard University Press;
Fig. 4–15   Hale Observatories;
Fig. 4–16   By permission of The Houghton Library, Harvard University;
Fig. 4–17   Lick Observatory;
Fig. 4–18   After Bok and Bok, *The Milky Way*, courtesy Harvard University Press;
Fig. 4–19   Lick Observatory;
Fig. 4–20   After Bok and Bok, *The Milky Way*, courtesy Harvard University Press, and a graph by Harold L. Johnson and Allan R. Sandage in the *Astrophysical Journal, 124*, 379. Reprinted by permission of the University of Chicago Press, © 1956 by the American Astronomical Society;
Fig. 4–21   Photograph courtesy Harvard College Observatory. X-ray data courtesy George W. Clark, Massachusetts Institute of Technology;
Fig. 4–23   Astronomy Netherlands Satellite data/Space Research Laboratory, Utrecht.

Chapter 5—Opener   The Kitt Peak National Observatory/William C. Livingston;
Fig. 5–2   Sacramento Peak Observatory;
Figs. 5–3 and 5–4   Hale Observatories;
Fig. 5–5   Sacramento Peak Observatory;
Fig. 5–6   The Aerospace Corp./David K. Lynch;
Fig. 5–7   M. Kanno, Hido Observatory, University of Kyoto;
Fig. 5–8   High Altitude Observatory/Robert M. MacQueen and Charles L. Ross;
Fig. 5–9   Based on a diagram by H. C. Van de Hulst;
Fig. 5–10A   University of Hawaii;
Fig. 5–10B   Jay M. Pasachoff;
Fig. 5–12   Big Bear Solar Observatory;
Figs 5–13, 5–16, 5–18, and 5–19   Jay M. Pasachoff;
Fig. 5–14   Williams College Expedition/Duncan H. Read;
Fig. 5–20   Dale R. Corson;
Fig. 5–21   Sacramento Peak Observatory/Air Force Cambridge Research Laboratories;
Fig. 5–23   M. Waldmeier, Swiss Federal Observatory;
Fig. 5–24   Big Bear Solar Observatory/Harold Zirin;
Fig. 5–25   Gustav Lamprecht;
Fig. 5–26   National Oceanic and Atmospheric Administration/Patrick S. McIntosh;
Fig. 5–27   Hale Observatories;
Fig. 5–28   Harvard College Observatory/NASA;
Fig. 5–29   Stephen H. Schneider and Clifford Mass, *Science, 190*, 741, copyright 1975 by the American Association for the Advancement of Science;
Fig. 5–30   Hale Observatories;
Fig. 5–31   Naval Research Laboratory/NASA;
Fig. 5–32   NASA;
Fig. 5–33   © 1975 Field Enterprises Education Corp.;
Fig. 5–34   Courtesy of The Archives, California Institute of Technology.

CHAPTER 6—Opener   Theodore R. Gull/The Kitt Peak National Observatory;
Fig. 6–2   George H. Herbig, Lick Observatory;
Figs. 6–10 and 6–11   Raymond Davis, Jr., Brookhaven National Laboratory.

CHAPTER 7—Opener   Lick Observatory;
Fig. 7–3   Hale Observatories;
Fig. 7–4   Lick Observatory;
Fig. 7–5   Hale Observatories;
Fig. 7–7   From C. R. O'Dell, in IAU Symposium No. 34, "Planetary Nebulae," eds. D. E. Osterbrock and C. R. O'Dell. Reprinted by permission of the International Astronomical Union;
Figs. 7–9 and 7–10   Lick Observatory;
Fig. 7–11   Hale Observatories;
Fig. 7–12   © Ben Mayer, Los Angeles.

CHAPTER 8—Opener   Lick Observatory
Fig. 8–2   By permission of The Houghton Library, Harvard University;
Figs. 8–3 and 8–4   Hale Observatories;

Fig. 8–5   Hale Observatories, courtesy Sidney van den Bergh. Photograph by R. Minkowski;

Fig. 8–6   William Miller and Museum of Northern Arizona;

Fig. 8–8   Jocelyn Bell Burnell;

Fig. 8–9A   National Radio Astronomy Observatory;

Fig. 8–10   Joseph H. Taylor, University of Massachusetts-Amherst;

Fig. 8–14   Lick Observatory;

Fig. 8–15   Herbert Friedman and Seth Shulman, Naval Research Laboratory.

CHAPTER 9—Opener   Hale Observatories/Jerome Kristian;

Fig. 9–5   Drawing by Charles Addams; reprinted with permission of *The New Yorker Magazine, Inc.* © 1974;

Fig. 9–7   Chapin Library, Williams College;

Fig. 9–8   Hale Observatories/Jerome Kristian;

Fig. 9–9   Lois Cohen—Griffith Observatory;

Fig. 9–10   Joseph Weber, University of Maryland.

PART III—Opener   *(right)* Lunar and Planetary Laboratory, University of Arizona; *(left)* NASA;

CHAPTER 10

Fig. 10–2   Lick Observatory;

Fig. 10–5B   Jay M. Pasachoff;

Fig. 10–7   George Lovi, Vanderbilt Planetarium;

Fig. 10–8   Burndy Library, photograph by Owen Gingerich;

Figs. 10–9, 10–12, 10–14, and 10–17   Chapin Library, Williams College;

Fig. 10–11   Courtesy of Owen Gingerich;

Fig. 10–13   Uppsala University Library, photographed by Charles Eames, reproduced courtesy of Owen Gingerich;

Fig. 10–15   University of Michigan Library, Dept. of Rare Books and Special Collections, translation by Stillman Drake, reprinted courtesy of *Scientific American;*

Fig. 10–16   New Mexico State University Observatory;

Fig. 10–24   B.C. by permission of John Hart and Field Enterprises, Inc.;

Fig. 10–25   United Press International photograph;

Fig. 10–28A   NASA;

Fig. 10–28B   © 1977 by The New York Times Company. Reprinted by permission.

CHAPTER 11—Opener   NASA;

Fig. 11–1   Lunar and Planetary Laboratory, University of Arizona;

Fig. 11–2A   Mount Wilson and Palomar Observatories;

Fig. 11–2B   Lick Observatory;

Fig. 11–2C   Base photograph: Lick Observatory;

Fig. 11–4   Lick Observatory;

Figs. 11–5, 11–6A, 11–8, and 11–10   NASA;

Fig. 11–6B   Drawing by Alan Dunn; reprinted with permission of *The New Yorker Magazine, Inc.* © 1971;

Fig. 11–7   General Electric Research and Development Center;

Fig. 11–9   Lick Observatory;

Fig. 11–11   Judith W. Frondel, Harvard University;

Fig. 11–12   Data from Gerald J. Wasserburg and D. A. Papanastassiou, courtesy of Gerald J. Wasserburg, California Institute of Technology.

CHAPTER 12—Opener   NASA;

Fig. 12–2   New Mexico State University Observatory;

Figs. 12–6 to 12–11   NASA.

CHAPTER 13—Opener   NASA;

Fig. 13–1   Hale Observatories;

Fig. 13–4   Jet Propulsion Laboratories/R. M. Goldstein;

Figs. 13–6 and 13–7   NASA;

Fig. 13–8   Tass from Sovfoto;

Figs. 13–9 and 13–10   Science Service.

CHAPTER 14—Opener NASA;
Fig. 14–1 U.S. Geological Survey, EROS Data Center;
Figs. 14–5 to 14–8 Jay M. Pasachoff;
Quotation from Paul A. Weiss (abridged from Rockefeller Institute Review 2, No. 6, pp. 8–14) with permission of Dr. Weiss and the Rockefeller University Press. Reprinted in *A Random Walk in Science*, R. W. Weber, ed. (London: The Institute of Physics, 1973).

CHAPTER 15—Opener NASA;
Figs. 15–1 and 15–2 Lick Observatory;
Figs. 15–4 to 15–14 NASA.

CHAPTER 16—Opener NASA/Lunar and Planetary Laboratory, University of Arizona;
Fig. 16–1 Lunar and Planetary Laboratory, University of Arizona;
Fig. 16–4A *Sky and Telescope;*
Fig. 16–4B Yerkes Observatory;
Fig. 16–5 Hale Observatories/Charles T. Kowal;
Figs. 16–6, 16–7, and 16–9 NASA;
Fig. 16–10 NASA/B. Roy Frieden, University of Arizona.

CHAPTER 17—Opener New Mexico State University Observatory;
Fig. 17–2 New Mexico State University Observatory;
Fig. 17–3 Lowell Observatory photographs;
Fig. 17–5 Smithsonian Astrophysical Observatory/Kaare Aksnes;
Fig. 17–6 NASA;
Figs. 17–7 and 17–11 Lunar and Planetary Laboratory, University of Arizona;
Fig. 17–9 James L. Elliot/Cornell University, painted by Rick Sternbach;
Fig. 17–12 Master and Fellows of St. John's College of Cambridge;
Fig. 17–13 Lick Observatory;
Fig. 17–15 Lowell Observatory photograph;
Fig. 17–16 Hale Observatories;
Page 269, A1, 2, 3 and B1 and 2 NASA; *B3* Lick Observatory.

CHAPTER 18—Opener Institut d'Astrophysique de Liège, Belgium with the Schmidt camera at the Observatoire de Haute Provence, France;
Fig. 18–1 William Liller, Harvard College Observatory;
Figs. 18–2, 18–4, 18–7, and 18–10 Hale Observatories;
Fig. 18–3 European Southern Observatory/Jean Pierre Swings;
Figs. 18–5 and 18–9 NASA;
Fig. 18–6 National Portrait Gallery, London, painted by R. Phillips prior to 1721;
Fig. 18–8A Leon Tadrick in *The Sciences*, Vol. 15, No. 8, Nov. 1975 © 1975 The New York Academy of Sciences;
Figs. 18–8B and 18–8C *Sky and Telescope* by Roger Sinnott from orbital elements by Joseph Brady and Edna Carpenter, Lawrence Radiation Laboratory, University of California;
Fig. 18–11 Smithsonian Astrophysical Observatory;
Fig. 18–12 Peter Bloomer, *Horizons West*, by permission of Meteor Crater Enterprises, Inc.;
Fig. 18–13A Herzberg Institute of Astrophysics, Ottawa;
Fig. 18–13B World Wide Photos;
Fig. 18–14 Harvard College Observatory.

CHAPTER 19—Opener Twentieth-Century Fox, Inc. © 1977;
Fig. 19–1 © 1974 United Features Syndicate, Inc.;
Fig. 19–2 Peter van de Kamp, Sproul Observatory;
Fig. 19–3 The Kitt Peak National Observatory;
Figs. 19–4 and 19–5 Cornell University photographs;
Fig. 19–6 Museum of Modern Art/Twentieth-Century Fox, Inc.;
Figs. 19–7 and 19–8 Robert M. Sheaffer;
Fig. 19–9 © 1975 United Features Syndicate, Inc.;
Fig. 19–10 Chapin Library, Williams College;
Fig. 19–11 Twentieth-Century Fox, Inc. © 1977;
Marginal Note, page 302 Written by Larry Rhine and Mel Tolkin © copyright 1975 Tandem Productions, Inc. All Rights Reserved.

PART IV—OPENER (*left*) Lick Observatory; (*right*) Hale Observatories;
PART IV—Introduction Reproduced courtesy of the Trustees, The National Gallery, London, painted circa 1578;

CHAPTER 20—Opener Hale Observatories;
Fig. 20–2 Hale Observatories;
Fig. 20–3 Lick Observatory;
Fig. 20–5 Data from W. Becker, photograph from Hale Observatories;
Fig. 20–6 Agris Kalnajs, Mt. Stromlo Observatory.

CHAPTER 21—Opener Jay M. Pasachoff;
Figs. 21–1 and 21–2 Courtesy of Bell Laboratories;
Fig. 21–3 Jay M. Pasachoff;
Fig. 21–13 Harvard University/E. M. Purcell;
Fig. 21–16A Gart Westerhout, University of Maryland and U.S. Naval Observatory;
Fig. 21–16B Gerrit Verschuur, University of Colorado;
Fig. 21–19 Graphic Films, Inc., Hollywood.

CHAPTER 22—Opener Eric Becklin, California Institute of Technology;
Fig. 22–1 Institut d'Astrophysique de Liège, Belgium;
Fig. 22–2 Hale Observatories;
Fig. 22–3A Dennis Downes, Max Planck Institut für Radioastronomie, Bonn, G.F.R.;
Fig. 22–3B Eric Becklin, California Institute of Technology;
Fig. 22–4A Lund Observatory, Sweden;
Fig. 22–4B Harvard-Smithsonian Center for Astrophysics/Riccardo Giacconi and Herbert Gursky; with permission of D. Reidel Publishing Co.;
Fig. 22–5 NASA.

PART V—Openers Hale Observatories;
PART V—Introduction Lick Observatory Archives;

CHAPTER 23—Opener Hale Observatories;
Figs. 23–1 to 23–4, 23–6, 23–7, and 23–11 Hale Observatories;
Fig. 23–5 Courtesy of the U.K. Schmidt Telescope Unit, Royal Observatory, Edinburgh;
Fig. 23–10 Lick Observatory;
Fig. 23–14 The Kitt Peak National Observatory/Hyron Spinrad, University of California at Berkeley;
Fig. 23–15 Middle photograph from Lick Observatory. Top left photograph from Hale Observatories. Radio map based on Phillip J. Hargrave and Martin Ryle, *Monthly Notices of the Royal Astronomical Society, 166,* 305, copyright 1974 by the Royal Astronomical Society;
Fig. 23–16 Top photograph from Hale Observatories;
Fig. 23–17 Courtesy of *La Recherche*;
Fig. 23–21 National Radio Astronomy Observatory and E-Systems, Inc.;
Fig. 23–22 Westerbork Synthesis Telescope, Wilkes, Strom, and Wilson, *Nature, 250,* 625, copyright 1974 by Macmillan Journals Limited, London;
Fig. 23–23 Courtesy of Richard G. Strom, George K. Miley, Jan Oort, and *Scientific American*;
Fig. 23–24 Westerbork Synthesis Telescope, G. K. Miley, G. C. Perola, P. C. van der Kruit, and H. van der Laan, *Nature, 237,* 269, copyright 1972 by Macmillan Journals Limited, London.

CHAPTER 24—Opener Hale Observatories/Maarten Schmidt;
Figs. 24–1 and 24–2 Hale Observatories;
Fig. 24–6 Hale Observatories/Halton Arp;
Fig. 24–7A Hale Observatories/montage by W. W. Morgan, Yerkes Observatory;
Fig. 24–7B Princeton University Observatory.

CHAPTER 25—Opener Lotte Jacobi;
Fig. 25–1B From the collection of Mr. and Mrs. Paul Mellon;
Fig. 25–6 Courtesy of Bell Laboratories;
Fig. 25–8 Based on Patrick Thaddeus, *Annual Review of Astronomy and Astrophysics, 10,* 305, copyright 1972 by Annual Review, Inc.;

Fig. 25–9   Based on D. P. Woody, J. C. Mather, N. S. Nishioka, and P. L. Richards, *Physical Review Letters, 34,* 1036, copyright 1975 by The American Physical Society;

Fig. 25–10   Robert V. Wagoner, Stanford University;

Fig. 25–11   Jay M. Pasachoff.

CHAPTER 26 — Opener   National Radio Astronomy Observatory and E-Systems, Inc;

Fig. 26–1   The Kitt Peak National Observatory;

Fig. 26–2   Jay M. Pasachoff;

Fig. 26–3   © 1973 *The Journal of Irreproducible Results, Inc.;*

Fig. 26–4   NASA;

Fig. 26–5   The Kitt Peak National Observatory;

Marginal Note, page 394   Copyright © 1973 by The Massachusetts Medical Society. Reprinted by permission of The Viking Press and Lewis Thomas.

I also thank Frederick P. Woodson and the Photo Laboratory of the Hale Observatories, Dennis di Cicco at *Sky and Telescope,* Martha Liller at the Harvard College Observatory, Mildred Shapley Matthews at the Lunar and Planetary Laboratory, Rita Beebe at New Mexico State University Observatory, Spencer R. Weart and Joan N. Warnow at the Center for the History of Physics at the American Institute of Physics, and John Lancaster at the VLA of the National Radio Astronomy Observatory for their assistance in obtaining photographs.

Index

Principal references are in **boldface** type.
References to figures or marginal notes are in *italic* type.
References to tables are followed by t.
References to Color Plates are prefaced by CP.

Entries are alphabetized by letter, ignoring spaces, rather than by word. For example, B stars immediately precedes Bubble Nebula rather than appearing at the beginning of the B's. Significant initial numbers are alphabetized under their spellings; for example, 21 cm is alphabetized as twenty-one. Less important initial numbers and subscripts are ignored in alphabetizing; for example, 3C 273 is at the beginning of the C's, M1 appears at the beginning of the M's, and CO_2 appears next to CO. Greek letters are alphabetized under their English spellings.